Glencoe

Teen Health

my.mheducation.com

Send all inquiries to:
McGraw-Hill Education
8787 Orion Place
Columbus, OH 43240

Glencoe Teen Health, Student Edition:
ISBN: 978-1-26-412840-2
MHID: 1-26-412840-1

Printed in the United States of America.

9 10 11 12 LWI 25 24 23 22

Program Authors

Mary H. Bronson, Ph.D. recently retired after teaching for 30 years in Texas public schools. Dr. Bronson taught health education in grades K–12 as well as health education methods classes at the graduate and undergraduate levels. As Health Education Specialist for the Dallas School District, Dr. Bronson developed and implemented a district-wide health education program. She has been honored as Texas Health Educator of the Year by the Texas Association for Health, Physical Education, Recreation, and Dance and selected Teacher of the Year twice by her colleagues. Dr. Bronson has assisted school districts throughout the country in developing local health education programs. She is also the coauthor of Glencoe Health.

Michael J. Cleary, Ed.D., C.H.E.S. is a professor at Slippery Rock University, where he teaches methods courses and supervises field experiences. Dr. Cleary taught health education at Evanston Township High School in Illinois and later served as the Lead Teacher Specialist at the McMillen Center for Health Education in Fort Wayne, Indiana. Dr. Cleary has published widely on curriculum development and assessment in K–12 and college health education. Dr. Cleary is also coauthor of Glencoe Health.

Betty M. Hubbard, Ed.D., C.H.E.S. has taught science and health education in grades 6–12 as well as undergraduate- and graduate-level courses. She is a professor at the University of Central Arkansas, where in addition to teaching she conducts in-service training for health education teachers in school districts throughout Arkansas. In 1991, Dr. Hubbard received the university's teaching excellence award. Her publications, grants, and presentations focus on research-based, comprehensive health instruction. Dr. Hubbard is a fellow of the American Association for Health Education and serves as the contributing editor of the Teaching Ideas feature of the American Journal of Health Education.

Professional Reviewers

Amy Eyler, Ph.D., CHES
Washington University in St. Louis
St. Louis, MO

Shonali Saha, M.D.
Johns Hopkins School of Medicine
Baltimore, MD

Roberta Duyff
Duyff & Associates
St. Louis, MO

Teacher Reviewers

Lou Ann Donlan
Altoona Area School District
Altoona, PA

Steve Federman
Loveland Intermediate School
Loveland, OH

Rick R. Gough
Ashland Middle School
Ashland, OH

Jacob Graham
Oblock Junior High
Plum, PA

William T. Gunther
Clarkston Community Schools
Clarkston, MI

Ellie Hancock
Somerset Area School District
Somerset, PA

Diane Hursky
Independence Middle School
Bethel Park, PA

Veronique Javier
Thomas Cardoza Middle School
Jackson, MS

Patricia A. Landon
Patrick F. Healy Middle School
East Orange, NJ

Elizabeth Potash
Council Rock High School South
Holland, PA

Table Of Contents

Your Health and Wellness

MODULE 1

Social Health

· ·

Table Of Contents

MODULE 2

Dating Relationships and Abstinence

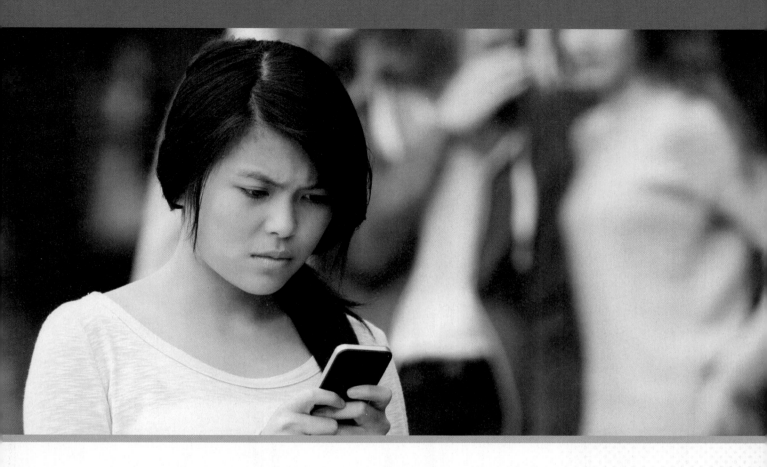

MODULE 3

Bullying and Cyberbullying

Table Of Contents

MODULE 4

Emotional Health

. . . .

MODULE 5

Mental and Emotional Disorders

· ·

Table Of Contents

MODULE 6
Conflict Resolution

MODULE 7

Violence Prevention

Table Of Contents

MODULE 8

Nutrition

MODULE 9

Physical Activity

. . . .

Table Of Contents

MODULE 10

The Life Cycle

MODULE 11

Personal Health Care

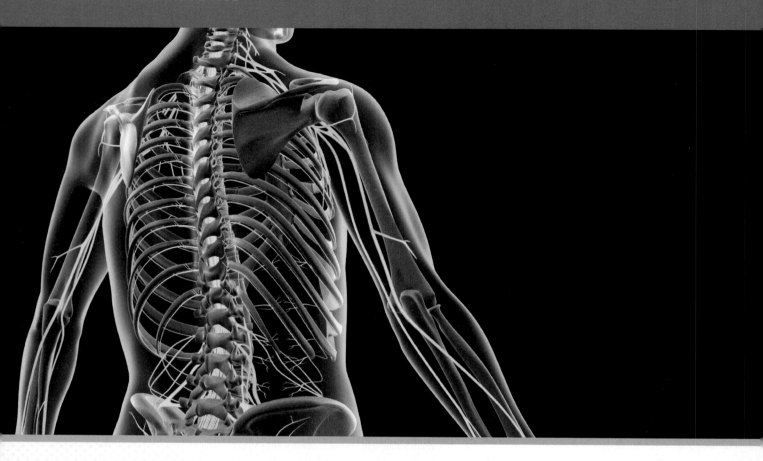

MODULE 12

Your Body Systems

MODULE 13

Tobacco

· · · · · · · · · · · ·

Table Of Contents

MODULE 14

Alcohol

MODULE 15

Drugs
· · · · · · · · · ·

MODULE **16**

Using Medicines Wisely

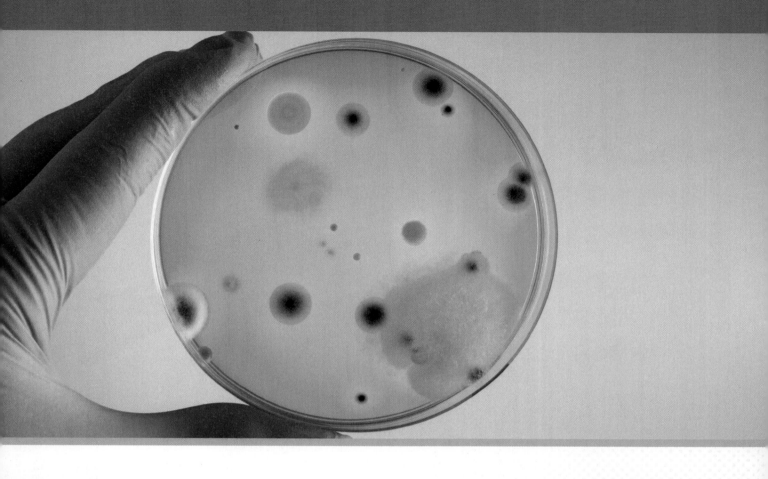

MODULE **17**

Communicable Diseases

· ·

MODULE 18

Noncommunicable Diseases

MODULE 19

Safety

. . . .

MODULE **20**

Green Schools and Environmental Health

Your Health and Wellness

LESSONS

1 Your Total Health

2 Health Influences and Risk Factor

3 Building Health Skills

4 Making Decisions and Setting Goals

5 Choosing Health Services

Your Total Health

Before You Read

Quick Write Write a short paragraph describing how a person with "total health" might look or act. What might this person's lifestyle be like?

Vocabulary
health
behavior
social health
wellness
habits
mind-body connection

BIG IDEA Total health includes healthy ways of thinking, feeling, and interacting with other people, as well as physical health.

What Is Health?

MAIN IDEA The three aspects of health include physical, mental/emotional, and social health.

Do you know someone you would describe as "healthy"? What kinds of traits do they have? Maybe they are involved in sports. Maybe they just "look" healthy. **Health** is the combination of physical, mental/emotional, and social well-being. Looking fit and feeling well are important, but there is more to having good health. Good health also includes getting along well with others and feeling good about yourself.

To maintain good health you need to think about all three parts of health. These are physical health, mental/emotional health, and social health. You must maintain each part of health equally to stay healthy.

Think of good health as a triangle that is made up of three equally-sized lines. Each line is labeled with one part of health. This triangle is balanced. This person thinks equally about physical, mental/emotional, and social health.

Now let's suppose we're looking at another person's triangle. This person thinks more about the social health side than the other two sides. This person's triangle has a longer line on the social side. The health triangle is out of balance. This person is not maintaining good health.

Physical Health

Physical health is one side of the health triangle. It involves the health of your body. Do you have a daily routine that includes physical activities? Engaging in physical activity every day will help to build and maintain your physical health. If you feel strong and have lots of energy, you probably have good physical health.

You can improve your physical health in different ways. Some of these ways include the following:

- **Eating healthy foods.** Choose nutritious meals and snacks.
- **Visiting the doctor regularly.** Get regular checkups from a doctor and a dentist.
- **Caring for personal hygiene.** Shower or bathe each day. Brush and floss your teeth at least twice every day.
- **Wearing protective gear.** When playing sports, using protective gear and following safety rules will help you avoid injuries.
- **Getting enough sleep.** Most teens need about nine hours of sleep every night.

> TO MAINTAIN GOOD HEALTH YOU NEED TO THINK ABOUT ALL THREE PARTS OF HEALTH. THESE ARE PHYSICAL, MENTAL/EMOTIONAL, AND SOCIAL HEALTH.

You can also have good physical health by avoiding harmful **behavior**, such as using alcohol, tobacco, and other drugs. Behavior is the way you act in the many different situations and events in your life. Smoking or using tobacco can harm your mouth, heart, and lungs. The use of tobacco has been linked to many diseases, such as heart disease and cancer. Alcohol and other drug use can damage your liver, brain, and other organs.

Internal factors can influence your behavior with regard to the use of tobacco, alcohol, and other drugs. For example, you may be curious about what it feels like to smoke a cigarette, so you might try one. Or, you may be afraid of the effects of drugs on your health, leading you to never use drugs. Your interests may place you in situations where other people are using tobacco, alcohol, or other drugs. You would need to use your decision-making skills to avoid harmful behaviors.

Mental/Emotional Health

Another side of the health triangle is your mental/emotional health. It involves the health of your mind and emotions. How do you handle your feelings, thoughts, and emotions each day? You can improve your mental/emotional health by talking and thinking about yourself in a healthful way. Share your thoughts and feelings with your family, a trusted adult, or with a friend.

If you are mentally and emotionally healthy, you can face challenges in a positive way. Be patient with yourself when you try to learn new subjects or new skills. Remember that everybody makes mistakes—including you! Next time you can do better.

Taking action to reach your goals is another way to develop good mental/emotional health. This can help you focus your energy and give you a sense of accomplishment. Make healthful choices, keep your promises, and take responsibility for what you do, and you will feel good about yourself and your life.

Reading Check

Analyze What are the three sides of the health triangle? Why should the three sides be balanced?

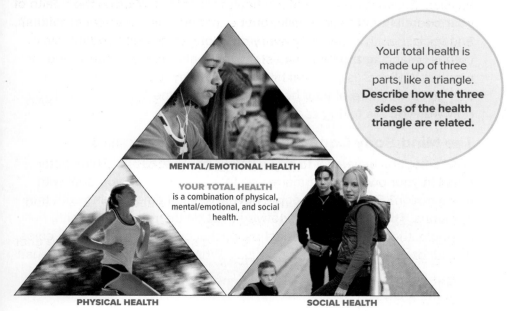

Your total health is made up of three parts, like a triangle. **Describe how the three sides of the health triangle are related.**

MENTAL/EMOTIONAL HEALTH

YOUR TOTAL HEALTH is a combination of physical, mental/emotional, and social health.

PHYSICAL HEALTH

SOCIAL HEALTH

Social Health

A third side of the health triangle is your **social health**, or your ability to get along with the people around you. It involves the health of your friendships. Social health means how you relate to people at home, at school, and everywhere in your world. Strong friendships and family relationships are signs of good social health.

Do you get along well with your friends, classmates, and teachers? Do you spend time with your family? You can develop skills for having good relationships. Good social health includes supporting the people you care about. It also includes communicating with, respecting, and valuing people. Sometimes you may disagree with others. You can disagree and express your thoughts, but be thoughtful and choose your words carefully.

Achieving Wellness

MAIN IDEA Your overall wellness means that you are healthy for a long period of time.

What is the difference between health and wellness? **Wellness** is a state of well-being or balanced health over a long period of time. It involves having balanced health over a period of time. Your health changes from day to day. One day you may feel tired if you did not get enough sleep. Maybe you worked very hard at sports practice. The next day, you might feel well rested and full of energy because you rested. Your emotions also change. You might feel sad one day but happy the next day. You can improve your wellness by developing good health **habits**. Habits are patterns of behavior that you follow almost without thinking. Good health habits include:

- eating healthy foods,
- participating in 60 minutes of physical activity on most days,
- managing stress, and
- developing healthy relationships with others.

Your overall health is like a snapshot of your physical, mental/emotional, and social health. Your wellness takes a longer view. Being healthy means balancing the three sides of your health triangle over weeks or months. Wellness is sometimes represented by a continuum, or scale, that gives a picture of your health at a certain time. It may also tell you how well you are taking care of yourself.

The Mind-Body Connection

Your emotions have a lot to do with your physical health. Think about an event in your own life that made you feel sad. How did you deal with this emotion? Sometimes people have a difficult time dealing with their emotions. This can have a negative effect on their physical health. For example, they might get headaches, backaches, upset stomachs, colds, the flu, or even more serious diseases. Why do you think this happens?

Reading Check

Define What is wellness?

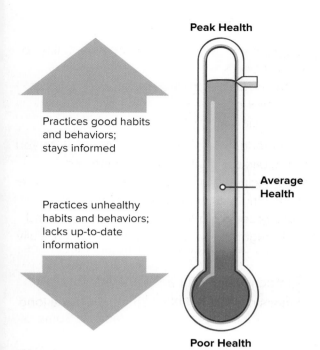

Peak Health

Practices good habits and behaviors; stays informed

Average Health

Practices unhealthy habits and behaviors; lacks up-to-date information

Poor Health

The Wellness Scale identifies how healthy you are at a given point in time. **Describe how you would rate your health on the wellness scale.**

Your mind and body connect through your nervous system. This system includes thousands of miles of nerves. The nerves link your brain to your body. Upsetting thoughts and feelings sometimes affect the signals from your brain to other parts of your body.

The **mind-body connection** is how your emotions affect your physical and overall health and how your overall health affects your emotions. If you become very sad or angry, or if you have other strong emotions, talk to someone. Sometimes talking to a good friend helps. Sometimes you may need the services of a counselor or a medical professional.

Myth vs. Fact

Myth The cause of mental illness is poor upbringing or lack of will power.

Fact Mental illnesses can be caused by some of the same factors as physical illnesses. For example, type 1 diabetes is caused by chemical imbalances in the body, and a mental illness such as depression is believed to be caused by chemical imbalances in the brain.

Lesson 1 Review

What I Learned

1. **VOCABULARY** Define the term *health*. Use it in an original sentence.

2. **LIST** What are the three sides of the health triangle?

3. **COMPARE AND CONTRAST** What is the difference between health and wellness?

4. **DESCRIBE** How can internal factors influence whether a person uses tobacco, alcohol, or other drugs?

Thinking Critically

5. **EVALUATE** Caitlin spends a lot of time with her friends, watching movies and listening to music. She goes to bed late at night and has trouble waking up for school. What part of her total health could be out of balance? What could she do to improve this area of her wellness?

6. **DESCRIBE** What are three traits that you might find in a person who has good social health?

Applying Health Skills

7. **PRACTICING HEALTHFUL BEHAVIORS** Write down three behaviors you could practice to improve and maintain your personal health. Choose one from each side of the health triangle. Keep a journal for one week and write down the steps you took to practice healthful behaviors.

Health Influences and Risk Factors

Before You Read

Quick Write List three influences that might affect your health.

Vocabulary

culture
heredity
chromosomes
genes
cultural background
media
lifestyle factors
attitude
risk
sedentary lifestyle
cumulative risk
prevention
abstinence

BIG IDEA Taking responsibility for your health means practicing healthful behaviors.

What Influences Your Health?

MAIN IDEA You can control some, but not all, of your health influences.

What are your favorite foods or activities? Your answers reflect your personal tastes, or likes and dislikes. Your health is influenced by your personal tastes, but it is also influenced by many other factors. Some of these factors include:

- heredity
- environment
- family and friends
- **culture**, or the collected beliefs, customs, and behaviors of a group
- media
- attitudes
- behavior

Heredity

Biological parents pass traits on to their children. A child can control some of these traits but not all of them. For example, you cannot control the natural color of your hair or eyes. **Heredity** is the passing of traits from parents to their biological children. It controls these and other physical traits, or parts of your appearance. You inherited, or received, half of your DNA from your mother and half from your father.

Each human has 46 **chromosomes**, which are threadlike structures that carry genes. Each parent contributes 23 chromosomes to a child. The chromosomes contain **genes**, or the basic units of heredity. Genetic information, such as hair and eye color, are stored in genes. Genes are made from chemicals called DNA. Each child in a family may inherit different chromosomes from parents. That's why siblings may look and act differently. Family history is an influence on your own health. Knowing your family history can help identify illnesses and diseases for which you might be at risk. In some cases, you can take action to reduce your risk.

Environment

Think about where you live. Do you live in a city, a suburb, a small town, or in a rural area? Where you live is the physical part of your environment. Environment is another factor that affects your personal health. Your *physical environment* includes the home you live in, the school you attend, the air and water around you.

Your *social environment* includes the people in your life. They can be friends, classmates, and neighbors. It is important to know that your friends and peers may influence your choices. Your peers are people close to you in age who are a lot like you. You may feel pressure to think and act like them. Peer pressure can also influence healthful choices. The influence can be positive or negative. Helping a friend with homework, volunteering with a friend, or simply listening to a friend are examples of positive peer influence. A friend who wants you to drink alcohol, for example, is a negative influence. Recreation is also a part of your social environment. Playing games and enjoying physical activities with others can have a positive effect on your health.

IT IS IMPORTANT TO KNOW THAT YOUR FRIENDS AND PEERS MAY INFLUENCE YOUR CHOICES.

Traits such as eye and hair color are inherited from parents. **Name two examples of inherited traits you can see in your family or in the family of a friend.**

Culture

Your family is one of the biggest influences on your life. It shapes your **cultural background**, or the beliefs, customs, and traditions of a specific group of people. Your family and your culture are two related influences on your health. Your family and their culture may influence the foods you eat as well as the activities and special events you celebrate with special foods. Some families fast, or do not eat food, during religious events. Ahmed's family observes the holiday of *Ramadan*. During this holiday, members of his family fast until sundown. Your family might also celebrate traditions that include dances, foods, ceremonies, songs, and games.

Your culture can affect your health. In cultures that encourage eating a healthy diet and exercising, the people are generally healthier. Culture can also influence how people behave toward each other, such as how males interact with females.

Media

What do television, radio, movies, magazines, newspapers, books, billboards, and the Internet have in common? The **media** is various methods of communicating information. Through the media you receive information that has a powerful influence on your health. The behaviors you see in the media might influence how you act. The way people look might influence how you think you should look. The products that people use in the media might influence what you think you should own. Some of those influences might be good for your health. Other influences by the media might harm your health. The purpose of the media is to influence your beliefs. In this way, the media suggests the ideal body type for a female or male.

Healthful information that you receive through the media might include helpful facts about health. Ads on the Internet and commercials on TV might alert you to new tools and services that can help you maintain your health. These might include a new health program offered at your local community center. Other ads might highlight new running shoes or healthful foods that are available in your area.

Unhealthful information is also found in the media. Ads featuring burgers and fries that look delicious might encourage you to adopt an unhealthful eating plan. Health information online may be posted by unreliable sources and may not be accurate.

The goal of advertising is to make you feel that you must have a product. The product may be good or bad for your health. You can make wise health choices by learning to evaluate the quality of everything you see, hear, or read.

Reading Check

Recognize How does the media influence your health?

Your Behavior and Your Health

MAIN IDEA The choices you make can help or harm your health.

Do you protect your skin from the sun? Do you get enough sleep so that you are not tired during the day? Do you eat healthful foods? Do you listen to a friend who needs to talk about a problem? Do you have a trusted adult you can talk to when you have questions or a problem? Your answers to these questions reflect your personal **lifestyle factors**. Lifestyle factors are behaviors and habits that help determine a person's level of health. Positive lifestyle factors promote your good health. Negative lifestyle factors promote poor health.

Your **attitude** is a personal feeling or belief. It plays an important role in your health. You will also have greater success in managing your health if you keep a positive attitude. Teens who have a positive attitude about their health are more likely to practice good health habits and take responsibility for their health. Having a trusted adult to talk to is also an important, and positive, part of practicing good health habits.

POSITIVE LIFESTYLE FACTORS PROMOTE YOUR GOOD HEALTH. NEGATIVE LIFESTYLE FACTORS PROMOTE POOR HEALTH.

The celebration of Kwanzaa is a tradition in many African American families. **Name two holidays celebrated by people of other cultures.**

Risk Behaviors

"Dangerous intersection. Proceed with caution." "Don't walk." "No lifeguard on duty." You have probably seen these signs or similar signs. They are posted to warn you about possible risks or dangers and to keep you safe.

Risk is the chance that something harmful may happen to your health and wellness. We all face risk every day. It's part of everyday life. Some risks are easy to identify. Everyday tasks such as preparing food with a knife or crossing a busy street carry some risk. Other risks are more hidden. You might not know that some foods you like have high-fat ingredients.

You cannot avoid every kind of risk. However, the risks you can avoid often involve risk behavior. Playing a sport can be risky, but if you wear protective gear, you may avoid injury. Wear a helmet when you ride a bike to avoid the risk of a head injury if you fall. Smoking cigarettes is another risk behavior that you can avoid. Riding in a car without a safety belt is a risk behavior you can avoid by buckling up. Another risk behavior is having a **sedentary lifestyle**, or a way of life that involves little physical activity. You can avoid many kinds of risk by taking responsibility for your personal health behaviors.

> YOU CAN AVOID MANY KINDS OF RISK BY TAKING RESPONSIBILITY FOR YOUR PERSONAL HEALTH BEHAVIORS.

Lifestyle factors affect your personal health. **Identify which positive lifestyle factors you practice.**

- Eating well-balanced meals, starting with a good breakfast.
- Getting at least 60 minutes of physical activity daily.
- Sleeping at least eight hours every night.
- Doing your best in school and other activities.
- Avoiding tobacco, alcohol, and other drugs.
- Following safety rules and wearing protective gear.
- Relating well to family, friends, and classmates.

Risks and Consequences

All risk behaviors have consequences. Some consequences are minor or short-term. You might eat a sweet snack just before dinner so that you lose your appetite for a healthy meal. Other risk behaviors have serious and sometimes life-threatening consequences. These are long-term consequences.

Experimenting with alcohol, tobacco, or other drugs have long-term consequences that can seriously damage your health. They can affect all three sides of your health triangle. They can lead to dangerous addictions, which are physical and mental dependencies. These substances can confuse the user's judgment and can increase the risks he or she takes. Using these substances may also lead to problems with family and friends, and problems at school.

A risk that affects your health is more complicated when it is part of a **cumulative risk** (KYOO-myuh-luh-tiv). Cumulative risk occurs when one risk factor adds to another to increase risk. For example, making unhealthy food choices is one risk. Not getting regular physical activity is another risk. Add these two risks together over time, and you raise your risk of developing diseases such as heart disease and cancer.

Many choices you make affect your health. Knowing the consequences of your choices and behaviors can help you take responsibility for your health.

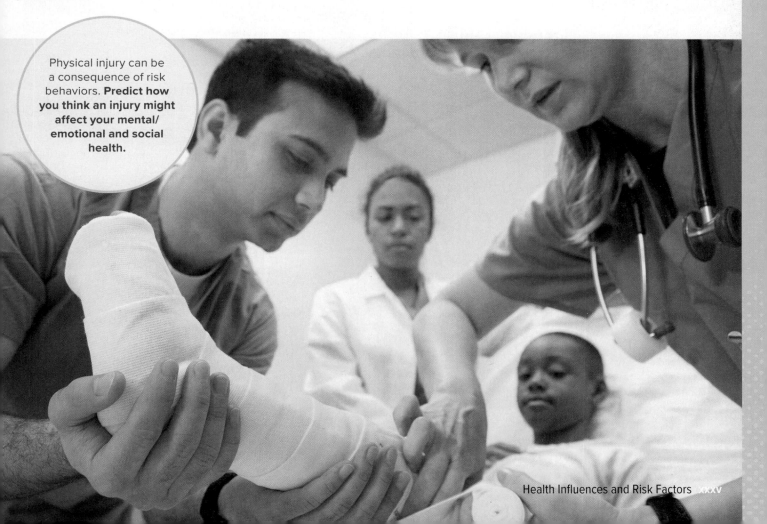

Physical injury can be a consequence of risk behaviors. **Predict how you think an injury might affect your mental/ emotional and social health.**

Reducing Risks

Practicing **prevention**, or taking steps to avoid something, is the best way to deal with risks. For example, wear a helmet when you ride a bike to help prevent head injury. Slow down when walking or running on wet or icy pavement to help prevent a fall. Prevention also means watching out for possible dangers. When you know dangers are ahead, you can take precautions to prevent accidents. Precautions are planned actions taken before an event to increase the chances of a safe outcome.

- **Staying Informed.** You can take responsibility for your health by staying informed. Learn about developments in health to maintain your own health. Getting checked by a doctor is another way to stay informed about your own health.

Reading Check

Explain How can avoiding risk behaviors benefit your physical health?

Getting regular checkups is one form of prevention. **Describe how staying informed promotes good health.**

- **Choosing Abstinence.** If you practice **abstinence** (AB-stuh-nunhs) you make a conscious, active choice not to participate in high-risk behaviors. You show that you care for your own health and others' health by preventing illness and injury. By choosing not to smoke, you may avoid getting lung cancer. By staying away from alcohol, illegal drugs, and sexual activity, you avoid the negative consequences of these risk behaviors.

Abstinence is good for all sides of your health triangle. It promotes your physical health by helping you avoid injury and illness. It protects your mental/emotional health by giving you peace of mind. It also benefits your relationships with family members, peers, and friends. Practicing abstinence shows you are taking responsibility for your personal health behaviors and that you respect yourself and others. You can feel good about making positive health choices, which will strengthen your mental/emotional health as well as your social health.

> PRACTICING PREVENTION, OR TAKING STEPS TO AVOID SOMETHING, IS THE BEST WAY TO DEAL WITH RISKS.

Developing Good Character

Self-Discipline
Developing a healthy lifestyle takes commitment. For example, you may need to remind yourself to turn off the TV or computer and participate in activities that build your good physical health.

Lesson 2 Review

What I Learned

1. **VOCABULARY** Define the term *lifestyle factors*. Use it in an original sentence.

2. **LIST** What are some ways teens can reduce risks related to health problems?

3. **EXPLAIN** How can knowing your family health history help protect your health?

Thinking Critically

4. **DIFFERENTIATE** How can factors, such as heredity, environment, family and friends, culture, media, attitudes, and behavior, influence you to make healthful or unhealthful choices?

5. **APPLY** Watch a TV commercial for a health-related product. What information does the commercial tell you about the product? Do you think the information is truthful? Explain.

Applying Health Skills

6. **ANALYZING INFLUENCES** For a week, identify and write down as many influences on your health choices as you can. Label each influence as being positive or negative. Explain why you chose the labels you did for each influence.

Building Health Skills

BIG IDEA Learning health skills now will benefit you throughout your life.

Skills for a Healthy Life

MAIN IDEA You can learn skills that will help you lead a healthy life.

Health skills help you become and stay healthy. Health skills can help you improve your physical, mental/emotional, and social health. Just as you learn math, reading, sports, and other kinds of skills, you can learn skills for taking care of your health now and for your entire life. Wearing a helmet and pads when skateboarding or playing football to prevent injury is an example of putting health skills into practice. Another example is washing your hands often to prevent illness, such as a cold. Health skills will also help you communicate well and develop healthy relationships.

Self-Management Skills

When you were younger, your parents and other adults decided what was best for your health. Now that you are older, you make many of these **decisions**, or choices, for yourself. You take care of your personal health. You are developing your self-management skills. Two key self-management skills are practicing healthful behaviors and managing stress.

These ten skills affect your physical, mental/emotional, and social health and can benefit you throughout your life. **Describe three ways that health skills can benefit you.**

Health Skills	What It Means to You
Accessing Information	You know how to find valid and reliable health information and health-promoting products and services.
Practicing Healthful Behaviors	You take action to reduce risks and protect yourself against illness and injury.
Stress Management	You find healthy ways to reduce and manage stress in your life.
Analyzing Influences	You recognize the many factors that influence your health, including culture, media, and technology.
Communication Skills	You express your ideas and feelings and listen when others express theirs.
Refusal Skills	You can say no to risky behaviors.
Conflict-Resolution Skills	You can work out problems with others in healthful ways.
Decision Making	You think through problems and find healthy solutions.
Goal Setting	You plan for the future and work to make your plans come true.
Advocacy	You take a stand for the common good and make a difference in your home, school, and community.

When you eat healthy foods and get enough sleep, you are taking actions that promote good health. Stress management is learning to cope with challenges that put a strain on you mentally or emotionally. Strategies for managing stress can help you deal with stress in a healthy way.

Practicing Healthful Behaviors

Healthful behaviors affect your physical, mental/emotional, and social health. You will see benefits quickly when you practice healthful behaviors. These benefits last as you grow and change. If you exercise regularly, your heart and muscles grow stronger. When you eat healthful foods and drink plenty of water, your body works well. Getting a good night's sleep will help you wake up with more energy. Respecting and caring for others will help you develop healthy relationships. Managing your feelings in positive ways will help you avoid actions you may regret later.

Practicing healthful behaviors can help you learn new skills. It can also help you meet challenges and enjoy life. Practicing healthful behaviors can help prevent injury, illness, and other health problems.

When you practice healthful actions, you can help your total health. Your total health means your physical, mental/emotional, and social health. This means you take care of yourself and do not take risks. It means you learn health-promoting habits. When you eat well-balanced meals and healthful snacks and get regular physical checkups you are learning good health habits. Staying positive is another good health habit.

MANAGING YOUR FEELINGS IN POSITIVE WAYS WILL HELP YOU AVOID ACTIONS YOU MAY REGRET LATER.

Studying for a test can cause stress. **Explain why it is valuable to learn stress management skills.**

Managing Stress

Gabrielle was a good softball player. When she played for fun with friends, she enjoyed it. She could hit, catch, and run the bases well. Before league games, however, Gabrielle often felt worried. She did not want to make a mistake. Her worrying gave her a headache. Her stomach got so upset she could hardly eat. These things made her play poorly. Gabrielle was showing signs of stress.

Learning ways to deal with stress is an important self-management skill. **Stress management** is identifying sources of stress and learning how to handle them in ways that promote good mental/emotional health. Relaxation is a good way to deal with stress. Exercise is another way to positively deal with stress.

Making Decisions and Setting Goals

The path to good health begins with making good decisions. These include making the decision to eat nutritious foods, get enough sleep and exercise, and begin studying for a test early. Perhaps you make more of your own decisions now than you did when you were younger. Some of those decisions might be deciding which clothes to buy or which classes to take. As you grow older, you gain more freedom, but with it comes more responsibility. For example, you will be challenged to make difficult decisions. You will need to understand the short-term and long-term consequences of decisions you make.

Another responsibility is goal-setting. A **goal** is something you hope to accomplish. Setting realistic goals can help you learn a new skill or task. Setting healthful goals can also help you maintain your health. An example is setting a goal to get regular physical activity. Decision-making and goal-setting are skills that you will use in many ways throughout your life.

Accessing Information

MAIN IDEA Learning to find reliable health information will protect your overall health.

Knowing how to get **reliable** health information is an important skill. Information that is reliable is trustworthy and dependable. Where can you find all this information? A main source of information is adults you can trust. Parents and guardians, teachers, and your school nurse are good sources. They can help you find books, articles, and web sites on a variety of health topics. Community resources give you other ways to get information. These include the library and government health agencies. Organizations such as the American Red Cross can also give you good information.

Reading Check

Define What are healthful behaviors?

Reliable Sources

You can find more facts about health and health-enhancing products or services through media sources such as television, radio, and the Internet. TV and radio interviews with health professionals can give you information about current scientific studies related to health. The Internet has information from government agencies, universities, and health care providers.

Web sites that end in .gov and .edu are often the most reliable sites. These sites are maintained by government organizations and educational institutions. Getting health information is important, but so is analyzing whether that health information is valid, or reliable. You may also want to carefully review web sites ending in .org. Many of these sites are maintained by organizations, such as the American Cancer Society or American Diabetes Association. However, some sites ending in .org may not be legitimate. They may be maintained by groups that are "sound-alike" organizations and may offer information that is not legitimate. Try to learn something about the person or organization providing the information. Ask questions such as:

- Is the author of a health article an expert on the subject?
- Does the author name scientific studies or other sources for the facts?
- Are there other books or articles that agree or disagree with the author?
- Is the report based on research done by a respected institution?
- Is the report simply one person's opinion, unsupported by evidence or facts?

When you check the reliability of a source, you can also go online to a site named Charity Navigator. Charity Navigator lists organizations that are legitimate. If your organization is not listed, it may be a fraud. You can also ask your health teacher, school nurse, family doctor, or other trusted adult about the source. Remember that any media source that sells products should be considered with caution.

> **GETTING HEALTH INFORMATION IS IMPORTANT, BUT SO IS ANALYZING WHETHER THAT HEALTH INFORMATION IS VALID, OR RELIABLE.**

The Internet can be a good source of health information. **Identify the most reliable web sites to go to when searching for health information.**

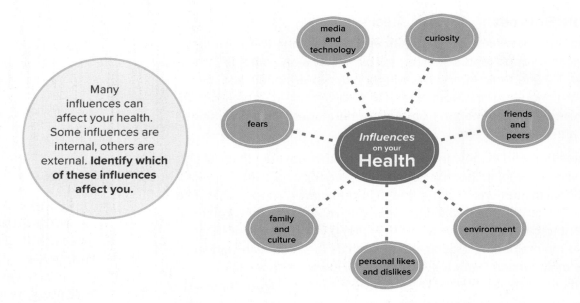

Many influences can affect your health. Some influences are internal, others are external. **Identify which of these influences affect you.**

Reading Check

Explain What questions would you ask to determine whether information is reliable?

Analyzing Influences

Learning how to analyze health information, products, and services will help you act in ways that protect your health. The first step in analyzing an influence is to identify its source. A TV commercial or an infomercial may tell you a certain food has health benefits. In this case, the source is an advertiser who is trying to get you to buy the food. Next, you should think about the motive, or reason, for the influence. Does the advertiser really take your well-being into consideration? Does the ad make you curious about the product? Does it try to scare you into buying the product?

Analyzing influences involves recognizing factors that affect or influence your health. Some of these factors are internal. Others come from outside sources. Your decisions have to do with more than just knowing facts. They also have to do with your own values and beliefs. The opinions of your friends and family members affect your decisions. Your culture and messages from the media also affect your decisions. Knowing what influences you will help you make responsible choices in the future.

Interpersonal Communications Skills

MAIN IDEA Good communication skills will help you maintain healthy relationships.

Three of the ten health skills involve the way you communicate with other people. Your relationships with others depend on good **communication** skills. This is the exchange of information through the use of words or actions. Good communication skills include telling others how you feel. They also include listening to others and understanding how others feel. Two types of communication exist. They are verbal and nonverbal communication.

Verbal communication involves a speaker or writer on one end, and a listener or reader on the other. Good communication involves speaking clearly and carefully. Speaking skills help you express your ideas and feelings in healthful ways. Good communication also involves good listening skills. Listening skills let you understand the messages other people send you. A speaker's message has meaning only if the listener receives it.

Nonverbal communication skills involve more than speaking and listening. Nonverbal features include tone of voice, body position, and using expressions. Often, how you say something is more important than what you say. Body language is also a form of communication. Body language includes facial expressions and gestures. You send messages through the words you choose and how you say them, through your facial expressions, and even through your posture. When you communicate effectively, you can prevent misunderstandings.

The idea of communicating may seem obvious to you. You may be saying to yourself "I already know how to communicate." To understand communication skills, consider the difference between hearing and listening. Imagine that someone is speaking to you while your attention is elsewhere. Maybe you are in the middle of watching a show, or perhaps your mind is just wandering. You may have heard the speaker's words without really listening to what is being said. Active listeners pay close attention to what the speaker is saying and ask questions to make sure they understand.

Refusal Skills

Mia worked in a fast-food restaurant after school. Her friend Erin came in one day and asked Mia to let her have a soft drink without paying. "Come on, no one will know," Erin said. Mia knew she was being asked to do something wrong. She could get into trouble. Even if she didn't get into trouble, her conscience would bother her.

An important communication skill is saying no when others want you to do something unhealthy. It may be something that is wrong. It may be something that you are not comfortable doing. When you are pressured to do something wrong, tension can build. You may worry what will happen if you don't go along with the group. Will your friends still like you? Will you still be a part of the group? It is at these times that **refusal skills** can help. Refusal skills are strategies that help you say no effectively.

Refusal skills are especially useful during your teen years. There may be times when you are asked to do something you do not want to do like Mia in the story above. There may be things that you are unable to do. Maybe you are not interested. Maybe the activity costs money and you do not have enough money to participate in the activity. Maybe it is an activity that goes against your values—something you feel is wrong or unhealthy. Using refusal skills can sometimes be challenging, but they can help you stay true to yourself and to your beliefs. Also, other people will respect you for being honest about your needs and wants.

> OFTEN, HOW YOU SAY SOMETHING IS MORE IMPORTANT THAN WHAT YOU SAY.

Reading Check

Describe What are some ways to say no to risky behaviors?

The S.T.O.P Strategy

The steps for practicing refusal skills are based on the letters in the word S.T.O.P. This makes it easy to remember.

- **Say no.** "No, I can't go with you today."
- **Tell why not.** "I would be breaking a promise."
- **Offer other ideas.** "What about tomorrow?"
- **Promptly leave** if you need to.

When you use refusal skills, show that you mean what you say by using strong body language. Strong body language includes eye contact, crossed arms, and a serious expression. You may need to walk away.

When you stand up for a decision you make, you also need to stand up for the values and beliefs that helped you make the decision. This is especially true when you choose to stay away from situations that could be harmful. Refusal skills are a great tool to use when you need to stay away from behavior that is unhealthy, unsafe, or goes against your values and beliefs.

Conflict Resolution

Conflicts, or disagreements with others, are part of life. Learning to deal with them in a healthy way is important. Imagine that your favorite TV show is about to begin when a family member comes along and changes the channel. At such times, **conflict-resolution skills** can help you find a way to satisfy everyone. Conflict-resolution skills provide the ability to end a disagreement or keep it from becoming a larger conflict. Also, by using this positive health behavior, you can keep conflicts from getting out of hand. Maybe the solution to this conflict is turning off the television and doing something else the family enjoys.

People have different wants, needs, and ways of looking at things. People often disagree over how to spend time, spend money, or share resources. Sometimes this causes a conflict between people. Dealing with conflict takes good conflict-resolution skills. Here are some conflict-resolution tips.

- Take a time-out to let everyone calm down.
- Allow each person to tell his or her side of the story.
- Let each person ask questions of the other person.
- Keep thinking of creative ways to resolve the conflict.

Advocacy

People with **advocacy** skills take action in support of a cause. They work to bring about a change by speaking out for something like health and wellness. When you speak out for health, you encourage other people to live healthy lives. You influence others to make good choices. Advocacy also means keeping others informed. By sharing health information, you allow others to make healthful choices.

Myth vs. Fact

Myth The way you react to stress cannot be changed.

Fact You can change the way you think about stressful situations. For example, you might consider a math test a challenge to be conquered rather than a stressful situation to suffer through.

Using refusal skills effectively can help you avoid a potentially dangerous situation. **Describe a time when you have used refusal skills.**

Ways to Take Action

Some people choose careers as advocates. Their job is to help other people learn about health issues and concerns. They warn people about risks and share information about good health behaviors. Helping others learn about health issues is part of growing into a mature, responsible adult. Advocates use a number of different ways to get the word out about their cause. Here are some ways you can practice getting the word out.

- Write letters to government leaders, blogs, and online and print news editors.
- Collect signatures from people in your community.
- Organize activities in your school or neighborhood.
- Volunteer with a group that shares your feelings. If no group exists, start your own group.
- Contact online news sites, blogs, local radio or television stations to see if they will give your cause attention.

Lesson 3 Review

What I Learned

1. **VOCABULARY** What are *health skills?*

2. **RECALL** What do the letters in the word S.T.O.P. stand for when practicing refusal skills?

3. **DESCRIBE** What are the ten health skills mentioned in this lesson?

Thinking Critically

4. **DESCRIBE** Write a paragraph describing a time when you used refusal skills.

5. **ANALYZE** How do your friends influence your health?

Applying Health Skills

6. **COMMUNICATION SKILLS** Practice having a conversation with a classmate. Think about ways to show you are listening. Why is it important to let the other person know you are listening?

Making Decisions and Setting Goals

Before You Read

Quick Write Describe a decision you made recently. List the steps you took when making that decision.

Vocabulary

decision making
ethical
values
criteria
goal setting
short-term goal
long-term goal

BIG IDEA Mental and emotional disorders can be treated.

Decisions and Your Health

MAIN IDEA All of your decisions have consequences that can positively or negatively affect your health.

Do you have more privileges and responsibilities now than you had last year? As you get older, you will be able to make more decisions on your own. Making responsible decisions will allow you to gain more privileges. For example, your parents or guardians may allow you more freedom to choose after-school activities. The choices and decisions you make affect each part of your health triangle. To make good decisions, think about two things: which choice is healthful and which choice fits my values.

When you make decisions, think about the consequences. Some decisions may help you avoid harmful behaviors. Not using tobacco is a decision that will have a positive long-term effect on your physical health. Deciding to report a bully can help to promote good mental/emotional health. Making the decision to form friendships with others who share your values will promote good social health.

The decision-making process has six steps. It includes analyzing the information that's available. For some decisions, you may also want to get help from your parents, guardians, or other trusted adults.

The Decision-Making Process

MAIN IDEA The decision-making process is an important tool to learn. It can be used in all aspects of your life.

You make decisions every day. You may decide what to eat for breakfast. You may decide what music to buy. You may decide to join a sports team. Some decisions are easy to make. Other decisions are more difficult. Understanding **decision making**, which is the process of making a choice or solving a problem, will help you make the best possible decisions. The decision-making process can be broken down into six steps. You can apply these six steps to any decision you need to make. Using the decision-making process can help keep you from making decisions without thinking them through.

Reading Check

Give Examples Give an example of a health decision that could affect your health now and in the future.

Tran is a member of the school chess club. They are preparing for a competition with another school. Tran has a club meeting at 3:30 tomorrow but his friends want him to go to a movie with them. He decides to use the decision-making process to help him decide what to do.

Step 1: State the Situation

The first step in the decision-making process is to identify the situation as you understand it. When you understand the situation and your choices, you can make a sound decision. Ask yourself these questions to help you identify the situation: What choice do you need to make? What are the facts? Who else is involved?

Tran wrote down answers to these questions. He decided that he must make a choice between going to chess club and going to a movie. The facts are that the chess club depends on him to participate and he needs to practice for the upcoming competition. He also likes to be with his friends. He realizes that his choices involve himself, the members of the chess club, and his friends.

You can learn the skill of making good decisions. **Describe two decisions you have made in the past 24 hours.**

WHEN YOU UNDERSTAND THE SITUATION AND YOUR CHOICES, YOU CAN MAKE A SOUND DECISION.

Step 2: List the Options

When you feel like you understand your situation, you need to think of your options. List all of the possibilities you can think of. Be sure to include only those options that are safe. It is never an option to risk your health or the health of someone else. It is also important to ask an adult you trust for advice when making an important decision.

Tran listed his options: Go to chess club. Go with friends to see a movie. Then he explained the situation to his mom and asked her for advice. His mom thought there might be a third option. She suggested that he go to chess club for part of the meeting and leave early to join his friends at the movie theatre.

Step 3: Weigh the Possible Outcomes

After listing your options, you need to evaluate the consequences of each option. The word H.E.L.P. can help you work through this step of the decision-making process.

- **H (Healthful)** What health risks will this option present to me? What health risks will this option present to others?
- **E (Ethical)** Does this choice reflect what you and your family believe to be **ethical**, or right? An ethical decision is choosing to take the right action. Does this choice show respect for you and others?
- **L (Legal)** Will I be breaking the law? Is this legal for someone my age?
- **P (Parent Approval)** Would your parents approve of this choice?

Tran used the questions in the H.E.L.P. formula to think through the consequences of each choice. He decided that none of the choices presented health risks to himself or others. He also felt that all of the choices were ethical. He was concerned, though, that the chess club members might feel like he did not respect them if he skipped club to go see a movie. Tran determined that all of the choices were legal. Then he considered how his parents might react. He thought they would probably not be too upset with any of the choices.

Step 4: Consider Your Values

When you have a decision to make, you should always consider your **values**, which are the beliefs that guide the way a person lives. Your values reflect what is important to you and what you have learned is right and wrong. Honesty, respect, consideration, and good health are values. What do you value? Values provide **criteria**, or standards on which to base decisions, that help you evaluate the possible outcomes of your decision. For example, if you value good health, you will exercise regularly and eat nutritious foods. If you value honesty, you will not make decisions that involve lying, cheating, or stealing.

Tran considered the values he was taught growing up. He realized that keeping commitments and friendship were both very important to him. He wanted to keep his commitment to the members of the chess club, but he also wanted to be with his good friends.

AN ETHICAL DECISION IS CHOOSING TO TAKE THE RIGHT ACTION.

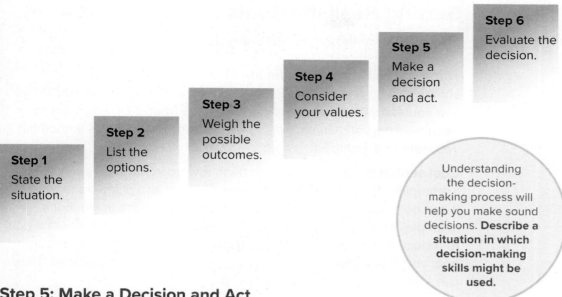

Step 1
State the situation.

Step 2
List the options.

Step 3
Weigh the possible outcomes.

Step 4
Consider your values.

Step 5
Make a decision and act.

Step 6
Evaluate the decision.

Understanding the decision-making process will help you make sound decisions. **Describe a situation in which decision-making skills might be used.**

Step 5: Make a Decision and Act

You've weighed your options. You've considered the risks and consequences. Now you're ready for action. Choose the option that seems best for you. Make sure you are comfortable with your decision and how it may affect others. Make sure the option supports your values. If you are not sure about your decision, ask a trusted adult for advice. Remember that this step is not complete until you take action.

Tran realized that he really wanted to participate in both activities. He wanted to go to chess club and he wanted to go to a movie with his friends. So he decided to take his mom's suggestion. He went to chess club and told the advisor that he would be leaving early. After practicing his chess game, he joined his friends at the movie theatre. Tran realized that some decisions are harder to make than others. He knew that it didn't take too much time to make this decision. He also knew that more difficult decisions can take much more time.

Step 6: Evaluate the Decision

After you've taken action, you need to evaluate the results. Evaluating the results can help you make better decisions in the future. To evaluate the results, ask yourself some questions. Was the outcome positive or negative? Were there any unexpected outcomes? Was there anything you could have done differently? How did your decision affect others? Do you think you made the right decision? What have you learned from the experience? If the outcome was not what you expected, try again. Use the decision-making process to find another way to deal with the situation.

Tran took some time to evaluate the decision he had made. He asked himself the questions listed above. Here is what he determined: He felt like the outcome was positive. There were no unexpected outcomes. He would not have done anything differently. His decision did not affect others negatively. He felt like he made the right decision. He learned that it is worth the time to use the decision-making process to make the best decisions possible.

Reading Check

List What are the six steps of the decision-making process?

Setting Realistic Goals

MAIN IDEA Goals can be short-term or long-term.

When you think about your future, what do you see? Do you see someone who has graduated from college and has a good job? Do you see yourself with a spouse or partner in the future? Do you want to become a parent? These are some of your goals. A goal is something you want to accomplish.

Goal setting is the process of working toward something you want to accomplish. It's a powerful tool. Setting a goal and planning to achieve it can help you focus on what you want to accomplish in life. For example, if you want to improve your eating habits you might set a goal to reduce the amount of high-fat foods you eat each day. Goals can help you make good decisions and give you a sense of purpose. They can also help you measure your progress.

Types of Goals

There are two basic type of goals—**short-term goals** and **long-term goals.** Short-term goals are those that you can achieve in a short length of time. Getting your homework turned in on time might be a short-term goal. Long-term goals are those that you plan to reach over an extended period of time. Getting a college education might be a long-term goal.

Often long-term goals are made up of short-term goals. For example, if your long-term goal is to finish your class project by the end of the semester, you might have several short-term goals to achieve. You might set a time when you will select a topic for your project. Then you might determine how you will deliver the project—electronically or otherwise.

Reading Check

Analyze What are the two types of goals? How are they related?

Jamie has set a goal to be chosen for the all-star team. **Describe a goal you have set recently.**

Then you can focus on the research. Finally, you might set a goal of having the project finished a few days before it is due. All of these short-term goals lead to the long-term goals of completing your class project.

Reaching Your Goals

To accomplish your short-term and long-term goals, you need a plan. A plan will help you identify the criteria for your goal. The S.M.A.R.T. goal criteria can help you develop an action plan. A goal that is S.M.A.R.T. is one that is:

- **Specific.** Identify your reasons for wanting to achieve this goal. Ask yourself: What is my goal? Why do I want to achieve this goal? Who can help me reach the goal? How long will it take to reach the goal? When should I begin?
- **Measurable.** Determine how long it will take you to achieve the goal. If it's a long-term goal, should you set short-term goals to identify your progress?
- **Attainable.** Be honest with yourself. Are you capable of attaining the goal? If you're currently about five feet tall, is it realistic to set a goal to grow to six feet in height?
- **Relevant.** Does the goal help you achieve something that is relevant to you? A person who sets a goal to drink more water may prefer soda, but drinking more water will make you healthier. If you choose to be healthier, drinking more water is a relevant goal.
- **Timely.** What is your timeframe for achieving the goal? What is your action plan to meet that deadline? What are the steps in your plan that will help you succeed?

Finally, it's important to acknowledge the completion of a goal. When you achieve a goal, reward yourself. Try to choose a reward that is healthful.

Lesson 4 Review

What I Learned

1. **RECALL** What are the benefits of using the decision-making process?

2. **VOCABULARY** Define *short-term* and *long-term* goals. Then give an example of each.

3. **EXPLAIN** How are short-term and long-term goals related?

Thinking Critically

4. **APPLY** Imagine that your long-term goal is to become a teacher. List three short-term goals you could set to help you achieve your long-term goal.

5. **PREDICT** How might priorities, changing abilities, and responsibilities influence setting health goals?

Applying Health Skills

6. **GOAL SETTING** Choose a personal health goal. Discuss strategies and skills needed to attain a personal health goal.

Choosing Health Services

Before You Read

Quick Write Imagine that you are choosing a health care specialist for an injury or disease. Make a list of questions you might ask him or her about your health.

Vocabulary

health care
health care system
primary care provider
allied health professionals
preventive care
specialist
hospice care
health insurance
managed care
health maintenance organization (HMO)
preferred provider organization (PPO)
point-of-service (POS) plans
Affordable Care Act (ACA)
public health
National Institutes of Health (NIH)
Food and Drug Administration (FDA)
Centers for Disease Control and Prevention (CDC)
famine

BIG IDEA Health care providers can help you live a healthy life.

What is Health Care?

MAIN IDEA The health care system includes many people whose job is to help keep you well.

The primary person who is responsible for taking care of your health is you. You can take care of your health in many ways. Some of those ways include:

- eating a healthful, well-balanced diet.
- getting eight hours of sleep each night.
- avoiding using alcohol, tobacco, and drugs.
- exercising regularly.
- bathing or shower and wash your hair daily.
- brushing your teeth twice a day.
- taking safety precautions when playing sports or riding in a car.

Living a healthful lifestyle can help you stay healthy. However, everyone needs **health care** at some point. Health care is any service provided to individuals or communities that promote, maintain, or restore health. The **health care system** includes all the medical care available to a nation's people, the way they receive the care, and the way the care is paid for.

The health care system has two goals. They are to prevent and cure disease and injury. An example of disease prevention is getting a checkup each year. Another example is getting a flu shot. Ways to prevent injuries include getting a checkup before joining a sports team and using safety gear while playing sports.

Preventive care is not typically provided in an emergency room. **Identify the purpose of preventive care.**

Health Care Providers

MAIN IDEA Health care providers includes many professions, including doctors, nurses, nutritionists, physical therapists, and more.

Many different professionals can help you with your health care. You may be most familiar with your own doctor who is your **primary care provider**. A primary care provider is someone who provides checkups and general care. Your doctor is probably a pediatrician. A pediatrician specializes in health care for people under 21.

Other providers are **allied health professionals**. They are medical professionals who perform duties which would otherwise have to be performed by doctors or nurses. These professionals provide care under the supervision of a doctor, nurse practitioner, or physician's assistant. They also answer questions and to give regular health checkups. These providers are nurses, pharmacists, health educators, counselors, mental health specialists, dentists, and nutritionists.

Preventive Care

Preventive care is aimed at revealing health problems while they are minor. It includes the steps taken to keep disease or injury from happening or getting worse. Getting regular checkups is one way to prevent health problems and maintain wellness. Regular checkups are recommended once each year for people of all ages. During a checkup, a health care provider will check a patient's height, weight, heart, lungs, and give any immunizations. Immunizations help prevent certain diseases, such as measles and chicken pox. For older adults, immunizations can help prevent pneumonia and shingles. Most immunizations are given according to a schedule based on age.

Preventive care also helps teach of ways to prevent injury and illness. They may learn about healthy lifestyle behaviors. These include healthful eating, being physically active, and avoiding risk behaviors.

Specialist	Specialty
Allergist	Asthma, hay fever, other allergies
Cardiologist	Heart problems
Dermatologist	Skin conditions and diseases
Oncologist	Cancer
Ophthalmologist	Eye diseases
Orthodontist	Tooth and jaw irregularities
Orthopedist	Broken bones and similar problems
Otolaryngologist	Ear, nose, and throat
Pediatrician	Infants, children, and teens

Different specialists treat different conditions. **Identify which specialists treats you and your peers.**

Specialists

Sometimes your primary care provider may send you to see a **specialist**. A specialist is a health care professional trained to treat a special category of patients or specific health problems. The chart lists some types of specialists and the people or conditions they treat. As part of preventative care, females should see a gynecologist, a doctor that specializes in female reproductive health. Usually females begin seeing a gynecologist once a year after they are 18 or when they become sexually active.

Health Care Settings

MAIN IDEA A health care setting includes many tools and equipment used to keep you well.

In the past, a clinic or hospital were the main sources of health care. Today, new types of health care delivery settings are open. People now can go to their doctors' offices, hospitals, surgery centers, assisted living communities, and hospices.

Doctor's Offices

Doctors' offices are probably the most common setting for receiving health care. A doctor, nurse practitioner, or physician's assistant do checkups. They may have some equipment in the office to help diagnose health problems. Many medical needs can be met in a doctor's office.

Hospitals

Hospitals offer care for more serious health issues. Surgery and emergency medical care are provided at a hospital. Hospitals provide patient care 24 hours a day, seven days a week.

Reading Check

Explain When might you go to a health care specialist?

Many medical needs can be met in a doctor's office. **Name other types of health care settings.**

Hospice care offers comfort to terminally ill patients. **Identify the expertise that hospice workers must have.**

Surgery Centers

For some minor surgeries, a doctor may send a patient to a surgery center instead of a hospital. Surgery centers offer outpatient surgical care. The patients go home after their surgery. Complex surgeries cannot be done in a surgery center.

Clinics

Clinics are similar to doctors' offices and often have a group of primary care physicians and specialists on staff. At a clinic, the patient may see any doctor who is available. This might make it more difficult for the doctor to get to know you and your health issues. However, for people who do not need to go to the doctor often, a clinic might be a good fit.

Assisted Living Communities

As people get older, they may not be able to take care of themselves as well as they used to. Assisted living communities offer older people an alternative to nursing homes. In nursing homes, medical staff take care of all of a resident's needs. In assisted living communities, the residents can choose which services they need. They may be unable to drive and need transportation. They may need reminders to take medications. They may need to have food prepared for them. In an assisted living community, the residents are able to live in their own apartments as long as they are able. Some nurses and other medical staff are available when the residents need help.

Hospice Care

Hospice care is care provided to the terminally ill that focuses on comfort not cure. Terminally ill patients have serious illnesses. They will not recover. Hospice workers are specially trained and are experts in pain management. They are also trained and skilled at giving emotional support to the patient and the family. Some terminally ill patients receive hospice care in their own homes. Others go to an inpatient hospice facility.

Paying for Health Care

MAIN IDEA Health insurance can help pay for various medical costs.

No one knows when they will need health care. If an accident occurs, health care costs can be expensive. A long illness can also be expensive to treat. To pay for these costs, people buy **health insurance**, or a plan which a person pays a set fee to an insurance company in return for the company's agreement to pay some or all medical expenses when needed.

In exchange for a monthly fee, an insurance company agrees to pay some of a person's health care costs. Insurance companies can pay for health care costs because they group people into coverage pools. A coverage pool usually includes people who are well and people who are sick. The fees paid by people who are well helps to pay for the health care costs of people who are sick. Some health insurance plans will also pay for part of the cost of prescription medicines.

> A COVERAGE POOL USUALLY INCLUDES PEOPLE WHO ARE WELL AND PEOPLE WHO ARE SICK.

> Health insurance plans can be complex. **Describe the goal of health insurance.**

Many people obtain health insurance for their families through an employer. About half of all employers in the U.S. offer a health insurance plan for their employees to purchase. An employer may pay for part of the cost of the insurance plan.

An employer may offer a few different health insurance plans. The employee is responsible for choosing the best plan for them. Employees need to study each plan and choose the plan that works best for them and their family.

Types of Health Care Plans

Health insurance companies offer types of plans. One is a **managed care** plan which is a health insurance plan that saves money by encouraging patients to select less costly forms of care. The person who purchases this plan agrees to see doctors who have agreed to participate in the plan. Three main types of managed care plans are:

- **health maintenance organization (HMO)** An HMO plan is a contract with selected physicians and specialists to provide medical services. HMO members pay a fixed monthly rate. They must use selected physicians and specialists who are part of the plan.
- **preferred provider organization (PPO)** is a health insurance plan that allows its members to select a physician who participates in the plan or visit a physician of their choice. Members may also visit a physician who is not on the plan, but they will usually have to pay more.
- **point-of-service (POS) plans** combine the features of HMOs and PPOs. Members see participating physicians at a reduced cost or see physicians who are not part of the plan for a higher cost.

Government Public Health Care Plans

About 50 percent of companies in the U.S. offer health insurance to their employees. People without employer-based health insurance must buy private health insurance. Some people cannot afford this insurance. For them, the government provides low-cost or free care. The care is provided through county hospitals or free clinics and is paid for with taxpayer money. In 2017, the U.S. spent $3.5 trillion on public health care.

The government also provides three health care plans. *Medicaid* is for people who are unemployed. *Medicare* is for people over the age of 65 or who are disabled. A third plan is the ***Affordable Care Act (ACA).*** The ACA is a law that was enacted in 2010 to ensure that all U.S. citizens have access to affordable health care. It offers health insurance plans to people who work in low-income jobs, or those who are self-employed.

People who earn a lower income can receive a subsidy to pay some of the cost of health insurance. The subsidies help to make the health insurance affordable. The ACA is intended to reduce the cost of health care for uninsured people.

· · · · · · · · · ·

Reading Check

Define What is health insurance?

· · · · · · · · · ·

What is Public Health?

MAIN IDEA Government and nongovernment agencies play an important role in public health.

The **public health** system in the U.S. is designed to monitor and promote the welfare of the population. It includes the hospitals and clinics offering low-cost or free care in counties. It also includes agencies that track disease and provides funding for research to treat or cure disease.

Federal Health Agencies

The healthcare system also includes federal health agencies. The Department of Health and Human Services (HHS) is one of those federal health agencies. Some agencies of the HHS are:

- **National Institutes of Health (NIH)** is an agency of the Federal government that funds medical research to enhance health, lengthen life, and reduce illness and disability. It supports more than 38,000 research projects nationwide.
- **Food and Drug Administration (FDA)** is a department of the Federal government that regulates foods, drugs, and other substances. It assures the safety of foods, cosmetics, medicines, and medical devices.
- **Centers for Disease Control and Prevention (CDC)** is an agency of the Federal government that protects the health, safety, and security of Americans, monitors health and disease outbreaks, and provides health information to the public.
- Indian Health Service (IHS) provides health services to American Indians and Alaska natives of federally recognized tribes.
- Health Resources and Services Administration (HRSA) provides access to health care for people who earn a lower income, are uninsured, or live in areas where health care is scarce.
- Substance Abuse and Mental Health Administration (SAMHSA) works to improve the quality and availability of substance abuse prevention, addiction treatment, and mental health services.
- Agency for Healthcare Research and Quality (AHRQ) supports research on health care systems, health care quality and cost issues, access to health care, and effectiveness of medical treatments.
- The Consumer Product Safety Commission (CPSC) works to reduce risks from unsafe products. The CPSC sends out recall information if a product is found to be unsafe.

THE HEALTHCARE SYSTEM ALSO INCLUDES FEDERAL HEALTH AGENCIES.

Myth vs. Fact

Myth The biggest problem facing the health-care system in the United States is a lack of new technology.

Fact The biggest problem facing the health-care system in the United States is the high cost of health care. Many people, including families with children, cannot afford basic health care or medical insurance.

Voluntary Health Organizations

Another group of health organizations are not linked to the government. They collect donations from the public. They also encourage members of the public to volunteer at events to raise funds. This money pays for health education and research programs. The American Heart Association, American Cancer Society, and American Diabetes Association are examples. These organizations provide important health information about research and new developments to treat these diseases. Other examples of organizations that help people with health issues include the National Domestic Violence Hotline (https://www.thehotline.org/), which helps people who are being abused by a partner, and Alcoholics Anonymous (https://www.aa.org/), which helps people who want to stop using alcohol. The LGBT National Help Center (https://www.glbthotline.org/) provides help for people with questions about sexual orientation and/or gender identity.

International Public Health

Other groups operate around the world. One is the World Health Organization (WHO). WHO is part of the United Nations. It operates in 200 countries and territories.

The WHO provides information and aid to struggling communities and nations. It offers vaccinations to prevent disease. It also provides **famine** relief. A famine is the widespread shortage of food. It can lead to the starvation and death of many people in a country. WHO relies on donations from the public to continue its work.

Reading Check

Explain How are organizations like the American Heart Association funded differently than government organization?

Lesson 5 Review

What I Learned

1. **EXPLAIN** Why is it important to have regular physical checkups?

2. **NAME** What are three types of health care settings?

3. **DESCRIBE** What are two ways people pay for health care?

Thinking Critically

4. **APPLY** How might seeing a health care provider for regular checkups keep a person's medical costs down?

5. **SYNTHESIZE** How might an assisted living community affect the health of a senior citizen?

6. **ANALYZE** What are some global influences on personal and community health?

Applying Health Skills

7. **PRACTICING HEALTHFUL BEHAVIORS** The company Matthew's dad works for does not offer health insurance. Matthew's family cannot afford to buy health insurance. They know that there are things they can do to reduce the number of times they need to go to the doctor. What tips would you give Matthew and his family for staying healthy? Make a list of at least five items.

Social Health

LESSONS

1

Building Character

. . . .

Before You Read

Quick Write Write a short description of a person who has been a positive role model for you.

Vocabulary

character
character trait
citizenship
integrity
tolerance
prejudice
hate crime
accountability
empathy
role model
constructive criticism
clique

. . . .

Reading Check

Identify Name two characteristics of trustworthiness.

BIG IDEA Character is the way a person thinks, feels, and acts.

Understanding Character

MAIN IDEA A person's character demonstrates his or her values and beliefs.

Do you tell the truth? Do you do what you say you will do? Do you show your family and friends that you care about them? These are some signs of good **character**, or the way a person thinks, feels, and acts. A person's character is demonstrated through core values such as trust, respect, and responsibility. Good character is an important part of a healthy identity.

Your character also affects your physical, mental/emotional, and social health. Physical activity, healthy eating, and safety habits show respect for your body and your physical health. Being honest and reliable builds strong friendships and family relationships. Showing kindness to and caring for others helps you get along with the people in your life.

Six Traits of Good Character

Groups function best when each member demonstrates good character. You are a member of many groups, like a family, a sports team, or friends. In order for everyone to get along, they need to have good **character traits**, or qualities that demonstrate how a person thinks, feels, and acts. The six basic traits of good character include: trustworthiness, respect, responsibility, fairness, caring, and **citizenship**. Citizenship is the way you conduct yourself as a member of a community.

Trustworthiness

Gabriel and Tuan are taking a quiz. Gabriel studied for the quiz and feels prepared. Tuan tries to look at Gabriel's quiz, but Gabriel covers his answers. Gabriel shows trustworthiness and integrity by not allowing Tuan to look at his answers.

If you are honest and truthful, other people will trust you. They will feel that you are reliable. Think about groups projects you have worked on. You were assigned to work on a project with other students. Each day you come to class on time with your part of the work complete. Your team members learn to trust you.

Trustworthy people also have integrity. **Integrity** is being true to your ethical values. Imagine you see a student leave her wallet on the lunch table at school. A person with integrity would return it immediately.

Being loyal also makes you trustworthy. A loyal friend will not say, or allow others to say, untrue or unkind things about you. You can be loyal to your school and your community by always acting with integrity and citizenship.

Respect

Lorinda asks Dipali to come a party at her house. When Dipali arrives, many people are drinking alcohol. Dipali doesn't want to feel pressured to drink, so she decides to leave. Lorinda begins teasing Dipali because she chooses not to drink alcohol. Soon other people at the party also tease Dipali. Dipali leaves the party and goes home. She shows respect for herself by choosing not to do something that might risk her health.

Showing respect begins with self-respect. This person avoids high-risk behaviors. This means avoiding sexual activity and the use of tobacco, alcohol, and other drugs.

A person who shows respect also considers the feelings of others. So, Lorinda is not showing respect for Dipali. A person who shows respect for others thinks about how they want to be treated and treat others the same way. You show respect with good manners. Another way to show respect is to listen to another person's point of view, even when it is different from yours.

Showing **tolerance**, which is the ability to accept other people as they are, is another form of respect. Some differences may be gender, race, culture, abilities, sexual orientation, or body size. In your school, some students are better than others at subjects like Math or Science. Some students may have been born in another state or even another country.

> **GOOD CHARACTER IS AN IMPORTANT PART OF A HEALTHY IDENTITY.**

> Community members show good citizenship by helping others. **Name two other ways to show good citizenship.**

Respecting Others

No two people are alike. Some people are interested to learn about the differences of others. Other people are less comfortable. Those who are less comfortable may show **prejudice**. They may dislike another person because of that person's gender, skin color, religion, cultural background, a disability, sexual orientation, or body size.

Prejudice is used to make a person feel ashamed or uncomfortable. Prejudice may also hint that physical violence might be possible. Violence that happens as a result of prejudice can be called a **hate crime**. It's a crime committed against another person or group based on racial, religious, or sexual background.

Discrimination can take many forms. One is to show prejudice against people of another race, religion, or culture. Another form is gender-based. Females are usually the target of this type of discrimination. Females may be told that they aren't good at certain subjects, such as Math or Science. Males may tell jokes about females that are intended to cause hurt feelings. These are examples of gender discrimination and harassment.

Prejudice creates a barrier to developing healthy relationships with others. Each person is unique. A person who is curious and tolerant of others gains the opportunity to learn about cultures other than their own. Also, discriminating against others is against the law. If a person commits a violent act because of prejudice, that person could go to jail for committing a hate crime.

> **PREJUDICE CREATES A BARRIER TO DEVELOPING HEALTHY RELATIONSHIPS WITH OTHERS.**

Responsibility

Yuri and his brother have a new puppy. They take turns walking the dog. One day, Yuri's friend, Jacob, asks him to go see a movie after school. Yuri wants to go, but it is his turn to walk the dog. Yuri calls Jacob and asks if they can go to a movie on another day. He is accepting responsibility for taking care of his pet.

What are some of your responsibilities? Do you complete your school work on time? Do you make your bed or put your laundry away? As you get older, you take on more responsibilities. This means that you will accept **accountability**, which is a willingness to answer for your actions and decisions. When you take responsibility for your actions, you do not blame others for your mistakes.

To accept responsibility means to be willing to take on duties and tasks. Showing responsibility means doing what you say you will do. You take credit for things done well and not done so well. When you have a responsibility, you follow through without being asked or reminded.

Taking Responsibility for Your Health

By making good decisions you take responsibility for your health. One way is to make healthful food choices. This allows you to maintain or improve your physical health. Another way is to avoid behaviors that can harm your health. This is practicing abstinence. Some behaviors to abstain from are sexual activity or using alcohol, tobacco, and other drugs.

Taking responsibility can be demonstrated in many ways. **Name two ways that you can demonstrate responsibility at home.**

Fairness

Abby and her friend Taylor are judging a school art contest. Abby wants to award the prize to Paul because his painting is very good. Taylor wants to award the prize to her friend Katy. "I know Katy is your friend, but Paul really deserves to win," says Abby. "It's only fair to give it to the person who deserves to win." Abby understands the importance of fairness.

Fairness demonstrates good character. It means treating people equally and honestly. Many things that you do, such as taking turns and sharing, demonstrate fairness. Fairness includes being a good sport, win or lose.

Caring A caring person treats others with kindness and understanding. Some characteristics of a caring person are:

- Considers the feelings of others.
- Shows gratitude to others who offer help.
- Offers forgiveness to those who have hurt them.
- Gives time, attention, and help to others.

A caring person is a kind person. A caring friend will show sympathy if someone is feeling sad. They also show **empathy**, which means that you identify with and share another person's feelings.

Citizenship Every one of us is the citizen of a community. A community is your neighborhood, your school, your city, or your country. The way you act as a member of a community is citizenship. Citizenship is a part of good character. Teens who obey the rules and follow the laws are good citizens.

One aspect of citizenship is doing what you can to help your communities. One way is to take a stand to prevent violence and bullying. Another way is to advocate or take a stand to help the community. Becoming a volunteer for a beach or forest clean-up is one way to be a good citizen. Caring and citizenship work together. You are a good citizen because you care about other people, the community, and the environment.

Reading Check

Name What are two traits of fairness?

Reading Check

Define What is empathy?

Reading Check

Identify What are the six traits of good character?

What Shapes Your Character?

MAIN IDEA Your life experiences and role models shape your character.

Many influences shape your character. From an early age, you learn values from your family members. As you mature, you make your own choices about what kind of character you will have. Do you value honesty in others? You may choose to always be honest. Do you see fairness around you? You may choose to be fair. You may learn that you appreciate kindness in others and choose to be kind. You choose and practice to have good character.

Life Experiences

Character is shaped by your family and the people around you. If you learned to share and respect your brothers or sisters, you learned fairness and respect. As you grow older and begin school, your teachers, other adults, and friends can shape your character. When you find someone you trust, you learn about being trustworthy. At school, you learn to be responsible for your schoolwork to follow the rules, and to care for others. As you learn about your community and environment, you learn about citizenship.

DO YOU VALUE HONESTY IN OTHERS?

Teens give their time to help the community. **Explain how you think it might feel to provide food to hungry people.**

Role Models

One way to learn how to show good character is to watch and listen to others. This behavior is called modeling. You learn to act a certain way by modeling the behavior of others. The people whose behavior you copy is a **role model**. A role model is a person who inspires you to think or act in a certain way. A role model may be someone your age, older, or even someone younger.

A teacher may be a good role model. A teacher is honest and fair. A coach should model trustworthiness and respect for others. A family member may be a model of caring and responsibility.

Role models help us learn new skills. They show us how to act under certain conditions. They also help us build positive relationships with people who are older or younger than us.

Reading Check

Explain Who are the first teachers of character?

Parents or Guardians	Stories	Life Experiences	Examples Set by Others
The earliest influence on your character was likely a parent or guardian. Parents and guardians are our first teachers of character.	Have you heard the expression, that's the moral of the story? Many stories contain a moral. The moral teaches about values or character traits.	You learn from things you do in your life. You may have done something that turned out to be a mistake. Did you learn from that mistake?	Role models inspire us to act or think in a certain way. They set good examples. Who do you look up to for inspiration?

Many factors can influence a person's character. **Describe one influence on your character.**

Developing Good Character

MAIN IDEA Good character means making good choices.

As you watch, listen to, and learn from others, you choose the traits you want to demonstrate your character. For example, if you do your family chores, you demonstrate responsibility. If you listen to different points of view, you demonstrate respect. When you help a person in need, you demonstrate caring and kindness.

Demonstrating good character is a choice. If you choose to be honest, you will tell the truth and you will not cheat. If you obey the rules and respect authority, you are choosing citizenship. You practice making choices to demonstrate good character. Good character is something anyone can choose. The more you practice the traits you value, the more they become a part of your character.

Reading Check

Analyze What are two ways to develop good character?

Character in Action

MAIN IDEA Good character helps you develop and maintain healthy relationships.

Good character is shown in actions and words. Good character improves a person's life and affects all of the other people in that person's life. When your actions and words show good character, you have many good and strong relationships. How many people do you talk to each day? When your communications with people show good character, you will have positive relationships.

Working together can build good character. **Explain what traits of good character can show when working together.**

Making a Difference at Home

Your first relationships are with your family members. Your actions and words show your character when you treat your family members with caring and respect. Another way is to show thanks for the things your family does for you. One way to show thanks is to help out with chores without being asked. Another way is to be patient and kind to your brothers or sisters.

For example, it's easy to get frustrated when trying to teach a younger sibling a new skill. One way to help a sibling may be to offer **constructive criticism**, which is using a positive message to make a suggestion. If a younger sibling is having trouble learning to tie shoes, tell the sibling that everyone has trouble learning new skills. Offer encouragement and think about ways you can make the steps easier to remember.

Finally, teens who are experiencing problems at home can still maintain good character. If the problem is with a parent, talk to another trusted adult. Ask for advice on how to solve the problem. If the problem is between you and sibling, try talking to a sibling to resolve the problem. If that doesn't work, ask a parent or guardian to help.

Reading Check

Name What are two ways to build good character at home?

GOOD CHARACTER IS SOMETHING ANYONE CAN CHOOSE.

These teens help out with a task at home without being asked. **How do these teens make a difference?**

. . . .

Reading Check

Describe How can cliques be harmful? How can they be helpful?

. . . .

Making a Difference at School

Displaying good character can make a difference at school. At school, you have relationships with friends, classmates, teachers, and other adults. You can build your character by accepting others and showing tolerance toward differences.

Another way is to be accepting of others who may be different. Most people feel most comfortable around others who are most like them. This can lead to the formation of a **clique**, or a group of friends who hang out together and act in similar ways. Being a part of a clique can provide a person with a sense of belonging. However, cliques can be negative if they leave out others or show prejudice toward those who are different.

To show good character at school, finish assignments on time, obey school rules, and treat all teachers and students with respect. Good character is also shown by taking a stand against violence, bullying, and conflict.

Good character can be demonstrated at home, at school, and in the community. **Describe one way you can show good character at school.**

YOU CAN BUILD YOUR CHARACTER BY ACCEPTING OTHERS AND SHOWING TOLERANCE TOWARDS DIFFERENCES.

Making a Difference in Your Community

Good character is also shown by being a good citizen. One way to make a difference in the community is to protect the environment. Throw away trash properly. Recycle bottles, cans, and papers. Find the time to volunteer your time or skills to help others.

Most communities have volunteer programs. Some volunteers make and serve food at shelters, collect clothes for donation, and work teach younger children new skills. When you volunteer your time and skills to help others, you make a difference in your community.

Being a volunteer is an act of citizenship. It shows that you advocate fairness, equality, caring, and giving. You are helping to improve the lives of others who live in the same community. In this way, you are advocating for a better community and better world.

> **BY VOLUNTEERING YOU ARE HELPING IMPROVE THE LIVES OF OTHERS WHO LIVE IN THE SAME COMMUNITY.**

Lesson 1 Review

What I Learned

1. **DEFINE** Define *character* and use it in a sentence.

2. **EXPLAIN** How does good character contribute to physical, mental/emotional, and social health?

3. **IDENTIFY** Where are three places that you can demonstrate good character?

Thinking Critically

4. **HYPOTHESIZE** Think of an act of citizenship you know about. It can be an act of someone you know, or someone you have read about or seen on TV. Tell how the act demonstrates citizenship. Does the example include any other traits of good character? If so, tell how.

5. **APPLY** Gary's friend, Jon, has cystic fibrosis, a chronic disease that affects lung function. As a result of the cystic fibrosis, Jon has a lung infection and will miss the school play. Jon is feeling frustrated and discouraged. How can Gary show empathy for Jon?

Applying Health Skills

6. **COMMUNICATION SKILLS** Work with a small group to make a poster that outlines rules for respectful behavior in your school community. Use what you learned in this lesson. When your poster is complete, share it with your class.

7. **ADVOCACY** Imagine that a clique of students is bullying another student at lunch and in the school halls. What steps can you take to advocate against bullying? What traits of good character does it take to advocate against bullying?

Practicing Communication Skills

BIG IDEA Healthy relationships depend on good communication.

What is Communication?

MAIN IDEA Communication involves your words, postures, gestures, and facial expressions.

You communicate with different people every day. Communication is the exchange of information through the use of words or actions. It involves three parts. These are the person who sends the message, the person who receives the message, and the message. You communicate on the phone, in writing, and with your actions.

You use the phone to communicate with people you may not see every day. You communicate in writing and online with your teachers. E-mails and text messages let you share quick information, such as dates or directions. When you communicate by phone, in writing, and online, you use only your words. When you communicate in person, you also use your facial expressions and body language.

Verbal and Nonverbal Communication

Verbal communication involves using words, either spoken or written, to express feelings, thoughts, or experiences with another person. **Body language** is the use of postures, gestures, and facial expressions. It is a form of **nonverbal communication**, which is getting messages across without using words. Posture is how a person sits or stands and the message the body sends. An open, friendly posture shows that the message is being received.

Gestures are motions made with hands or the head to express a meaning. Nodding is a gesture that shows agreement or understanding. Facial expressions can also communicate feelings to others. An open, friendly facial expression communicates interest in what the speaker is saying.

Tone of voice, or intonation, provides clues as to how the speaker wants the listener to feel about the information. Intonation can communicate thoughtfulness, happiness, and even anger.

Verbal communication is one way to find meaning in what someone else says. Body language is another. It is nonverbal communication, or messages sent with expressions and gestures.

When you talk to another person, do you look right at the other person? Do you show you care about the conversation? Your facial expressions show that you are truly interested in talking with the other person. Take the time to watch people as they talk. What does their body tell you about them as they talk?

Mixed Messages

Sometimes, a person who is communicating might send a **mixed message**. It's a situation in which your words say one thing but your body language says another. If you feel shy or nervous, you might talk quietly and fidget with your hands. You might not look directly at the person to whom you are speaking. An angry person might talk with his or her arms folded. If you are feeling embarrassed, you might look down instead of at the person you are talking to.

If your body language does not match your voice, you may send a mixed message. For example, saying "I'm fine" in an angry tone might make others think something is wrong. A good communicator will look directly at someone and speak in a caring voice. Sometimes it can be hard to understand a person who says one thing but who displays body language that sends a different message.

Relationships

A relationship occurs when two or more people have a connection or are associated for some reason. You will have different types of relationships with people throughout your life. If you have brothers or sisters, you have a sibling relationship with them. Your friendships with other people are relationships. When you get a job, your relationship with your manager will be one of a boss and an employee. People also have romantic relationships.

AN OPEN, FRIENDLY FACIAL EXPRESSION COMMUNICATES INTEREST

Talking and using body language are two ways to communicate with others. **Describe the message sent by the body language of the boys in the photo.**

Because each individual person is different and unique, including you, every relationship you have will be different and unique. For example, you probably do not have the same relationship with each of your siblings or your parents. You may get along better with one sibling compared to another. You probably do not have the same relationship with each of your friends. You may have your friends from the basketball team and your friends from the video production group. These friendships are based on different interests. If you are friends with someone from a different culture, there may be cultural differences in your relationship.

Power in Relationships

Everyone has several types of relationships. Some are with parents or guardians, siblings, other family members, your peers, and people in the community. Power plays a role in each of these relationships. Power refers to who has the greatest amount of influence in the relationship.

Your Family When you were a young child, your parents or guardians told you what to do. Other family members who were older than you also had power over you. You may remember being told to mind your grandparents, or other older adult family members. If your parents left you in the care of a neighbor, that person had power over you. During your teen years, you will begin making more decisions for yourself. As you assume more power over your life, your relationship with your parents will change.

Your Friends Among your friends, there may be one person who takes the leadership role. This person might make most of the decisions about which activities you engage in. The person who makes the most decisions has the power. With your friends, you may develop new interests and want to try different activities. You may begin to try to influence your friends to try new activities too. This will change the power in your relationships with friends.

Your Dating Partners Dating couples also experience power in relationships. One person in the relationship may make more decisions than the other. When teens go on group dates, the group decides what they will do. When two teens go on a one-on-one date, one person may have more power than the other.

Abuse in Relationship Power

In general, the power in relationships is determined by factors such as age, status, and position. In general, older people who have more experience have power over younger people. A wealthy or prominent person in the community may use his or her status to gain power over others. A person's position might give him or her power. Teachers, parents, police officers, and other authority figures have status that gives them power.

Power in relationships can be abused. If one person uses their age, status, or position to gain power and hurt or demean another person, the relationship is abusive. Abuse in family relationships can take the form of neglect, violence, child abuse, or elder abuse. Among peers, an abuser may become a bully. In a dating relationship, an abuser may become force the other person to engage in activities that go against their values.

Communications and Gender Stereotypes

A person's gender can affect the way they communicate with people of the opposite gender. This behavior creates a **stereotype**, or belief about people who belong to a certain group. A gender stereotype assumes that a person must act in a certain way just because the person is male or female.

According to one gender stereotype, females are expected to be passive while males are expected to be assertive. An example of this stereotype might occur when male and female students are working together on a class project. The males in the group would be expected to be the leaders of the group. It might be expected that the boys communicate what needs to be done and assign roles to the group. The behavior of the males might be considered assertive. A female who behaves the same way as the boys might be considered aggressive. Both the male and female are fulfilling the same role in the same way, but the female is viewed negatively.

Gender stereotypes can have a harmful effect. These stereotypes do not acknowledge that each person is an individual with their own thoughts and feelings. Think again about the example of the group project. A female in the group might want to develop skills as a leader. She may choose not to volunteer if she feels that the role is for males only. A male in the group may feel that he is required to take a leadership role, even if he does not want that role. Your gender does not define how you behave and what goals you set for yourself.

Reading Check

Explain Why is it wrong to use gender stereotypes?

Good Communication Skills

MAIN IDEA Good communication includes listening and showing that you understand what the other person is saying.

Interpersonal communication is sharing thoughts and feelings with other people. It requires sending and receiving information that is understood by everyone involved. You send messages and receive messages by speaking, listening, and writing. It takes skills to be a good communicator. You need to be a good speaker, listener, and writer to send and receive messages.

Speaking and listening in person is the most direct way to communicate. In person, the other person hears your words, and also sees your face and body. Your facial expressions and body language can tell a lot. If you look directly at the other person, you show attention and respect. If you look around or past the other person as you speak, you may look like you don't care about the conversation. If you are smiling, you look like you care. If you look bored, you look like you don't care. If you hold your arms folded and stiff, you may appear firm and disrespectful.

Reading Check

Name What are three examples of body language?

When you communicate with someone in person, this can be a good time to discuss your personal boundaries. Suppose you do not feel comfortable being hugged. You may tell your friends that you value their friendship, but you do not like being hugged. You and your friends can suggest a different way for them to express their happiness or warmth for you. This is an example of setting your personal boundaries. Setting your personal boundaries in person may be more effective than another method. A conversation in person helps you communicate the meaning of your message through your tone, facial expression, and body language.

Body language can also be important when using refusal skills. When you use refusal skills, show that you mean what you say by using strong body language. Strong body language includes eye contact, crossed arms, and a serious expression. You may need to walk away.

Body language is part of communication. **What message does this teen send with body language?**

SPEAKING AND LISTENING SKILLS

Outbound ("Sending")	Inbound ("Receiving")
Think, then speak. Avoid saying the first thing that comes to mind. Plan what you're going to say. Think it through.	**Listen actively.** Recognize the difference between hearing and listening. Hearing is just being aware of sound. Listening is paying attention to it. Use your mind as well as your ears.
Use "I" messages. Express your concerns in terms of yourself. You'll be less likely to make others angry or feel defensive.	**Ask questions.** This is another way to show you are listening. It also helps clear up anything you don't understand. It prevents misunderstandings, which are a roadblock to successful communication.
Make clear, simple statements. Be specific and accurate. Stick to the subject. Give the other person a chance to do the same.	**Mirror thoughts and feelings.** Pay attention to what is being said. Repeat what someone says to show that you understand.
Be honest with thoughts and feelings. Say what you really think and feel, but be polite. Respect the feelings of your listener.	**Use appropriate body language.** Even if you disagree, listen to what the other person has to say. Make eye contact, and don't turn away.
Use appropriate body language. Make eye contact. Show that you are involved as a speaker. Avoid mixed messages. Beware of gestures, especially when speaking with people of different cultural backgrounds. Some gestures, such as pointing, are considered rude in certain cultures.	**Wait your turn. Avoid interrupting.** Let the person finish speaking. You'll expect the same courtesy when it's your turn.

Different skills are involved in sending and receiving messages. **Explain how these skills are related to one another.**

Speaking Skills

If you want to get your message across, you will need to develop good speaking skills. You can learn several important traits that will help you become a better speaker.

- **Think before speaking.** When you speak without thinking, you risk being misunderstood. Think about what you will say. You do not want to start talking without thinking about your words first.
- **Make clear, simple statements.** Be specific. Focus on the subject and your message. Use examples if necessary.
- **Use "I" messages.** "I" messages are statements that present a situation from the speaker's personal viewpoint. For example, rather than saying, "You're not making sense," instead try saying "I'm not sure what you mean."
- **Be honest.** Tell the truth to describe your thoughts and feelings. Be polite and kind.
- **Use appropriate body language.** Make eye contact. Think about your expressions and gestures. Show that you are paying attention.

Listening Skills

Listening skills are as important as speaking skills. Be an active listener. **Active listening** means hearing, thinking about, and responding to another person's message. You can develop skills to become a good listener.

- **Pay attention.** Listen to what the speaker is saying. Think about that person's message.
- **Use body language.** Face the speaker. Look at the speaker. Focus on the words you hear. Use your body to show that you are listening.
- **Wait your turn.** Before you respond to what someone is saying, let the other person finish speaking. Then you will have your chance to ask any questions or respond.
- **Ask questions.** If it is appropriate, ask questions to make sure you understand the other person. Use "I" messages when you ask questions.
- **Mirror thoughts and feelings.** After the other person finishes speaking, repeat back in your own words what you believe that person said. This will show that you are listening. It will also help you better understand the message.

Writing Skills

We also communicate through writing. You may write e-mails, text messages, notes, or letters to family members and friends. When you write, remember that the other person cannot see you or hear your tone of voice, which may lead to misunderstandings. For example, a person reading a message that is written in a direct style may interpret the tone of the message as angry when the writer did not intend to communicate in an angry tone. You can also practice guidelines for good communication in writing.

- **Write clear, simple statements.** Be sure to state your thoughts and feelings clearly.
- **Reread your words before sending your message.** Remember, the other person cannot see your body language or hear your tone of voice. Do your best to make sure your thoughts will be understood.

Make sure your written words say what you mean. **Explain how you can be sure your message is clear.**

Your Communication Style

MAIN IDEA Communications styles include assertive, aggressive, and passive.

Once you know how to communicate clearly, you can choose your communication style. There are three styles of communication. One is **assertive** communication, the ability to stand up for yourself in a firm but positive way. Another form is **aggressive**, meaning overly forceful, pushy, hostile, or otherwise attacking in approach. A third approach is **passive**, a tendency to give up, give in, or back down without standing up for ones rights and needs. Each of these styles has its own traits which make communication more, or less, effective.

An assertive communicator is friendly but firm. An assertive communicator states his or her position in a firm but positive way. For example, you tell a friend, "I need to be home on time because I promised my parents." An assertive communicator shows respect for himself or herself and others.

An aggressive communicator is overly forceful, pushy, hostile, or otherwise attacking in his or her approach. An aggressive communicator may think too much about himself or herself and not show respect to others. For example, you tell a friend, "You'd better get me home on time." An aggressive communicator can hurt other's feelings or make them angry.

A passive communicator has a tendency to give up, give in, or back down without standing up for his or her rights and needs. A passive communicator who has a firm curfew might tell a friend, "It doesn't matter when I get home." A passive communicator may care too much about what others think of him or her. Sometimes a passive communicator may not feel confident or have self-respect. You can change from a passive to an assertive communicator by reminding yourself that your thoughts and opinions have value.

Reading Check

Assess Which communication style is the most effective?

Cultural Perspectives

Differences in Body Language Body language is used in different ways in different cultures. For example, in many cultures of Southeast Asia, it is not considered appropriate to show facial expressions of sadness or anger. People from these cultures might smile to mask negative feelings or a negative statement. In other cultures, it is considered impolite to make eye contact with communicating with a member of the opposite gender. Certain hand and arm gestures often have different meanings in different cultures.

Lesson 2 Review

What I learned

1. **DEFINE** What does *communication* mean?

2. **EXPLAIN** What is a mixed message?

3. **EXPLAIN** Which of the three communication styles do you think is most effective? Explain your answer.

Think Critically

4. **DEMONSTRATE** Imagine you have a friend who sends you a text message asking to copy your homework. You say no and the friend posts a negative message about you on a social media site. Use "I" messages to respond.

5. **SYNTHESIZE** What are some nonverbal ways to show consideration for others?

Applying Health Skills

6. **ANALYZING INFLUENCES** Pay attention to conversations you see in school or on TV.

   Watch for an example of *mixed messages* in conversation. Describe the example.

Family Relationships

Before You Read

Quick Write Name the people you can talk to if you have a problem.

Vocabulary

relationships
family
extended family
nurture
role
abuse
physical abuse
sexual abuse
neglect

Reading Check

Define What is a family?

BIG IDEA Healthy dating relationships involve healthy boundaries and healthful ways of showing affection.

What Makes a Family?

MAIN IDEA Family members support one another.

Your **relationships** are the connections you have with other people and groups in your life. One of your most important relationships is with your **family**. Family is the basic unit of society and includes two or more people joined by blood, marriage, adoption, or a desire to support each other. Your family is the greatest influence in developing your values and beliefs. The way you relate with your family prepares you for how you relate to other people for the rest of your life.

Many types of families can be found in the U.S. It is common for a nuclear family to live together. However, in other cultures, it is common for many members of a family to live together. This is called an **extended family**, which is a person's immediate family plus other relatives such as grandparents, aunts, uncles, and cousins.

Families Meet Needs

The main job of a family is to meet the needs of its members. Families provide for basic needs, such as food, clothing, and shelter. They should also provide support and comfort. It is within a family that many of us also learn life skills. These are skills that are necessary to live a healthy, successful life. They include making connections to others, communicating with others, taking on challenges, the use of self-control and focus, and critical thinking. Healthy families also **nurture**, or fulfill physical, mental, emotional, and social needs of other family members. A healthy family nurtures all sides of each family members health triangle.

- **Physical health.** Families care for their members by providing food, clothing, and shelter.
- **Mental/Emotional health.** Family members give each other love, acceptance, and support. They also pass along traditions, values, and beliefs.
- **Social health.** Families teach their members how to get along with each other and with people outside the family.

> **EVERY FAMILY MEMBER HAS A UNIQUE ROLE IN THE FAMILY.**

Reading Check

Explain What does *nurture* mean?

Strong families provide support for each of the members. **Name two ways that families can show support for each other.**

It is within your family that you learn the character, values, and beliefs that shape your decisions throughout life. You may learn these from your parents or other extended family members. Your character, values, and beliefs may be influenced by your family's culture, traditions, and religious beliefs. These influence the way you treat other people and the decisions you make during your life. They can also influence the activities in which you participate, the foods you eat, and the health habits you choose. Your values and beliefs can even help you choose not to participate in risky behaviors, such as using tobacco, alcohol, or drugs.

Roles and Responsibilities in the Family

Every family member has a unique role in the family. A **role** is a part you play when you interact with another person. Each role has its own set of responsibilities. Parents and other adults in a family have the responsibility of meeting the basic needs of the family. Parents also have the responsibility of teaching and modeling good health and communication skills.

The role of children in the family is to follow family rules, learn good communication skills, and show respect and appreciation by accepting or sharing responsibility. For example, a teen can show appreciation by helping out more around the house if both parents are busy. Helping older family members, watching younger siblings, or finishing household tasks are all ways to show your parents that you appreciate them.

Family roles can change. As you get older, you will take on more responsibilities. You may be asked to help with older family members, such as grandparents. You may take on more responsibilities in the home or with younger siblings. Each time you accept a new responsibility, you gain the opportunity to learn a new skill while showing respect, love, and support for your family.

> **AS YOU GET OLDER, YOU WILL TAKE ON MORE RESPONSIBILTIIES.**

Family responsibilities can change. **How is this teen showing a willingness to take responsibility as part of a family?**

Building Strong Families

MAIN IDEA A strong family is built on good relationships.

People with strong family relationships feel connected. They believe in the abilities of each member of the family, as well as feel safe and secure. The list below provides guidelines for making and keeping strong family relationships.

- **Support other family members.** Knowing that your family believes in you adds meaning when you succeed. Support can also help you when you don't succeed.
- **Show appreciation for each other.** Families grow stronger when each member shows appreciation to the other members. For example, a child may say "Thank you for dinner" to parents or guardians. Another way to show appreciation is to help with tasks such as cleaning the dishes or folding laundry.
- **Follow family rules.** Many families have rules. Some are related to responsibilities, such as when to do your homework, or when to take out the trash. Other rules may include what time to be home, when to use the computer, or when the TV can be on. Following the rules at home builds trust and respect among family members.
- **Spend quality time together.** Take time for activities that include the whole family. Some families always have dinner together to talk about events of the day. Others plan evenings at home or weekend outings together.
- **Use good communication skills.** Talking openly helps everyone solve problems and disagreements. Communication helps to develop trust and respect.
- **Show and share responsibility.** Do your tasks and chores without being asked. Taking on additional responsibility during busy times is a way of modeling respect for your parents.
- **Show respect.** Speak to family members in a respectful tone of voice. Show respect for differences among family members. Show respect for the rights of others as well as their privacy and personal belongings.

Reading Check

Assess How do the criteria listed above help build strong family relationships?

Changes in the Family

MAIN IDEA Families deal with change.

Changes and challenges affect every family. Having strong family relationships makes it easier to cope with changes. Changes within the family will occur as the children grow and accept increased responsibilities. If one sibling goes to college, another sibling may take over the responsibility of walking a younger sibling home from school.

Some changes may be minor and end after a period of adjustment. Other changes may affect a family's structure. The birth of a baby, a marital separation, becoming part of a blended family, and divorce are all permanent changes to the family that will require a period of adjustment.

Reading Check

Name What are two changes that can occur in a family?

Change	Positive Ways To Cope
Moving to a new home	Before the move, look at a map of your new neighborhood. Find your new house, your school, and nearby parks. When in a new neighborhood, try to meet other teens.
Separation, divorce, or remarriage of parents	Tell both parents you love them. Talk to them or to another trusted adult about how you feel. A separation or divorce is not your fault.
Job change or job loss	If the family needs to limit spending for a while, ask how you can help.
Birth or adoption of a new sibling	Spend time with your new sibling. Ask your parents how you can help. Imagine what your relationship might be like in the future.
Illness or injury	Show that you care about a sick or injured family member by spending time with him or her and asking how you can help.
Death, loss, and grief	Accept the ways family members express grief. Don't expect their ways of coping to be the same as yours. Pay extra attention to younger members of the family.

Changes that happen in a family affect all family members. **Name two serious changes that can affect families.**

Change may also occur in the family situation. Some of these include the loss of a job, illness or injury, military service, or moving to a new home. If a family moves so that a parent can take a better job in another city, the adjustment to a new city is difficult. However, the family may benefit after adjusting to the new city by gaining new places to explore, meeting new friends, taking on new challenges, and feeling that your family has become closer because of the move.

Changes that affect your whole family may feel out of your control. Sometimes family change can cause you to worry, feel stress, or sadness. When families experience change, it becomes more important for you to communicate openly and truthfully with other family members about your feelings. During your talks with family members, you can all explore solutions to family issues that are affecting you. Sometimes outside help may be needed to deal with serious change. Family members may seek help from family counselors who are specially trained to help family work through issues. Others that may help include health care workers, religious leaders, lawyers, or law enforcement.

CHANGES AND CHALLENGES AFFECT EVERY FAMILY.

Serious Family Problems

MAIN IDEA Serious family problems may require help from counselors or others.

Family problems can sometimes be serious and require outside help. Drug or alcohol addiction is a serious problem. While it may be that only one family member is addicted to drugs or alcohol, the whole family suffers, requiring outside help to deal with the addiction.

A serious family situation that may require the help of police or other authorities is **abuse (uh-BYOOS),** or the physical, emotional, or mental mistreatment of another person. Abuse can affect children or adults. **Physical abuse** is the use of physical force, such as hitting or pushing. A person who is physically abused may show bruises, scratches, burns, or broken bones. Emotional abuse can also be serious. Emotional abuse occurs when someone always yells or puts down another person. Emotional abuse can damage a person's self-esteem.

Sexual abuse is sexual contact that is forced upon another person. Sexual abuse is unwanted use of forced sexual activity including touching private body parts or being forced to touch body parts. Sexual abuse also includes showing sexual materials to a child. Abuse often includes secrets, and threats to keep secrets.

A parent or guardian is responsible for caring for a child's physical, mental/emotional and social needs. If a parent or guardian fails to meet these needs, the parent can be charged with **neglect**, or failure to provide for the basic physical and emotional needs of a dependent. Physical neglect can include not providing food, shelter, clothing, or medical care. Emotional neglect means not giving love and respect.

All forms of abuse and neglect are against the law. Such family problems may also require outside help. Any person who feels abused or neglected must find someone who can help. The process of getting help can start by talking to another trusted adult, a teacher, a school counselor, or a medical professional.

Reading Check

Define What is abuse, and what forms can it take?

Developing Good Character

Taking Responsibility
When change happens in the family, you can show your support by taking on more responsibility. Jamar's mother has taken on more work hours to help support the family. Now she has to be at work at 6:00 a.m. Jamar offered to help his younger sister get breakfast, get ready for school, and get on the bus safely each morning.

Lesson 3 Review

What I Learned

1. **VOCABULARY** Define the word *neglect*.

2. **IDENTIFY** What are three ways to build and keep strong family relationships?

3. **IDENTIFY** What are some family problems that might require help from outside the family?

Thinking Critically

4. **SYNTHESIZE** Give some examples of ways you can use good communication skills with your family.

5. **COMPARE AND CONTRAST** Explain the difference between nurture and neglect.

Applying Health Skills

6. **ADVOCACY** Imagine you have a friend who you think may have a problem at home. Your friend is behaving in an unusual way. He or she seems sad, quiet, and withdrawn. How can you communicate your concern to your friend?

Peer Relationships

Before You Read

Quick Write List the people you consider your friends.

Vocabulary

peers
acquaintance
friendships
loyal
sympathetic
cooperation
peer pressure
negative peer pressure
assertive response

BIG IDEA Strong relationships will have a positive effect on your physical, mental/emotional, and social health.

Who Are Your Peers?

MAIN IDEA Your peer group is made up of people who are close in age and have things in common with you.

As a teen, you spend a lot of time among your **peers**, who are people close to you in age who are a lot like you The students at your school and other teens you know from your outside activities are your peers. Peers will be an important part of your life throughout your lifetime. Think about all the peers you encounter in your daily life. Your peers may be your friends, classmates, teammates, or neighbors.

A peer can also include an **acquaintance**, or someone you see occasionally or know casually. Sometimes your acquaintances become friends. A peer can be someone you have never met but with whom you have something in common. For example, if you volunteer for a national agency that works for a clean environment, you are peers with other teens who may volunteer for that agency. They may live in other cities or other countries, but they are still your peers.

> **PEERS WILL BE AN IMPORTANT PART OF YOUR LIFE THROUGHOUT YOUR LIFETIME.**

Influences on Your Friendships

Your family is one of the biggest influences on your health. Many times, we learn how to behave in relationships by watching others. Parents can show you ways to behave in a relationship if they express love and caring for each other.

Your family also influences who you choose as friends, and how you behave with your friends. Family members who show respect for themselves will probably show respect for friends. A person who respects himself will try to make good choices to avoid health and other risks.

Reading Check

Identify Who are your peers?

Friendships During the Teen Years

During your teen years, you develop many **friendships** which are relationships with people you know, trust, and regard with affection. Friendships usually begin with a common interest, such a sport, a class in school, or conversations on the bus to school.

Your peers and acquaintances are the people around you. Your friends are the relationships you choose. Your friends may have a shared interest or the same values as you do. Having friends is an important part of your social health and growth. In strong friendships, you appreciate the values of loyalty, honesty, trust, and respect.

What Makes a Good Friend

Friendships share many qualities and usually grow stronger with time. The qualities of strong friendships apply to all friends, including you as a friend. Strong friendships have a number of qualities in common.

- **Shared Values** Friendships can begin with shared values. If you are an honest, responsible person, you will appreciate honesty and responsibility in your friends. For example, if you work hard at your schoolwork and are eager to learn, you share interests with other good students. If you take responsibilities for younger siblings at home, your friends may also have strong family relationships and responsibilities.

- **Reliability** A good friend is reliable. Reliable friends do what they say they will do. They do not talk negatively about one another to others. It feels good to be reliable and to have a friend you can rely on.

Reading Check

Compare and Contrast What do peer relationships and friendships have in common? How are they different?

Reading Check

Name What are three qualities of a good friend?

Your peers are close to you in age and have a lot in common with you. **Describe some common interests that teens may share.**

- **Loyalty** A good friend is **loyal**, or faithful, to his or her friends. In friendships that are strong, both friends are loyal to each other. Good friends will respect and honor each other's interests, values, beliefs, and differences.

- **Sympathy** Good friends share sympathy for one another. For example, if you studied hard for a test but did not do as well as you had hoped, you may feel sad or disappointed. A **sympathetic** friend is aware of how you may be feeling at a given time. A friend will understand and respect your feelings. A friend who has empathy is able to identify and share your feelings.

- **Caring** A good friend cares about you and shows it. A friend can show caring by being interested in your feelings, your values, and your beliefs. A caring friend is a good listener. A caring friend gives time and attention to you and your interests.

- **Trust** Good friends trust each other. They learn through their friendship that trust is important. Trust in friendships goes both ways, and is proven in reliability and loyalty.

- **Cooperation** means working together for the common good. Helping each other and offering support are ways to show friends you care.

- **Respect** A good friend has self-respect. A good friend also has respect for his or her family, school, and friends. Good friends show respect by giving their time and attention to each other. You can show respect by displaying all the traits of friendship: reliability, loyalty, sympathy, caring, and trust. Good friends also show their respect each other's values and differences.

Friendships grow when two or more people have qualities in common. **Name the eight qualities that are common in friendships.**

Making New Friends

Teens typically find most of their friends at school, in their neighborhood, or in shared activities. Sometimes it can be hard to find new friends, such as when you change grades, schools, or move to a new neighborhood. However, you can develop skills to help make new friends.

- **Be yourself.** Identify your values, beliefs, and special interests. What would make you a good friend? You want to make friends who value you for who you are.
- **Break the ice.** Start a conversation with a compliment or a question. Show your interest. If the other person shares the same interest, you may begin a friendship.
- **Seek out teens who share your interests.** Join a club, sports team, or community group. There you will find peers who share some of the same interests as you.
- **Join a group that works for a cause you support.** You can show your citizenship and giving qualities to people who share your values. You will also help your community.

FRIENDSHIPS SHARE MANY QUALITIES AND USUALLY GROW STRONGER WITH TIME.

Strengthening Friendships

Some friendships last a long time. The friends may share the same interests and develop similar interests as they grow older. Other friendships may require some work to maintain. If both friends are willing, a friendship can be strengthened to make it long-lasting. A number of strategies can help to make a friendship strong.

- **Spend time together.** The more time you spend with someone, the better you get to know each other. Do your homework together, share a special interest, practice a sport, or work on a project together.
- **Communicate openly and honestly.** Open and honest communication will build trust and respect—qualities you want in your friendships.
- **Help each other through hard times.** Good friendships aren't only about the fun times. Good friends also share their time and sympathy when a friend has a problem and needs some extra care and support.

Reading Check

Describe How can power differences affect a friendship?

An important part of keeping friendships strong is identifying problems and working to solve them. **Name what these teens can do to resolve disagreements in a positive way.**

- **Respect each other's differences.** People are not exactly alike; they have some differences. Friends whose families represent a different culture from your own may have different customs. Show respect for ways your friends are different from you. Learning about other cultures is interesting. Be accepting of others.
- **Encourage each other to reach goals.** A part of friendship is sharing the interests and goals of others. Be giving of your time, attention, and support in your friend's goals.
- **Identify problems and work to solve them.** A part of communication includes discussing problems and expressing your interest in solving a problem. It could be a problem your friend shares or a problem between you and a friend.

In order to communicate your interest in a friend, remember to think before speaking. Be a good listener. Be honest and truthful with your feelings and opinions. Talking about a problem with a friend can help both you and the other person understand how you are different. It can also help you both understand the issue and help to find a solution.

In a strong friendship, the friends are equal. One friend does not use his or her age, economic status or position to influence a friend to do something against his or her values. In unequal friendships, power differences may be used by influence the other friend. Power differences can occur if one person is older and has more life experience. They can also occur if one friend has a higher economic or social status, or if the friends work together and one has a higher position.

Peer Pressure

MAIN IDEA Peer pressure can affect you in different ways.

Teens spend much of their time among peers at school and in other activities. Peers can use **peer pressure**, which is the influence that your peer group has on you. Teens want to fit in and be accepted. Sometimes, without even knowing it, they are influenced by their peers. For example, if you notice that almost everyone at school wears zippered sweatshirts, you may want a zippered sweatshirt too. You are influenced by what you see your peers do. This is called indirect peer pressure. No one is making you get a zippered sweatshirt, but you want one because you see everyone around you wears one.

At other times, you may feel direct peer pressure. A peer might tell you what you should do to fit in or be accepted. If you choose not to do what that person suggests, you may worry whether you will fit in and be part of the group. Remember that while your peers have a big influence on your life, you always have a choice to say no if you believe that a behavior or action will be harmful.

Reading Check

Define Explain what indirect peer pressure is and give an example.

Positive examples set by others can have a major influence on your choices and decisions. **Name two ways that peers can have a positive influence on you.**

Positive Peer Pressure

Like all influences, peer pressure can have a positive or negative effect. Positive peer pressure helps you make healthful choices. For example, if someone says "You are such a good dancer, we wish you would join the dance team," that is a positive suggestion. Dance may strengthen your physical health. You may also make new friends who have a shared interest.

Imagine that many of your peers volunteer at a food bank. They enjoy the sense of citizenship and caring. You may choose to volunteer too, based on your peers' positive experiences. Volunteering may make you feel good about yourself because you are helping others. You may make new friends. You are using an example you see in others to make a positive choice for yourself. This is positive peer pressure, or positive influence.

Negative Peer Pressure

Negative peer pressure is pressure you feel to go along with harmful behaviors or beliefs of others your age. Your peers may urge you to do something you do not agree with or do not want to do. When you face negative peer pressure, you have a choice to make. It helps to think about your values. If you feel you have to choose between making a healthful choice for yourself or fitting in with a group, think about the consequences. Remember, true friends will respect your decision.

PEER PRESSURE CAN HAVE A POSITIVE OR NEGATIVE EFFECT.

Negative peer pressure can take many forms. Encouraging a person to act in a way that is harmful or illegal is one form of negative peer pressure. Others may include dares or threats. Negative peer pressure can also come in the form of teasing or name-calling. You can learn to recognize negative peer pressure by using the H.E.L.P. guidelines. H.E.L.P. stands for Healthful, Ethical, Legal, and Parent-approved.

If what your friends tell you to do does not meet the H.E.L.P. guidelines, you can refuse. All of your actions are your own choices, but you can learn ways to resist negative peer pressure.

Reading Check

Explain Tell why negative peer pressure can be harmful or hurtful.

Developing Good Character

Caring Angelle noticed a new girl in her classroom. The new girl did not know anybody in school. At lunch, Angelle saw the new girl eating by herself. Angelle sat beside her and introduced herself. They discovered they both had seen the same movie. They talked about the movie and soon were laughing about the funny parts. Two of Angelle's peers sat down to join the conversation.

Resisting peer pressure is a skill you can learn. **List two ways to avoid negative peer pressure.**

- **Avoid the situation.** If you can tell that a situation might be unsafe, harmful, or against rules, do not participate.
- **Use assertive responses.** If your peers suggest a dangerous behavior or situation, say no. Use an **assertive response** which is a response that declares your position strongly and confidently.
- **Focus on the issue.** State your reasons for your choice. Avoid responding if your peers tease you. Avoid trading insults.
- **Walk away.** It is best to try to talk things out with peers who try to pressure you. If anyone gets angry, walk away.

Lesson 4 Review

What I Learned

1. **VOCABULARY** Define *friendship*.

2. **IDENTIFY** What are three ways to strengthen friendships?

3. **DEFINE** What is negative peer pressure? Give an example of it.

Thinking Critically

4. **COMPARE AND CONTRAST** What are some positive and negative peer influences?

5. **ASSESS** Write a short essay analyzing how power differences in intimate relationships can have a positive or negative effect on one or both people. Give examples.

Applying Health Skills

6. **ANALYZING INFLUENCES** Do you think adults experience as much peer pressure as teens? Write a brief paragraph explaining your opinion.

Dating Relationships and Abstinence

LESSONS

. . . .

1 Beginning to Date

2 Healthy Dating Relationships

3 Abstinence and Saying No

2

Beginning To Date

Before You Read

Quick Write Write a short paragraph that describes how your friendships with boys and girls have changed over the last few years.

Vocabulary

commitment

affection

BIG IDEA Your friendships become more important during adolescence.

Changing Friendships

MAIN IDEA Changes during the teen years include making new friends and forming new types of friendships.

You have learned that a friendship is a relationship between two people that is based on trust, caring, and consideration. People often choose friends with similar interests and values. Good friends can benefit your health in many ways. They can have a positive influence and help you resist harmful behaviors.

The teen years bring many physical, mental/emotional, and social changes. Adolescence is the stage of life between childhood and adulthood, usually beginning somewhere between the ages of 11 and 15. During puberty, your body starts to develop the physical traits of an adult. Your brain also changes, and you start to see the world in more complex ways. You may begin to try to understand yourself and how you fit into society. Teens will also discover new interests, including making new friends. You may also start to develop different types of friendships.

Thinking about Dating

Relationships become more important during the teen years than they were when you were younger. As you grow and mature, your friendships may change. You still share good times and have fun with your friends and acquaintances, but you also begin seeking deeper qualities in the people you choose as friends. These qualities may include loyalty and trust. Your group of friends may also begin to grow.

During the teen years, you may also develop feelings of attraction. For example, it may be that the girl across the street used to seem like just another neighbor. Now you might pay more attention to her when she is around. When you were younger, maybe your brother's best friend used to annoy you. Now you might find yourself worrying about how you look when he visits.

These kinds of new feelings cause some teens to begin to think about dating. Dating is a way to get to know people better. There is no specific time when you are supposed to start dating. Some people feel ready to date while in their teens. Others do not feel ready to date until much later.

Group Dating

Choosing to spend time with a group of your peers is one way to date and get to know others. A mixed group of teens may get together to watch a movie or play a game. Other fun group activities might include dancing, skating, or playing sports. Plus, these activities have the benefit of keeping you physically active. Some other advantages to going out with a group include:

- **Conversation** It is easier to keep conversation going when there are several teens in your group instead of only two.
- **Less Pressure** You will feel less pressure to engage in sexual activity and other risk behaviors when you are in a group.
- **Less Expense** A group date can be less expensive for each individual because everyone in the group can share the cost of an activity.

Group dating offers chances to grow and mature. You can learn how to communicate with different types of people. Going out with a group can help you learn more about activities you enjoy. It may help you discover new interests as you learn about other people. Dating in a group setting may even lead you to meet someone you would like to get to know better as an individual.

Teens may become attracted to just one person. **Describe when it is the right time for a teen to begin dating.**

. . . .

Reading Check

Explain Why is spending time in a group a good alternative to individual dating?

. . . .

Spending time in a group activity is one alternative to individual dating. **Tell why you think these teens enjoy one another's company.**

Individual Dating

As you become more mature, you may want to go out on a date with just one person. Individual dating is a big step that should be taken for the right reasons. Peer pressure is not a good reason to begin individual dating. You should wait until you are ready to date. Your parents may also let you know when they feel you are ready for individual dating.

You may find that you enjoy being around another person. You may also find that you share common interests and values. At some point, you may agree that you would like to spend time together by going on a date. You can think of dating as a special form of friendship. Like any other friendship, a healthy dating relationship should be based on caring and respect.

Individual dating is not always stress-free. You may feel nervous going out for the first time. You may worry about what your date will think of you. These thoughts and feelings may be new to you, but many people think and feel the same way before a date. Dating as a teen involves developing a different type of friendship than those you had when you were younger. A date does not have to mean the start of a lifelong **commitment**, which is a pledge or promise. Going on an individual date may turn out to be just the first step in making a special new friend.

Showing affection in a healthy way lets someone know that you care. **Explain why you think it is important to show someone that you care.**

Healthful Ways to Show Affection

Another change that happens during the teen years is the development of feelings of **affection**. Most people want to find someone special to care about deeply. It is one of the great gifts and joys of life. Showing affection, or feelings of love for another person, can take many different forms. One form of showing affection is sexual intimacy. However, it is more healthful to postpone sexual activity until adulthood and marriage.

However, teens can show affection in healthful ways. Holding hands and hugging are physical ways to show affection, but you can also do something thoughtful for another person. You might give a friend a card or a small gift as a way to show affection. For example, Bethany has a big soccer game coming up. Her friend Justin likes to create healthful meals and snacks, so he makes Bethany a tray of fruits and vegetables to share with her teammates. Bethany knows Justin has an important test later this week, so she sends him an encouraging note.

> LIKE ANY OTHER FRIENDSHIP, A HEALTHY DATING RELATIONSHIP SHOULD BE BASED ON CARING AND RESPECT.

You can show that you are a good friend by listening and being sympathetic to the other person's thoughts and ideas. These kinds of actions deepen the bonds of affection. They also display good character, which is a sure sign that you are maturing.

Before beginning to date and show affection for others you should set your personal boundaries. These are boundaries that tell what you will and will not do. Setting personal boundaries can include how you will show affection. They also include how you want someone else to show affection for you. It also includes being mature enough to discuss your personal boundaries with a person you are dating.

Lesson 1 Review

What I Learned

1. **VOCABULARY** What does the term *commitment* mean?

2. **IDENTIFY** Name three advantages of dating in a group.

3. **LIST** Name two healthful ways for teens to show affection.

Thinking Critically

4. **APPLY** If you were thinking of dating someone, what characteristics might you look for in the other person?

5. **EXPLAIN** What are some reasons why dating may not be right for everyone?

Applying Health Skills

6. **COMMUNICATIONS SKILLS - EVALUATE** Adam and Emily have been dating for a while. Adam is expressing that they should engage in more physical affection. What would you advise Emily to tell him?

Healthy Dating Relationships

Before You Read

Quick Write Identify some risks involved in sexual activity. Write a couple of sentences explaining them.

Vocabulary
consequences
limits
sexting
age of consent
dating violence
date rape

BIG IDEA Engaging in unhealthful dating behaviors, such as sexual activity, carries consequences that can have serious impact on your life.

Healthy Dating Relationships

MAIN IDEA Having healthy dating relationships involve healthful boundaries and healthful ways of showing affection.

Remember that a dating relationship is a unique kind of friendship. Qualities of good friendships include reliability and loyalty. Good friends support you and keep their promises. Good friends are also trustworthy and sympathetic. They allow you to share your thoughts and emotions. When you decide to date individually, it is important that you and your dating partner establish healthful boundaries. These boundaries have much in common with the qualities of good friendships.

Thinking About Your Future

As you grow, you will become more independent. Adults will not always be present to set limits and make sure that you stay within them. Knowing your own limits will become very important. The places you go and the people you spend time with can have **consequences**, which are the results of actions. Consequences can affect your personal safety, health, and plans for your future. You will need to be able to evaluate all situations, along with internal and external influences, and avoid people who could be negative influences. Remember to use refusal skills when pressured to do something that you are not ready to do.

Reading Check

Explain What is one way to help establish limits in your own life? How would setting limits help you?

> **WHEN YOU DECIDE TO DATE INDIVIDUALLY, IT IS IMPORTANT THAT YOU AND YOUR DATING PARTNER ESTABLISH HEALTHFUL BOUNDARIES.**

One way to establish limits in your own life and avoid these consequences is to write down the goals you want to achieve. Once you understand what you want to achieve, you will have a better idea of what limits, or boundaries, you need to set to help you reach your goals.

Setting Limits

Imagine playing a game or sport that had no rules. The activity would seem confusing, and it could be dangerous. Rules bring order and purpose to games, and they serve a similar purpose in daily life. Rules can take the form of **limits**, or invisible boundaries that protect you.

Dating couples should set limits for their relationship. Limits let one person know what the other person is willing to do, or not do. A dating partner may set a limit on the types of parties they will attend. He or she may want to avoid parties where alcohol or drugs might be available. Another limit is to avoid sexual activity and to limit how technology is used in a relationship. One important limit is to avoid sending sexually explicit messages online. Remember that messages and photos posted on the Internet are outside your control. You cannot stop them from reaching people you may not want to see them.

Sexting is the practice of sending someone sexually explicit photographs or messages. It is a dangerous activity. Once you send a photo or message online, you cannot control who sees them. The person who receives the photo or message can share them with anyone. If a couple breaks up, a former partner who is hurt or angry may post photos or messages online. This type of revenge is illegal in many states. However, the best way to avoid the possibility is to never send photos or messages that might be used to embarrass or humiliate someone. For this reason, all couples should set limits regarding the use of technology. To prepare for the discussion, ask yourself the following questions:

- What types of pictures or videos do I agree to have taken?
- Should any pictures and videos of me be shared online?
- Should my dating partner text, instant message, or call me while I'm in class?
- Should my dating partner share private information about me with others?

Dating couples should set limits for their relationship. Limits let one person know what the other person is willing to do or not do. Some examples of limits one teen might set for a dating relationship are:

- Say no to going to parties where there may be alcohol and other drugs,
- Refuse to participate in activities that could cause injury, and
- Avoid sexual activity.

One limit on sexual activity is set by each state. It's called the **age of consent** and is a law that defines the age at which a person is considered mature enough to become sexually active. The age of consent protects younger teens who might be manipulated by older teens or adults. Before reaching the age of consent, a teen cannot legally agree to become sexually active. To find the age of consent in your state, go to https://www.ageofconsent.net/states and click on your state.

Healthy dating relationships include limits or personal boundaries. **Describe why it is important to set boundaries for personal relationships.**

Reading Check

Define What is the *age of consent*? Write down the age of consent in your state.

Respect Both you and your date deserve to be treated with consideration and respect.

Communication When you are with someone you are dating, you should be yourself and communicate your thoughts and feelings honestly.

No Pressure You should never feel pressured to do anything that goes against your values or your family's guidelines.

Good communication skills are one characteristic of healthy dating relationships. **Describe how these teens are demonstrating good communication skills.**

Respecting Yourself and Your Date

Healthy friendships bring out the best in each person. Healthy dating relationships should do the same. Qualities of healthy friendships and dating relationships include mutual respect, caring, honesty, and commitment.

You can also use communication, cooperation, and compromise to help build a healthy dating relationship. For example, you may want to go for a hike, while your date may want to play disc golf. A compromise might involve a trip to a park that has a disc golf course. Sometimes compromise is not the best choice to resolve a disagreement. You should not compromise if the result would be harmful or unlawful. You should also avoid compromising on things that really matter to you, such as your values and beliefs. Values and beliefs are internal factors that can influence your dating relationships.

Practicing abstinence also reflects the respect you have for yourself and for others. Remember, if you want to show someone you care for him or her, you can do so in ways that do not involve sexual activity. You can also offer support by talking and listening to the other person.

Dating Violence

Healthy dating relationships are built on respect. Violence of any type is a sign of an unhealthy relationship and shows a lack of respect. When a person in a dating relationship says mean things, hits, or tries to control his or her partner, he or she is committing dating violence. **Dating violence** occurs when a person uses violence in a dating relationship to control his or her partner. It can include physical, emotional, or psychological abuse.

Teens who are just beginning to date are not always sure how to recognize a healthy dating relationship. If a dating relationship feels uncomfortable or becomes violent, it is unhealthy. If your dating partner becomes violent or abusive, find help to get away from the person. Ask a parent or other trusted adult for help if you are concerned about breaking off the dating relationship.

HEALTHY DATING RELATIONSHIPS SHOULD BRING OUT THE BEST IN EACH PERSON.

Another form of dating violence is **date rape**, which occurs when one person in a dating relationship forces the other person to take part in sexual activity. A dating partner might use force, teasing, or intimidation to convince another person to become sexually active. Remember, you always have the right to say no to sexual activity. You can say no at any time to anyone. You do not have to give a reason. In other cases of date rape, a dating partner might add date rape drugs to food or a soft drink while the victim is not looking or has left the area for a short time. After eating the food or drinking the drink, the victim may become unconscious.

Alcohol might also be used as a date rape drug. Using alcohol makes it harder to think clearly, to make good choices, to say no, or to resist an assault. Dating violence and date rape are crimes. To protect yourself, follow some simple rules:

- Stay with your drink at all times. If your drink has been left unattended, throw the drink out and get a new one.
- Avoid drinking from open containers or punch bowls. Drugs may have been added to the bowl.
- If your drink smells or tastes odd, throw it out.
- If you feel drugged or drunk after drinking a soft drink, get help immediately.

Sometimes, a victim who was drugged may not be sure that a crime has been committed. If you suspect that you were drugged and raped, tell someone. To learn more about how to prevent date rape, go to www.womenshealth.gov and search for "date rape" or "date rape facts."

Victims of date rape can suffer physical as well as emotional injuries. The victim needs help for physical and emotional injuries. The person should be tested for STDs. A victim may feel depressed or develop post-traumatic stress disorder (PTSD). Date rape victims should talk to a counselor to help them work through these feelings.

When a Dating Relationship Ends

Most dating relationships formed during the teen years do not last. One or both dating partners may simply change or outgrow the relationship. Whatever the reason for a breakup, the loss of a special relationship can be difficult to cope with. Breaking up can result in stress and depression. When a dating relationship ends, it is natural and normal to feel lonely and hurt. However, these feelings fade with time.

If it is the other person who decides to break up, it can be even more painful. However, the healthiest thing you can do is to respect the other person's wishes. Accept that person's decision and find a way to move on. It may not be healthful to start another dating relationship right away. Eventually, though, you will find another person with similar interests, values, and goals who you would like to get to know better.

Reading Check

Identify What are two examples of limits for teens?

Consequences of Early Sexual Activity

MAIN IDEA Choosing abstinence is a way to avoid the physical consequences of sexual activity.

As teens begin individual dating, they may face new pressures. One is the pressure to engage in sexual activity. The Internet, movies, TV, and magazines may show sexual activity among young people. This may make it look like normal behavior. In truth, most teens avoid sexual activity. The Centers for Disease Control and Prevention (CDC) conducts a Youth Risk Behavior Surveillance System (YRBSS) every two years. In 2015, the survey showed that less than four percent of teens under the age of 13 have engaged in sexual activity.

You may read and hear about sexual activity among teens in magazines, music, and online. Maybe you feel pressure from your friends. These are all external factors that may influence you. However, you can choose not to engage in sexual activity. You can practice sexual abstinence.

The choice to be sexually abstinent as a teen promotes good health. It helps teens avoid the risks that accompany sexual activity. Sexual abstinence shows that you are focusing on your current goals and your plans for the future. It also shows respect for the physical and emotional well-being of others. It is important to understand that being sexually active can have serious consequences. Sexual activity can affect all three sides of your health triangle—physical, mental/emotional, and social.

Physical Consequences

Teens who become sexually active expose themselves to several risks. Becoming sexually active brings the risk of being infected with a sexually transmitted disease (STD). STDs can damage the reproductive system and prevent a person from ever having children. Some STDs remain in the body for life—even after they are diagnosed and treated. Other sexually transmitted diseases, especially HIV/AIDS, can result in death. Any type of sexual activity can result in an STD.

Another risk is unplanned pregnancy. Most teens do not have the emotional maturity to be parents. Teens usually do not have the financial resources to take care of a baby. The teen years are a time for thinking about what you want to do with your life. Teens who become parents usually must put their own education and career plans on hold. When people wait until adulthood to become parents, they are better able to achieve their long-term goals.

Reading Check

Identify What are some specific physical consequences of early sexual activity?

Teen parenthood can be difficult both emotionally and financially. **List some other challenges that this teen might face.**

Mental/Emotional Consequences

Teens who become sexually active may also experience other consequences. Some of these consequences may affect a teen's mental/emotional health. The consequences can include:

- **Emotional distress** because one or both partners are not committed to each other.
- **Loss of self-respect** because sexual activity may go against their personal values and those of their families.
- **Guilt** over concealing their sexual activity from their parents and others.
- **Regret and anxiety** if sexual activity results in an unplanned pregnancy, an STD, or the breakup of the relationship with the partner.

Social Consequences

Sexual activity can affect a teen's social health. Becoming sexually active can limit a teen's interest in forming new friendships. The teen years are a time to meet new people and explore new interests. A teen who is involved in an exclusive relationship with one other person may not be open to meeting new people.

> WHEN A DATING RELATIONSHIP ENDS IT IS NATURAL AND NORMAL TO FEEL LONELY AND HURT.

Reading Check

Explain Why might teens experience regret and anxiety from a sexual relationship?

Myth vs. Fact

Myth People with good character and values never get STDs.

Fact Anyone can get an STD. If you do not know your partner's history of sexual activity, you cannot know whether that person has an STD.

Lesson 2 Review

What I Learned

1. **IDENTIFY** What are three possible physical, social, and mental/emotional consequences of teens engaging in sexual activity?

2. **EXPLAIN** What does the right to say "no" mean?

3. **EXPLAIN** What are three benefits of abstaining from sexual activity?

Thinking Critically

4. **ANALYZE** What are some ways you could practice personal boundaries while in a dating relationship or after one ends?

5. **DESCRIBE** What are two internal and two external factors that can influence your dating relationships? How can they influence them?

Applying Health Skills

6. **EVALUATE** How can sexually transmitted diseases (STDs) affect a teen now and in the future?

Abstinence and Saying No

Before You Read

Quick Write List three ways of saying no when someone pressures you to do something dangerous or unhealthy.

Vocabulary

risk behaviors

BIG IDEA Practicing refusal skills will help you deal with peer pressure.

Choosing Abstinence

MAIN IDEA Choosing abstinence involves communication with your dating partner, self-control, avoiding risky situations, and using refusal skills.

One of the most important limits you can set for yourself as a teen is choosing abstinence from **risk behaviors**, or actions or behaviors that might cause injury or harm to you or others. This includes avoiding tobacco, alcohol, drug use, and sexual activity. Choosing abstinence is the healthful choice. Tobacco, alcohol, and drug use can all cause health problems, addiction, and even lead to death. Abstinence from sexual activity protects teens from contracting sexually transmitted diseases (STDs), HIV/AIDS, and unplanned pregnancy. Most teens make the choice to wait to become sexually active. Among high school students, less than 30 percent reported being sexually active.

> MAKE SURE YOUR DATING PARTNER UNDERSTANDS AND RESPECTS YOUR LIMITS.

Committing to Abstinence

Choosing abstinence is not a decision you can make once and never think about again. It is a choice you will have to recommit to each time you face pressure to engage in sexual activity. Even if you have been sexually active before, you can still choose abstinence. It is important to talk about your decision with the person you date. The following tips may help the conversation go more smoothly:

- Choose a relaxed and comfortable time and place.
- Begin on a positive note, perhaps by talking about your affection for the other person.
- Be clear about your reasons for choosing abstinence.
- Be firm in setting limits in your physical relationship.
- To stay firm in your decision, continue to remind yourself of your limits and the reasons you are choosing abstinence in the first place.

Dealing with Sexual Feelings

Practicing abstinence requires planning and self-control. Sexual feelings are normal and healthy. You cannot prevent them, but you do have control over how you deal with them. Teens can learn ways to manage these feelings. Here are some tips to help you maintain self-control and practice abstinence:

- **Set limits on expressing affection.** Think about your priorities and set limits for your behavior before you are in a situation where sexual feelings may develop.
- **Communicate with your partner.** Make sure your dating partner understands and respects your limits.
- **Talk with a trusted adult.** Consider asking a parent or other trusted adult for suggestions on ways to manage your feelings and emotions.
- **Avoid risky situations.** Choose safe, low-pressure activities, such as a group date.
- **Date someone who respects and shares your values.** A dating partner who respects you and has similar values will understand your commitment to abstinence.

Reading Check

Analyze Why is it important to discuss your commitment to abstinence with your dating partner?

Communication and respect are signs of a good relationship. **Explain how setting limits with your dating partner can help strengthen your relationship.**

Avoiding Risky Situations

Compare the following two situations. You meet with a group of friends for an afternoon picnic. You eat, talk, and then go for a hike. The other situation is a late-night party at a home where the parents are away. Some couples are kissing. In which situation do you think you would be more likely to be pressured to participate in risk behaviors?

Where you go and what you do can have a big impact on your health and safety. Here are some basic precautions:

- **Before you go on a date, know where you are going and what you will be doing.** Find out who else will be there, and discuss with your parents or guardians what time they expect you home.
- **Avoid places where alcohol and other drugs are present.** These can impair a person's judgment. People under the influence of these substances are more likely to take part in high-risk behaviors.
- **Avoid being alone on a date.** You may find it more difficult to maintain self-control when you and your date are alone together. These situations also increase the risk of being forced or pressured into sexual activity.

Teens who are beginning to date must learn to avoid risky situations. **Name two ways to avoid risky dating situations.**

Know where you are going, what you will be doing, and when you will be home.

Avoid places where alcohol and other drugs are present.

Avoid being alone on a date at home or in an isolated place.

Using Refusal Skills

MAIN IDEA You can use and practice refusal skills to help you keep your commitment to abstinence.

Think about some of the reasons you decided to commit to abstinence. Memorize those reasons so that you remember them if you are ever pressured. You can use refusal skills, which are strategies that help you say no effectively, to help maintain your decision. Refusal skills take practice. The S.T.O.P. strategy can be an effective way to say no to risk behaviors. The letters in S.T.O.P. represent the four steps of the strategy. You can use one or all of the following steps:

- **Say no in a firm voice.** State your feelings firmly but politely. Say, "No, I don't want to." Make your no sound like you really mean it. Use body language to support your words. Make eye contact with the person.
- **Tell why not.** If the other person keeps up the pressure, explain why you feel the way you do. You do not need to use phony excuses or make up reasons. Just say, "No, thanks, I care about my health." Do not apologize. You have done nothing wrong.
- **Offer other ideas.** If the person pressuring you to do something is a friend, you might suggest alternatives. Suggest an activity that is safe and fun.
- **Promptly leave.** If all else fails, just walk away. Let your actions match your words. If you need a ride home, phone a parent or other trusted adult to come and pick you up.

Taking basic precautions to avoid high-risk situations can help you enjoy dating in your teens. **Explain why it is important to know where you are going and what you will be doing on a date.**

. . . .

Reading Check

Identify What is a refusal strategy you might use when being pressured to do something you do not want to do?

. . . .

By using refusal skills, you stand up for your values and build self-respect you also show others you have strength and character. **Identify the S.T.O.P. strategy step this teen appears to be taking.**

Dealing with Pressure

MAIN IDEA Knowing how to respond to people who want you to do something you do not want to do will help you resist that pressure.

Most of your friends are probably your peers—the people close to you in age who have a lot in common with you. Sometimes teens worry about what their friends think about them. Your friends' opinions can affect how you act. The influence that your peer group has on you is called "peer pressure." Peer pressure can be negative or positive.

Negative Peer Pressure

Friends should not pressure you to do something that is risky, unhealthful, or unsafe. They should also not pressure you to do something that goes against your values or your family's values. For example, friends should not pressure you to use tobacco, alcohol, or other drugs. They should not ask to copy your homework or ask you to break the rules of your school or community. True friends will respect your choices.

Negative peer pressure can take many forms. Encouraging a person to act in a way that is harmful or illegal is one form of negative pressure. Other forms include bribes, dares, or threats. Negative peer pressure can also come in the form of teasing or name-calling (or bullying). When you experience this kind of pressure, you can use your refusal skills and have responses ready for those who try to persuade you to do something negative.

> THE INFLUENCE THAT YOUR PEER GROUP HAS ON YOU IS CALLED "PEER PRESSURE."

Positive Peer Pressure

Your true friends give positive peer pressure when they suggest you do the right thing. They may encourage you to study more. They might suggest you to work on a group project or become a volunteer. They may ask you to join the science club at school or welcome new people into the group. Friends can help you say no to risk behaviors, such as using tobacco. Positive peer pressure can be good for you. It can improve your health and safety and help you feel better about yourself.

Reading Check

Define What is peer pressure?

Fitness Zone

Physical Fitness Plan One way a friend can give positive peer pressure is by promoting healthful activities such as exercise. You might set up a regular time to play sports or go for a run or hike with your friends. Sharing a fitness routine can help build stronger relationships and improve everyone's physical and emotional well-being.

Lesson 3 Review

What I Learned

1. **VOCABULARY** Explain what the term *risk behavior* means. Use it in an original sentence.

2. **LIST** List four ways to help keep a commitment to abstinence.

3. **IDENTIFY** Give one example of positive peer pressure and one of negative peer pressure.

Thinking Critically

4. **EVALUATE** Jay has told Ron twice that he does not want to sneak into the school basketball game. Jay has even offered to pay for Ron's ticket, but Ron insists on sneaking in. What should Jay do next?

5. **ANALYZE** Do you think adults experience as much pressure from their peers as teens do? Write a paragraph explaining your position.

Applying Health Skills

6. **COMMUNICATION SKILLS** Write a skit about a situation in which a teen uses each step of the S.T.O.P. strategy to refuse to participate in a risk behavior.

Bullying and Cyberbullying

LESSONS

3

Bullying and Harassment

Before You Read

Quick Write Write about three factors you think are responsible for bullying. Share one factor with the class.

Vocabulary

bullying
labeling
intimidation
harassment
sexual harassment
gender discrimination

BIG IDEA Anyone can experience bullying and harassment, but there are effective ways to stop bullying.

What Is Bullying?

MAIN IDEA Most students have been bullied at one time or another.

Josh waits in line in the school cafeteria. Suddenly, he feels someone shove him, almost causing him to fall over. Josh turns and sees Rick, who was six inches taller than him and more muscular. Rick laughs and says, "Get out of my way, I'm next," as he moves in front of Josh. Josh is angry but feels helpless. A teacher who noticed what happened comes over and tells Rick to go to the back of the line and wait his turn.

Josh has experienced **bullying**. It's a type of behavior in which the person uses threats, taunts, or violence to intimidate another again and again. Bullies believe they have power over the people they target. They use their power to hurt others. If you have ever been bullied, you are not alone. Three out of four students have been bullied at one time or another. The 2017 Youth Risk Behavior Surveillance System reports that 19 percent of students in grades 9 to 12 nationwide experienced bullying. Bullying behavior often affects students in grades 7 to 9.

Types of Bullying

Bullies can target another person using two methods. One is direct bullying. The other is indirect bullying. Many bullies use a combination of both methods.

Direct bullying takes place face-to-face. It includes name-calling, taunting and teasing, hitting or pushing. Direct bullying can also include sending mean notes to another person. Indirect bullying is spreading rumors, leaving a person out of a group, or sending anonymous notes and messages about another person online.

What are Bullying Behaviors?

Bullying behaviors are different than two friends teasing each other. Bullies usually target a person who may not be in their group of friends. The behavior can include taunting another student to physical violence. They use three types of behaviors that include:

- **Physical bullying**—hitting, kicking, pinching, spitting, tripping/pushing, or taking personal belongings.
- **Verbal bullying**—teasing, **labeling** or name-calling, taunting, making sexual comments, or making threats to physically harm.
- **Psychological bullying**—**intimidation**, spreading rumors, isolating a person, threatening to use force, or embarrassing a person in public.

TEENS WHO ARE BULLIED CAN EXPERIENCE NEGATIVE PHYSICAL AND MENTAL/EMOTIONAL EFFECTS.

Who Becomes a Bully?

Bullying occurs when one teen feels that he or she has power over another teen. Remember the example of Josh and Rick at the beginning of the lesson? Rick is six inches taller than Josh and Rick is more muscular. Rick's size creates a power imbalance between the two boys. A study by the U.S. Department of Health and Human Services (HHS) identified several types of power imbalances that can lead to bullying. They are:

- **Physical size.** Bullies may be older, larger, and stronger than the person targeted.
- **Popularity.** The bully may be part of a popular group at school or has friends in a popular group.
- **Status.** The target of a bully may be a member of a minority racial or ethnic group, is a member of the LGBTQ+ community, or is overweight or underweight.
- **Abilities.** Bullies may target students with learning disabilities. They may also target students who are involved in the arts or who are not part of a sports team.
- **Being outnumbered.** Bullies target students who have few friends.
- **Presence of weapons.** Having access to weapons may lead to bullying.

Both boys and girls can be bullies. However, boys become bullies a little more often than girls. Boys are also bullied a little more often than girls.

Bullies also use different methods based on their sex. Girl bullies are more likely to spread rumors, use name-calling, exclude another girl from a group, and bully others online. Boy bullies are more likely to use physical bullying.

Reading Check

Identify Federal law protects against harassment based on what protected classes?

Bullies want to gain power over someone else. **List the types of power imbalances that can lead to bullying.**

What are the Effects of Bullying?

MAIN IDEA Anyone involved with bullying can be affected in negative ways.

Everyone involved with bullying can be affected—those being bullied, those who watch the bullying, and even the bully. Teens who are bullied are more likely to miss school to avoid the bullying. Some may drop out of school to escape it. Teens who are bullied may also have the following negative physical and mental/emotional effects, including:

- feelings of fear, helplessness, depression, loneliness, and thoughts of suicide
- low self-esteem
- problems sleeping
- headache, stomachache, or poor appetite

Among teens who witness bullying, most are afraid to tell school authorities. They fear that they may become the bully's next target. Teens who see others being bullied are more likely to tell parents, although they may fear that adults will do nothing to stop the bullying. Teens who witness bullying when it happens are more likely to:

- use tobacco, alcohol, or other drugs.
- have mental health problems, including depression and anxiety.
- miss or skip school.

Even the teens who become bullies are affected by their behavior. Some continue their violent behavior into adulthood. They are more likely to participate in risk behaviors. The HHS study learned that bullies are more likely to use alcohol and drugs as they get older. They are more likely to get into fights with others and drop out of school.

What is Harassment?

MAIN IDEA Harassment isn't teasing or joking. In some cases, it's a crime.

Harassment is the ongoing conduct that offends another person by criticizing his or her race, color, religion, physical disability, or gender. Bullying and harassment may include some of the same behaviors. Harassment, though, is defined by Federal civil rights law. This law refers to harassment that is severe, persistent, or creates a hostile environment. It protects against behaviors that target another person because of their:

- national origin
- race
- skin color
- gender
- age
- religious beliefs

Reading Check

Explain What are some negative effects of bullying on the health of the bullied person and of the bully?

Even bullies are affected by their behavior. **Name two negative effects that a bully might experience.**

Another form of harassment is unwanted touching. Sometimes, touching can be used as a way to show power or bully another person. An unwanted touch can be a forced hug. Always ask another person before touching them.

Sexual harassment is a third type. It's the uninvited and unwelcome sexual conduct directed at another person. Sexual harassment can be physical or verbal. Sending e-mails, text messages, or voice mails with a sexual meaning can be sexual harassment. It can also include mean comments about a person's gender. Sexual harassment is illegal.

A type of conduct related to harassment is **gender discrimination**. It's singling out or excluding a person based on gender. It can be directed at a male or a female. In either case, gender discrimination is wrong. You should not judge a person based upon his or her gender.

How Can I Respond to Harassment?

Harassment can be difficult to handle on your own. If you are a target of harassment, here are some strategies you can use to stop it:

- Tell the person to stop. Tell the person that the harassment is unacceptable and is against the law. Say that you will report it.
- Use an assertive communications style. Use a firm speaking voice, looking the person in the eye.
- Tell your parents or guardians and ask for advice on handling the harassment.
- If the harassment continues, report it to a teacher, school counselor, or principal. Remind that person that harassment is illegal.

Reading Check

Explain What is one main reason that bullies pick on others?

Fitness Zone

Avoiding Trouble Teens who participate in positive, healthy activities are better able to avoid the effects of bullying, such as depression, low self-esteem, and drug use.

Lesson 1 Review

What I Learned

1. **VOCABULARY** Define the term *bullying*. Use it in an original sentence.

2. **IDENTIFY** Name the forms of bullying.

3. **DESCRIBE** How could different forms of sexual harassment be reported?

Thinking Critically

4. **APPLY** Your cousin writes to tell you about a "really funny kid" who just came to his school. He explains that this new person gets a laugh by knocking other students' books out of their hands. How would you explain to your cousin that this action is inappropriate behavior?

5. **DESCRIBE** Write a short paragraph about a teen being intimidated and how the teen feels as a result. End the story with steps that you think a teen might take to end the bullying.

Applying Health Skills

6. **ACCESSING INFORMATION** Harassment is considered a hate crime in 46 of the 50 states. Find out what the laws are in your community regarding harassment. Make a poster explaining the penalties for this behavior.

Cyberbullying

Before You Read

Quick Write Write a poem or short story about a cyberbully. Give your poem or story a positive ending.

Vocabulary

cyberbullying

Cyberbullies target others with mean messages, photos, and social media posts. **Tell why cyberbullies are difficult to trace and punish.**

BIG IDEA Cyberbullying through technology is a growing problem among teens that causes harm and humiliation.

How is Technology Used to Bully?

MAIN IDEA Cyberbullying is more difficult to avoid than face-to-face bullying.

Cyberbullying is the electronic posting of mean-spirited messages about a person often done anonymously. It's bullying by using technology to send hurtful messages about another person. Cyberbullies want to harass, threaten, or spread rumors about another person. Cyberbullies post messages, rumors, and hurtful photos on social media sites, forums aimed at teens, and gaming sites.

Cyberbullying vs. Bullying

In some ways, a cyberbully is just like an in-person bully. Both types target one person to harm or humiliate that person. There are a few differences between cyberbullying and bullying. They include:

- Cyberbullies are persistent, and can post messages from mobile devices that target another person 24 hours a day.
- Information posted by cyberbullies is difficult to remove from social media and other online sites.
- Attacks by cyberbullies are not witnessed by others and so are hard to notice.

Effects of Cyberbullying

Cell phones and computers do not cause cyberbullying. Neither do social media sites. These sites are generally used to for positive reasons. Some positive reasons are to keep in contact with friends and family, get help with school work, and to find entertainment. Unfortunately, these technology tools are also used to hurt other people.

The use of technology by adolescents and teens is increasing. That means that more cyberbullying is occurring. All bullying has a negative effect on a person's physical, mental/emotional, and social health. Teens who are cyberbullied are more likely to:

- use alcohol and drugs.
- skip school.
- experience in-person bullying.
- be unwilling to attend school.
- receive poor grades.
- have lower self-esteem.
- have more health problems.

Reading Check

List What are at least three effects of cyberbullying on a teen who is being bullied?

Preventing and Stopping Cyberbullying

MAIN IDEA Several strategies can help you end cyberbullying.

Some teens believe that if they avoid using technology they can avoid cyberbullies. As you get older and go to college or start a job, your use of technology will probably increase. Rather than avoiding technology use, the answer may be to take steps that prevent cyberbullies from targeting you.

How Can I Prevent Cyberbullying?

So, what are some ways that you can prevent cyberbullying? Avoid putting personal information in digital messages, emails, or on social media sites. Set limits with friends about the types of messages and photos you want posted online. Any photo that's posted online will stay online forever. A person with good technology skills can retrieve a photo even after it's deleted. This includes photos that are sent via email, posted to a social media site, or sent via text.

How Can I Stop Cyberbullying?

Many states have laws against cyberbullying. To learn about the laws in your state, go to www.stopbullying.gov and search for "State Anti-Bullying Laws & Policies. You can take these steps to avoid being cyberbullied and to stop the cyberbullying cycle:

- Do not respond to cyberbullying messages.
- Do not forward cyberbullying messages.
- Block the person who is cyberbullying. Visit social media safety centers to learn how to block users.

Teens spend much of their social lives online, and so cyberbullying is more of a threat. **List two ways that a teen who is being cyberbullied can respond.**

- Keep evidence of cyberbullying. Write down the dates, times, and descriptions of incidents. Save and print screenshots, e-mails, text messages, etc.
- Report cyberbullying to your social media site so it can take action against users abusing its terms of service. Use your evidence.
- Report cyberbullying to Internet and cell phone service providers so they can take action against users abusing their terms of service. Use your evidence. Tell your parents or guardian that you are being cyberbullied.
- Report cyberbullying to law enforcement. Cyberbullying can be considered a crime. Some state laws also cover off-campus bullying that creates a hostile school environment. Use your evidence.
- Report cyberbullying to your school. Cyberbullying can create a disruptive environment at school and is often related to in-person bullying. The school can use your evidence to help stop the behavior and develop its anti-bullying policy.

Reading Check

Describe What are two strategies to prevent cyberbullies from targeting you?

STOP CYBERBULLYING

Do Keep a record and evidence of attacks.	**Don't** Respond to the cyberbully.
Do Tell an adult if you receive harassing messages.	**Don't** Forward messages or images sent by a cyberbully.
Do Block messages from the cyberbully, if possible.	**Don't** Share or post personal information with strangers or people you do not know well.
Do Report the incident to your social media site, Internet provider, cell phone service, school, and/or local law enforcement agency.	**Don't** Visit websites that are unsafe.
Do Keep your passwords safe and do not share them with anyone except your parent(s).	**Don't** Post any text or images online that could hurt or embarrass you or those you know.

Cyberbullies will search online for photos and messages they can use to target another person. **Describe why sharing photos online can make you the target of a cyberbully.**

Lesson 2 Review

What I Learned

1. **VOCABULARY** Define the term *cyberbullying*. Use it in an original sentence.

2. **EXPLAIN** Why it is important to report cyberbullying to police.

3. **EXPLAIN** What are at least three negative effects of cyberbullying on the victim?

Thinking Critically

4. **COMPARE AND CONTRAST** What is the difference between cyberbullying and traditional bullying?

5. **APPLY** Find ads in magazines or newspapers or on the Internet, and/or watch television commercials designed to stop bullying. How are the ads effective? How might the ads be more effective?

Applying Health Skills

6. **COMMUNICATION** A friend of yours has been receiving humiliating text messages about her weight. What advice would you give her on handling this cyberbullying? Write a "To Do" list of strategies for your friend.

Strategies to Stop Bullying

• • • • • • • • •

Before You Read

Quick Write What steps would you take to keep yourself safe from bullying? Write a short paragraph about these strategies.

Vocabulary

bullying behavior

• • • • • • • • •

BIG IDEA Every person can take steps to stop bullying behavior

How Should I Stop a Bully?

MAIN IDEA You can stop bullying now and in the future by using strategies.

Stopping bullying once it has started can be difficult. A teen who is bullied should tell adults. Keep in mind that anyone can become a bully. Remember to treat everyone you meet with respect. Teens model the behavior of parents and other trusted adults. But teens can also model good behavior for other teens too.

On-the-Spot Strategies

No one deserves to be bullied. If you are a target, here are some ways to stop the bullying when it is happening:

• • • • • • • • •

Reading Check

List What are three ways to handle bullying on the spot?

• • • • • • • • •

- **Tell the bully to stop.** Look at the person and speak in a firm, positive voice with your head up. Say that if it continues, you will report the bullying.
- **Try humor.** This works best if joking is easy for you. Respond to the bully by agreeing with him or her in a humorous way. It could catch the bully off guard.
- **Walk away and stay away.** Do this if speaking up seems too difficult or unsafe.
- **Avoid physical violence.** Try to walk away and get help if you feel physically threatened. If violence does occur, protect yourself but do not escalate the violence.
- **Find an adult.** If the bullying is taking place at school, tell a teacher or a school official immediately.

Strategies for the Future

There are also several ways to avoid being bullied in the future:

- **Talk to an adult you trust.** A family member, teacher, or other adult can help. Telling someone can help you feel less alone. They can also help you make a plan to stop the bullying.
- **Avoid places where you know bullies may wait to target other students.** Stairwells, hallways, courtyards without supervision, and playground areas can be risky locations.
- **Stay with a group or find a safe place to go.** Bullies are less likely to target a student who is not alone.

Do's	Don'ts
Do keep control of yourself.	Don't let your emotions get the better of you.
Do stay calm and speak softly.	Don't let the other person force you into a fight.
Do walk away if necessary.	Don't try to get even.
Do apologize if necessary.	Don't tease.
Do try to turn the other person's attention somewhere else.	Don't be hostile, rude, or sarcastic.
Do use your sense of humor.	Don't threaten or insult the person.
Do give the other person a way out.	
Do try to understand how the other person thinks or feels.	
Do tell an adult.	

These Do's and Don'ts can reduce the likelihood of bullying. **Explain how you think trying to get even can make bullying worse?**

Stop Bullying Behavior

At times, a teen may not realize that his or her actions or words may actually hurt feelings. If you are someone who likes to joke with your friends, watch how the person responds to your jokes. If he or she seems hurt, stop the behavior.

Sometimes, a bully may not intend to hurt the feelings of another person. Some teens may need help recognizing **bullying behavior**, which are actions or words that are designed to hurt another person. If you have already been called a bully or you think you might have bullied another person, follow these steps:

- Stop and think before you say or do something that could hurt someone.
- If you feel like being mean to someone, think about why you want to be mean to that person.
- Talk to an adult you trust. Describe what upsets you about the other person.
- Remember that everyone is different, and that our differences make us interesting and unique.
- If you think you have bullied someone in the past, apologize to that person.

Lesson 3 Review

What I Learned

1. **VOCABULARY** Define the term *bullying behavior*.
2. **NAME** What are two ways to stop bullying behavior?
3. **IDENTIFY** How might a bully try to correct his or her behavior?

Thinking Critically

4. **ANALYZE** Shayna is being teased repeatedly by Dejon. His sexual remarks bother her. She doesn't know what to do. What advice do you have for Shayna?
5. **DESCRIBE** What would you do to keep yourself safe walking home if you were worried about being bullied outside of school?

Applying Health Skills

6. **COMMUNICATION SKILLS** Your new classmate, Seth, is having trouble with a student who is bullying and teasing him. Seth feels uncomfortable facing the bully. What strategies would you offer Seth to help him deal with this problem? Explain why.

Promoting Safe Schools

. . . .

Before You Read

Quick Write Write a short paragraph about ways you can stay safe from bullying in school.

Vocabulary

revenge

. . . .

BIG IDEA Everyone, including students, teachers, and parents, can promote schools that are safe from bullying.

How Can Teens Promote Safe Schools?

MAIN IDEA By recognizing the signs of bullying, students can take a stand against it.

Keeping schools safe takes effort by students, parents, teachers, and school officials. You can help by being aware of bullying and recognizing the signs of bullying. Then, if you are confronted by a bully or witness bullying, you can take a stand against it.

What Are the Warning Signs?

Bullying and harassment are not always obvious. Those being bullied may try to hide the problem from friends. They may be afraid or embarrassed to talk about it. However, there are many warning signs that can indicate someone is being affected by bullying—either being bullied or bullying others. Recognizing these warning signs is an important first step for acting against bullying. Warning signs of bullying include:

- injuries that cannot be explained
- damaged or stolen belongings
- faking illness to avoid going to school
- trouble sleeping or having nightmares
- loss of friends or avoiding friends
- hurting oneself

These warning signs can also point to other issues or problems, such as depression or substance abuse. It is important to talk with friends who show changes in behavior. Talking to the person can help identify the cause of the problem. Many teens who are bullies or are bullied do not ask for help, so if you know somoene in serious distress or danger, do not ignore the problem. Talk to a trusted adult right away.

Take a Stand Against Bullying

Anyone can be affected by bullying, even a person who is not the target of a bully. Bullying can cause anxiety. It can also cause depression. If you see someone else being bullied you may not know what to do to stop it. You may not feel safe enough to step in at the moment, but there are many ways to help the person who is being bullied.

Helping a Friend When Bullying Occurs

Make sure that you're safe before trying to help a friend who is being bullied. If the bullying may become violent, find an adult. To help a friend escape, try these techniques:

- Help the person escape the situation. Say that a teacher, parent, or other adult is looking for him or her.
- If you know the bully, say "He's okay. Leave him alone." Only do this if it feels safe to do so.
- Distract the bully or offer an escape to the person being bullied by saying something like, "Mr. Smith needs to see you right now," or "Come on, we need you for our game," if it feels safe to do so.
- Avoid using violence or insults. It takes a lot of courage for someone to step up on behalf of a bullied person. Using threats and insults is the same as the bully.
- Encourage the person being bullied to tell an adult.

Helping A Friend After the Bullying

A friend who is being bullied needs support. You can help the person rebuild their self-esteem and model good behavior for other students. Try these tips:

- Invite the person who was bullied to join your friends at lunch. Showing the bully that the person is part of a group can reduce bullying.
- Send a text message or go up to the person who was bullied and say something like, "That wasn't cool and I'm here for you."
- Tell the person being bullied that you do not like the bullying, and ask what you can do to help.

Reading Check

Predict When should you ask for help from an adult or a health professional to deal with bullying?

Encouraging a friend who is being bullied is a way of showing respect for others. **List two ways you can help a friend who has been bullied.**

- Spend time with the person being bullied at school. Talk and sit with the person at lunch, or after school.
- Tell an adult you trust, like your teacher, parent or guardian. You can do this in person or leave a note.
- Set a good example and do not bully others.
- Look for ways to contribute to the anti-bullying programs at your school through creating posters, stories, or films.

WARNING SIGNS OF BULLYING BEHAVIOR

Signs of Being Bullied	Signs of a Bully
Unexplained injuries	Getting into physical or verbal fights
Lost or destroyed clothing, books, electronics, or jewelry	Having friends who bully others
Frequent headaches or stomach aches, feeling sick or faking illness	Increased aggressive behavior
Changes in eating habits, skipping meals or binge eating. Coming home from school hungry because the student did not eat lunch.	Sent to the principal's office or to detention frequently
Difficulty sleeping or having frequent nightmares	Unexplained extra money or new belongings
Declining grades, loss of interest in schoolwork, or not wanting to go to school	Blaming others for his or her problems
Sudden loss of friends or avoiding social situations	Not accepting responsibility for his or her actions
Feelings of helplessness or decreased self esteem	Competitiveness and worry about his or her reputation or popularity
Self-destructive behaviors, such as running away from home, harming him- or herself, or talking about suicide	

Warning signs can help identify teens who are being bullied or who may be bullies themselves. **List three signs of a bullying problem.**

When Should I Report Bullying?

MAIN IDEA At times, the best option for dealing with a bullying is reporting the behavior to an adult.

Reporting a bully can be difficult. If you are the target, you may fear that the bullying will get worse. If you are a bystander, you may dislike the idea of telling on someone else. You may fear that the bully will seek **revenge** which is to avenge oneself or another. You may worry that adults won't take action. Unless a bully is reported, the behavior will continue. The bully might escalate and bully more students.

If a bully threatens to become violent or becomes violent, immediately tell a parent, teacher, or the police. If the bully has a weapon, get away from the bully as quickly and safely as you can. Then, tell an adult that the bully has a weapon.

Reporting bullying to an adult might help it stop. **Describe how you should respond if a bully brings a weapon to school.**

At home, tell a parent or guardian about the bullying. At school, tell a teacher, school counselor, nurse, or principal. Many schools in the U.S. are required to respond to bullying. Every state in the U.S. has laws against bullying and harassment. Many states have policies on how bullying and harassment should be handled by school officials. If teachers and the principal at your school won't take action call the school district office or the state education office.

Lesson 4 Review

What I Learned

1. **VOCABULARY** Define the term revenge.

2. **IDENTIFY** Who can help prevent bullying in schools?

3. **EXPLAIN** What are three warning signs of being a bully?

Thinking Critically

4. **CRITIQUE** Review your school's policies against bullying and write a short paragraph describing what your school does well. Write down one change or addition you would make to the school policy. Why would you make the change or addition?

5. **DESCRIBE** What are three things you can do to take a stand to stop the bullying if you do not feel safe being face-to-face with the bully?

Applying Health Skills

6. **ACCESSING INFORMATION** You want your parent or guardian to learn more about bullying and how he or she can help prevent it. What would you explain to him or her? Write down some basic information on what parents can do to prevent bullying in schools.

Emotional Health

LESSONS

4

Your Mental and Emotional Health

Before You Read

Quick Write Think about a time when you were disappointed. Write down how you dealt with your feelings.

Vocabulary

mental/emotional health
optimistic
adapt
personality
self-concept
self-esteem
confidence
resilience

BIG IDEA Good mental/emotional health includes having a positive view of yourself and being resilient.

What is Mental and Emotional Health?

MAIN IDEA Mental/emotional health is the ability to handle the stresses and changes of everyday life in a reasonable way.

Raul thinks of himself as a pretty normal teen. He isn't the most popular guy in his class, but he has a few good friends. They spend a lot of time together hanging out and playing basketball. He likes most of his classes in school, especially science. He's even thought about going to college and becoming a scientist some day. Most days, he's in a good mood, laughing and joking with his friends. Sometimes, though, he feels sad or stressed out. When that happens, he shuts himself in his room listening to music until he feels better. Raul understands that feeling this way is just a normal part of being a teen.

Raul shows many signs of good **mental/emotional health**. It's the ability to handle the stresses and changes of everyday life in a reasonable way. He gets along well with others, is **optimistic**, meaning that he has a positive attitude about the future, and takes a strong interest in school. He has his problems, but they never bother him for long. Some signs of good mental and emotional heath are:

- Having a good attitude and a positive outlook on life.
- Recognizing your strengths and working to improve your weaknesses.
- Setting realistic goals for yourself.
- Acting responsibly.
- Being able to relax and have fun, both on your own and in a group.
- Being aware of your feelings and expressing them in a healthy way.
- Accepting constructive feedback from others without becoming angry or defensive.
- Being able to accept yourself and others.
- Knowing how to **adapt**, or adjust to new situations.
- Showing empathy—the ability to identify with and share other people's feelings.

Reading Check

Give Examples What are two signs of good mental and emotional health?

Having good mental and emotional health is an important part of your total health. Your mental and emotional heath can affect the other sides of your health triangle, and it can also be affected by them. For example, if your physical health is poor, you might often feel tired and sick. This could make it hard for you to feel good about yourself and your life. If your social health is good, by contrast, that means you feel connected to people you trust, such as friends and family. That makes it easier for you to deal with your problems in a positive way, because you always have someone to talk to about them.

How You See Yourself

MAIN IDEA Your personality and self-concept are two factors that determine your mental/emotional health.

One of the most important elements of mental/emotional health is the way you see yourself. Mentally and emotionally healthy teens can accept themselves as they are. This does not mean that they think of themselves as perfect, but they are aware of their own strengths and know how to put them to good use. At the same time, they recognize their weaknesses and can work to improve them without feeling down on themselves.

MENTALLY AND EMOTIONALLY HEALTHY TEENS CAN ACCEPT THEMSELVES AS THEY ARE.

Having a positive outlook on life is a sign of good mental/emotional health. **Name two other signs of good mental/emotional health.**

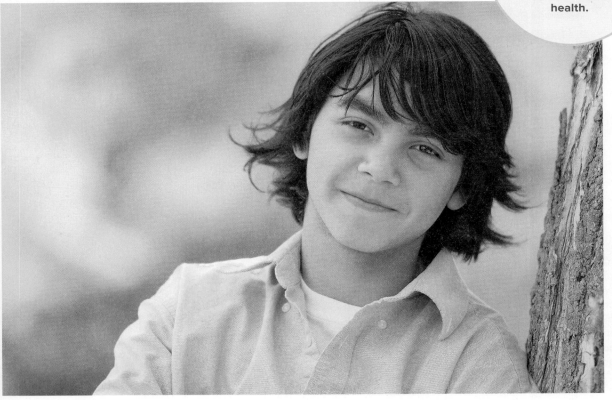

Self-Concept

Suppose you had to write an essay describing yourself. How would you begin? You might say something like, "I'm a good student," or "I'm a big sports fan." Many qualities go into making you yourself.

As a teen, you are in the process of learning about your own personality. Many factors shape your **personality**, which is a combination of your feelings, likes, dislikes, attitudes, abilities, and habits. Heredity is one of them. For example, if your parents are outgoing, you may be the same. Your environment and life experiences also help form your personality. You learn some behaviors from other people, such as family members, friends and peers. Your culture, your background, and your role in your family also affect your personality.

Your **self-concept** is the way you view yourself overall. It's not the same as your personality, but the two are related. Your personality is the way others see you, while your self-concept (also known as self-image) is the way you see yourself. Your self-concept is an important part of your mental and emotional health. It affects the way you relate to yourself and the world. For example, suppose you think of yourself as a person who plays by the rules. Now think about what would happen if external factors conflicted with how you think of yourself. For example, suppose you became friends with some teens who are popular at school, but are often getting into trouble. Trying to fit in with this group might go against your self-concept. This might cause you to feel anxious and unhappy.

Being able to accept constructive criticism is one sign of good mental/emotional health. **Describe one situation in which you have received constructive criticism and tell how you responded.**

Self-Esteem

A major part of your self-concept is **self-esteem**, or how you feel about yourself. When you have high self-esteem, you like and value yourself. You take pride in your achievements, and you face new challenges with **confidence.** Confidence is the belief in your ability to do what you set out to do. Having high self-esteem can improve all three sides of your health triangle. When you value yourself, you are more likely to take care of your body—for example, by eating right and being physically active. Self-esteem can also boost your social life. When you feel confident about yourself, you are more likely to be friendly and outgoing, which will help you make friends. Also, when you have confidence in yourself, you have a positive attitude about the future. You are more willing to take on new challenges and to stick with them if you don't succeed at first. This means that improving your self-esteem can make you more likely to succeed in all areas of your life.

Many factors can influence your self-esteem. The external messages you receive from your friends and family play an important role. Supportive and loving friends and family members can build your self-esteem. Critical or hurtful messages, on the other hand, can damage it. The media, another external factor, can also affect your self-esteem. It can influence your ideas about how you should look, what you should buy, and how you should act. However, these messages may not always be realistic. Finally, your own attitude affects your self-esteem. Your own attitude is an internal factor that can influence your self-esteem. When you think about yourself in a positive way, you boost your own self-esteem.

> WHEN YOU HAVE CONFIDENCE IN YOURSELF, YOU HAVE A POSITIVE ATTITUDE ABOUT THE FUTURE.

Building Self-Esteem

Whether your self-esteem is high or low, it can always be improved upon. You can develop skills to build your self-esteem and feel good about yourself. These skills can increase your overall level of mental/emotional health. The skills you can practice include:

- **Set realistic goals.** These are goals that are reasonable to accomplish. Divide your larger goals into smaller goals. You can build on each smaller success to reach your overall goal.
- **Focus on your strengths.** Find something you like to do, such as a hobby, school activity, or sport. Work to improve your skills. Try to enjoy yourself, even when you make mistakes. This will help you develop more confidence as well as self-esteem.

Reading Check

Identify What are two factors that influence your self-esteem?

Developing Good Character

Caring Good friends show both sympathy and empathy. These two words don't mean the same thing, but they are related. *Sympathy* means being aware of other people's feelings and showing concern for them. *Empathy* is a bit more than that. It means actually sharing in another person's feelings. If a friend's dog has died, you might show sympathy by knowing that your friend is sad and offering a shoulder to cry on. However, if you have ever lost a pet yourself, then you can feel empathy as well.

- **Ask for help.** Recognize and accept when you might need help. This is especially true when you are learning something new. Find someone who can help guide you.
- **Remember that no one is perfect.** Everyone has different abilities. You may be better at writing or swimming than some of your friends. Recognize that there is always room for improvement. Identify your weaknesses without judging yourself and make a solid effort to improve them. If others give you constructive feedback, try to learn from it. Be proud of yourself when you succeed, but know that sometimes failure is out of your control. Mistakes can teach you what doesn't work and push you to grow.
- **Think positively.** Even when you're not entirely sure of yourself, a positive attitude can help you be more confident. Being positive also helps you relate better to others. You're likely to be more honest and honorable and to respect others' life experiences.

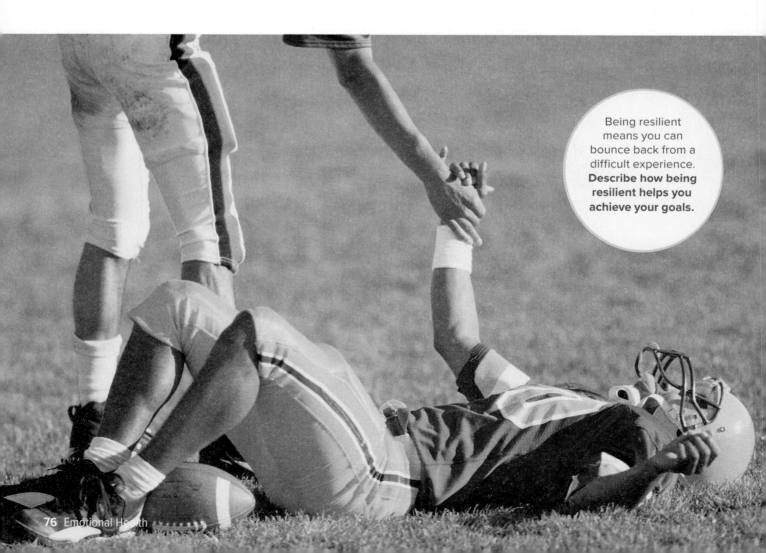

Being resilient means you can bounce back from a difficult experience. **Describe how being resilient helps you achieve your goals.**

Building Resilience

MAIN IDEA Developing resilience can build your self-esteem and improve mental/emotional health.

When you stretch a rubber band, it snaps back into shape as soon as you let it go. Some people are like that too. After a disappointment, they get right back on their feet and keep moving forward. This quality is called **resilience**, or the ability to recover from problems or loss. Being resilient does not mean that you never have problems. It simply means that you can face your problems and seek positive solutions. Resilience also isn't a quality that you are born with. It is a skill you have to develop and practice throughout your life. Here are some steps you can take to build self-esteem and develop resilience:

Reading Check

Define What is resilience?

- **Focus on your strengths.** Everyone has special talents or skills. If you are good with your hands, you might try activities such as crafts or woodworking. If you are a good student, consider tutoring someone who needs help.

- **Connect with others.** Friends, family members, and others can be an important source of support in difficult times. Take the time to build good relationships. Be there for others so that they will be there for you.

- **Motivate yourself.** Set realistic goals and plan out the steps needed to achieve them. Each time you reach a goal, you gain a sense of accomplishment.

- **Keep a positive attitude.** If you are facing a problem, try not to see it as an impossible obstacle. Focus on what you want to happen, not what you are afraid might happen.

- **Remember that no one is perfect.** It is normal to make mistakes, and it's okay to ask for help when you need it. Try not to be hard on yourself if you get something wrong. Instead, try to find a way to learn and grow from the experience.

> **FRIENDS, FAMILY, AND OTHERS CAN BE AN IMPORTANT SOURCE OF SUPPORT IN DIFFICULT TIMES.**

Lesson 1 Review

What I Learned

1. **VOCABULARY** Define *personality* and use the term in an original sentence.

2. **EXPLAIN** What can happen if you behave in a way that goes against your self-concept?

3. **LIST** What are three steps you can take to build positive self-esteem and resilience?

4. **IDENTIFY** What are some external and internal factors that can affect your self-esteem?

Thinking Critically

5. **INFER** Give an example of how your mental and emotional health might influence your physical or social health.

6. **ANALYZE** Explain how self-concept and self-esteem are related.

Applying Health Skills

7. **ANALYZING INFLUENCES** What do you think has had the most influence on your personality: your heredity, environment, or behavior? Explain your answer.

Understanding Your Emotions

Before You Read

Quick Write Think about times when you were upset or angry over something. Try to recall how you dealt with these feelings.

Vocabulary

emotions
mood swings
anxiety
emotional needs

Reading Check

Define What are mood swings?

BIG IDEA Learning to deal with emotions in healthy ways is important during your teen years.

What Are Emotions?

MAIN IDEA It is normal to experience a variety of emotions.

For two years, Tanesha has played only minor roles in school plays. Now, for the first time, she is being considered for a lead role. As she walks toward the cast list, she feels nervous and excited. When she sees her name at the top, she is thrilled. She can't resist jumping and clapping her hands for joy.

Have you ever been in a situation like Tanesha's? Even if you have, you can probably imagine her **emotions**, or feelings such as love, joy, or fear. In fact, nearly every life experience goes hand in hand with some kind of emotion.

It is normal to experience many different emotions, sometimes in just a short period of time. During your teen years, mood swings are common. They can happen because of changing hormone levels in your body, worries over the future, or concerns over relationships. **Mood swings** are frequent changes in emotional state. They occur during the teen years, especially when a teen is going through puberty.

During the teen years, mood swings become more common. **Identify the cause of mood swings during the teen years.**

Expressing Your Emotions

MAIN IDEA The way you express emotions affects your mental/emotional, physical, and social health.

Some emotions, such as happiness and love, can be very pleasant. Others, such as fear and sadness, can be very unpleasant. However, no emotion is good or bad in itself. All emotions, even unpleasant ones, are normal and healthy. What matters is learning to express them in healthy ways that don't cause any harm to yourself or others. Learning to manage your emotions is an important part of mental and emotional health.

Dealing with Anxiety

Have you ever had "butterflies in your stomach" before a big test or an important sports match? This is a normal reaction to **anxiety**, which is an overwhelming feeling of dread, much like fear. It is a form of stress. Mild anxiety can be actually useful because it gives you extra energy. When it builds up too much, however, it can be harmful. It can interfere with normal and necessary functions like eating and sleeping.

When you are feeling anxious, talking with others can sometimes help calm you down. For example, you might try talking through your problems with a family member, friend, or counselor. You can also try writing about your feelings in a journal. Physical activity can also help you feel less tense. One approach that will not help is to run away from the problem. If you meet a challenge directly, you will learn skills that can help you deal with similar challenges in the future. You may also find that the anxiety is not as bad the next time around.

Dealing with Fear

Fear can be a useful and important emotion. It helps us to sense and escape from danger. For example, suppose you begin to cross the street at a busy intersection. A car horn honks loudly and you look up to see a speeding car coming right at you. You escape getting hit by the car by jumping back quickly onto the curb. The emotion you felt, fear, at seeing the speeding car caused you to jump back on the pavement.

Sometimes, however, the things you fear are not real threats. For example, you might be afraid of meeting new people, even though they are unlikely to harm you. In this case, your fears could hold you back from living a full life. Likewise, fear of failure can hold you back from doing what you want in life. Fear becomes a problem when it stops protecting you and instead keeps you from reaching your goals.

Like anxiety, fear can be easier to handle if you talk about it. Support from friends and family may help you learn to face your fears. As you grow more used to situations, they can gradually come to seem less scary.

IT IS NORMAL TO EXPERIENCE MANY DIFFERENT EMOTIONS, SOMETIMES IN JUST A SHORT PERIOD OF TIME.

Dealing with Anger

Think about the last time you felt really angry. Did you feel like screaming and throwing things? Feelings like these are natural. In some cases, they can even be helpful. If a mugger attacks you, your anger might give you the strength to fight the attacker off and break away. However, if you're angry with your little sister for messing up your room, yelling, throwing things, and hitting will only make the problem worse.

This doesn't mean that anger is a bad emotion that should be hidden from others. You just need to express it in positive ways that do not hurt others. It is important to understand that there is a difference between feeling angry and expressing anger in a negative way that could hurt other people. One helpful approach is to think about what is causing your anger and look for a solution. Sometimes, though, you just feel angry for no good reason. In that case, it may help to find ways to relax, such as writing or listening to music. Physical activity can also help you release feelings of anger.

Defense Mechanisms

Defense mechanisms are strategies that people use to cope with strong emotions. Everyone uses them sometimes, usually without even realizing that they are doing it. Defense mechanisms are your mind's way of shifting focus away from an emotion that you do not want to face.

Some defense mechanisms are more harmful than others. Denial or projection can help protect you from pain in the short term. In the long term, however, they prevent you from dealing with your problems. Sublimation, on the other hand, can turn unwanted emotions into something positive.

Reading Check

Identify Name two defense mechanisms.

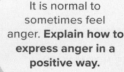

It is normal to sometimes feel anger. **Explain how to express anger in a positive way.**

Your Emotional Needs

MAIN IDEA All human beings share certain emotional needs.

What do people need in life? You might think first of physical needs, such as food, water, and shelter. These are *needs* because people must have them in order to survive. They are much more important than *wants*, things people like to have, such as books and games. However, once people have met their basic physical needs, they begin to notice their emotional needs. All people share several **emotional needs**, or needs that affect a person's feelings and sense of well-being. They feel a need to love and be loved by others. They also feel a need to belong. For instance, you might think of yourself as a part of a family, a student at your school, or a member of a team. Finally, people feel a need to make a difference in the world. They want to do something that matters to them and to be valued for their achievements.

Meeting Emotional Needs in Healthy Ways

During your teen years, you can meet your emotional needs in many ways. Unfortunately, some teens try to meet these needs in ways that can harm their health. For example, one teen might try to meet her need for love by becoming involved in a sexual relationship. This would put her at risk for pregnancy and disease. Another might try to meet his need to belong by joining a gang. This would put him at risk for violence, injury, and trouble with the law.

Fortunately, you have the power to choose healthy ways of meeting your emotional needs. For example, you could meet your need for love by spending time with your family and friends. You can meet your need to belong by joining a group or a club. As for your need to make a difference, helping out in your community can be a great way to help others. You will earn the respect of your neighbors, and you will boost your self-esteem as well.

Learning a craft such as working with clay can help to deal with difficult emotions in a healthful way. **Name two other activities that can help you deal with emotions in a positive way.**

- - - - - - - - - - -

Reading Check

Identify Which emotional needs can be met by your friends and family members?

- - - - - - - - - - -

Lesson 2 Review

What I Learned

1. **VOCABULARY** Define the word *emotions* and give two examples.

2. **DESCRIBE** What are two strategies for dealing with emotions in a healthy way?

3. **IDENTIFY** What emotional needs do all people share?

Thinking Critically

4. **INFER** How could the way you express emotions affect your social health?

5. **EVALUATE** Bradley's teacher assigns a presentation to the class. Bradley is very nervous. He is having trouble sleeping. Is Bradley's fear useful or not? Explain why.

6. **EXPLAIN** What is the difference between expressing anger in a positive or negative way? Give examples.

Applying Health Skills

7. **CONFLICT RESOLUTION** Sophia has asked her mom to pick her up at soccer practice. However, her mom doesn't show up until half an hour after practice is over. Sophia is angry. Write a dialogue between Sophia and her mother showing how she expresses her anger in a healthy way.

Managing Stress

Before You Read

Quick Write Write a few sentences briefly describing a time when you felt stress. How did you manage your feelings?

Vocabulary

stress
positive stress
distress
stressor
fight-or-flight response
adrenaline
fatigue
time management

Reading Check

Explain How can stress be positive?

BIG IDEA Keeping stress under control will improve all aspects of your health.

What is Stress?

MAIN IDEA Both positive and negative events can cause stress.

Chandra is feeling nervous. In just a few minutes, she has to make a presentation for her history class that will count as a major part of her grade. Her heart is pounding, her palms are sweating, and her stomach feels like it's tied up in knots. These symptoms are all normal responses to **stress**, which is the body's response to real or imagined dangers or other life events.

Stress is a normal part of life, and it isn't always bad. Many events that can cause stress are also rewarding. You may feel tense while watching a close football game, but that tension is part of the fun. A little bit of **positive stress**, or stress that can help you reach your goals, can motivate you to do your best. However, too much stress is called **distress**. It is negative stress, that prevents you from doing what you need to do or stress that causes you discomfort. It can cause health problems. You might feel tense or irritable, have headaches, or have trouble sleeping.

Causes of Stress

Many different events can be sources of stress in your life. A **stressor** is anything that causes stress. Some stressors are just everyday problems, like an argument with a friend. Others are serious threats, like having an earthquake strike your town. Sometimes, even positive events can be stressful. Moving to a new house or adding a new member to your family is a happy event, but it can still cause stress. Also, different people react to stressors in different ways. One person, for example, might find driving in traffic very stressful, while another might find it enjoyable.

> A LITTLE BIT OF POSITIVE STRESS CAN MOTIVATE YOU TO DO YOUR BEST.

How Your Body Responds to Stress

MAIN IDEA Long-term stress can damage your physical, mental/emotional, and social health.

Suppose that you are riding your bike down the road. Suddenly, a car pulls out right in front of you. As you squeeze the brake, you feel your heart speed up and your stomach clench. What you are experiencing is the **fight-or-flight response**, which is the body's way of responding to stress. The fight or flight response prepares a person to react to a stressful situation. In some scenarios, a person may run from the threat. An example of this is a person leaving a party early because they are feeling anxious. Sometimes, a person may remain and fight the threat. In a third scenario, a person's brain may perceive the threat to be something that cannot be outrun or fought. In this case, a person may freeze or their mind may go blank. This may happen when a teacher calls on a student in class. When you encounter a stressor, your brain signals your body to begin producing **adrenaline.** It's a hormone that increases the level of sugar in the blood, giving your body extra energy. The release of adrenaline triggers a process that affects nearly every part of your body.

The body feels stress when it occurs. **Describe what happens when the fight-or-flight response is triggered.**

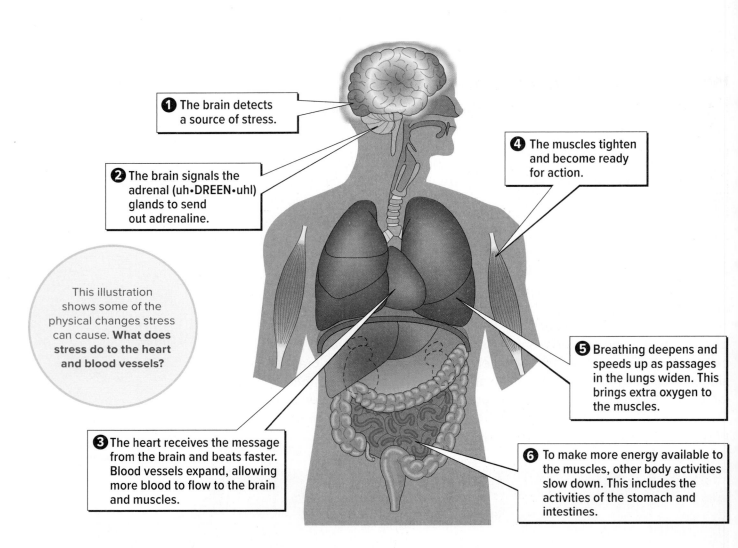

① The brain detects a source of stress.

② The brain signals the adrenal (uh·DREEN·uhl) glands to send out adrenaline.

④ The muscles tighten and become ready for action.

This illustration shows some of the physical changes stress can cause. **What does stress do to the heart and blood vessels?**

⑤ Breathing deepens and speeds up as passages in the lungs widen. This brings extra oxygen to the muscles.

③ The heart receives the message from the brain and beats faster. Blood vessels expand, allowing more blood to flow to the brain and muscles.

⑥ To make more energy available to the muscles, other body activities slow down. This includes the activities of the stomach and intestines.

How Stress Affects Your Health

The fight-or-flight response is useful because it helps you respond to threats. However, in some cases, you feel stress even when there is nothing to "fight or flee." In other cases, the stress continues for so long that your body can't handle it. When your stress levels are high, all three sides of your health triangle can suffer.

- **Physical effects** of stress can include headaches, back pain, **fatigue** or tiredness, and upset stomach. Stress can also cause sleep problems. Over time, it can lead to problems like weight gain, high blood pressure, or heart disease. It can also reduce the body's ability to fight off infection.

- **Mental and emotional effects** include feelings of anxiety, anger, or sadness. Stress can affect your mind as well, making you forgetful and unable to focus. As a result, your schoolwork might suffer. You may also feel "burned out" and lose interest in activities you enjoy.

- **Social effects** of stress are related to the mental and emotional effects. Because stress can make you irritable with others, your anger may burst out in harmful ways, leading to conflicts in your relationships. In some cases, high levels of stress can lead people to withdraw from friends and family members.

Reading Check

Identify Give an example of how stress can affect your social health.

Managing Stress

MAIN IDEA You can manage your stress level partly by avoiding stressors and partly by finding ways to cope with stressors you cannot avoid.

In some cases, you can take steps to avoid stress. For example, you can learn to recognize and stay away from stressful situations. However, it isn't possible to remove all sources of stress from your life. When you can't escape from stress, you can learn to manage your response to it.

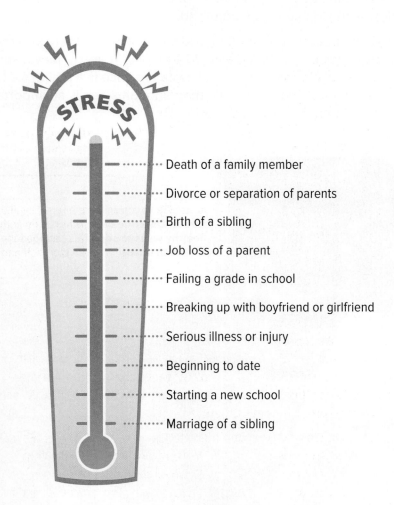

- Death of a family member
- Divorce or separation of parents
- Birth of a sibling
- Job loss of a parent
- Failing a grade in school
- Breaking up with boyfriend or girlfriend
- Serious illness or injury
- Beginning to date
- Starting a new school
- Marriage of a sibling

The more of the items on this list you are facing, the more likely it is that stress will affect your health. **Name two other stressors you would add to this list.**

WHEN YOU CAN'T ESCAPE FROM STRESS, YOU CAN LEARN TO MANAGE YOUR RESPONSE TO IT.

Strategies for Reducing Stress

In order to reduce stress in your life, it helps to know what is causing it. Once you know situations are most likely to be stressful for you, you can sometimes find ways to avoid them. For example, if riding on a crowded bus makes you feel tense, you could try walking instead.

Another important strategy for reducing stress is **time management**, which are strategies for using time efficiently. If you're like many teens, you may have trouble finding enough time for everything you need to do. Planning ahead can help. If you set aside regular blocks of time for schoolwork and chores, you won't have to rush to get them done at the last minute. Making a day planner—a detailed list of your daily activities—can help you stay organized.

Of course, no matter how carefully you plan your time, there's a limit to how many activities you can cram into your day. Trying to play a sport, act in the school play, and sing in a choir all at once is a recipe for too much stress and too little sleep. Getting enough sleep is an important part of reducing or managing stress.

Reading Check

Identify Give three strategies for reducing stress.

You can learn strategies to cope with stress. **Describe two healthy ways to cope with stress.**

Learn to set priorities. Decide which activities are the most important to you, and then give yourself the time to focus on them. Learn to say no when you are asked to take on more than you can handle. If you feel that you need help handling the stress in your life, talk to a parent or other trusted adult. Ask them to suggest other ways that you might reduce the stress in your life.

Strategies for Coping with Stress

Good health habits can help you deal with the stressors you can't avoid. When you eat healthy meals and snacks and get enough sleep, you have the energy to deal with your problems. It may also help to try some specific stress-relieving strategies, such as:

- **Relaxation.** Taking deep breaths is one good way to help yourself relax. Doing stretches or taking a hot shower can help loosen the tension in your muscles. You can also relax by just doing something you enjoy, such as reading a book or listening to music.
- **Physical activity.** Stress can make you feel all wound up. Physical activity can be a good way to use up that extra energy. Being physically active releases chemicals in your body that help you feel happy and calm. You can try lifting weights, playing sports, or just going for a walk.
- **Talking it out.** Talking about your feelings, or writing about them in a journal, can help get them off your chest. It may also help you think of new ways to deal with your problems.
- **Keeping a positive outlook.** A good laugh can be the best medicine. Laughing out loud relieves tension and leaves you feeling more relaxed. Even if you cannot see any humor in your situation, you can still try to think positively. For example, if you have a tough test coming up, you can tell yourself, "I know this subject, and I know I'll do well."

Reading Check

Explain How can budgeting your time help you manage stress?

Lesson 3 Review

What I Learned

1. **DEFINE** Explain how the words *stress* and *stressor* are related.

2. **GIVE EXAMPLES** Name one positive event and one negative event that could cause stress.

3. **IDENTIFY** What are two healthful strategies for dealing with stress?

Thinking Critically

4. **ANALYZE** How does stress affect your physical, mental/emotional, and social health? How are these effects related?

5. **HYPOTHESIZE** Do you think life is more stressful for teens today than it was for your parents? Why or why not?

Applying Health Skills

6. **STRESS MANAGEMENT** Give an example of a situation that might cause stress. Then outline strategies you might use for managing stress in this situation.

Coping with Loss

Before You Read

Quick Write Write a paragraph describing how you would help a friend who has lost a loved one.

Vocabulary

grief
grief reaction
coping strategies

BIG IDEA Keeping stress under control will improve all aspects of your health.

Understanding Grief

MAIN IDEA Grief is a normal response when a person suffers a painful loss.

Can you recall the first time you ever lost something important to you? Perhaps as a young child, you lost or broke a favorite toy. Maybe you had a close friend move away. Losses like these are a sad but unavoidable part of life. In some cases, the emotional pain they cause can be more intense than physical pain.

Perhaps the most painful loss of all is to have someone close to you die. At such times, it is normal for people to experience **grief** which is the sorrow caused by loss of a loved one. People may grieve over the death of a relative, a friend, or even a family pet. Grief can cause feelings of sadness, loneliness, or anger. How people experience grief, and how long it lasts, differs from person to person.

Each person deals with grief in their own way. **List the five stages of grief that most people experience.**

The Grief Reaction

Any kind of loss or serious disappointment can lead to a **grief reaction.** It's the process of dealing with strong feelings following any loss. For example, you might have this reaction to moving or having your parents get divorced. There is no right or wrong reason for feeling grief, and there is no right or wrong way to grieve. The process is different for each person. However, many people experience several stages of grief.

Reading Check

Describe How are loss and grief related?

- **Denial.** Immediately after a loss, people may be in a state of shock. They may feel numb or unable to believe that the loss has really happened.
- **Anger.** Some people feel angry with their loved one for leaving them. Others may blame other family members or even themselves.
- **Bargaining.** People think about what they could have done to prevent the loss. They may feel guilty for being alive when their loved one is dead. They may also wish they had done more for the person when he or she was alive.
- **Depression.** This stage brings deep sadness. People may cry often. They may have feelings of hopelessness or physical symptoms, such as trouble sleeping.
- **Acceptance.** At last, people begin to move on. They gradually regain some of their energy and find it easier to go about their daily lives.

People may not experience all the stages, and they may not happen in the order presented here. Some doctors describe the grief reaction less as a process than as a cycle, with the same feelings coming back again and again. Although these feelings may be most powerful early on, they can last for a long time. Even years after a loss, people may occasionally feel strong emotions about it.

Coping with Grief

MAIN IDEA Coping strategies help people deal with grief.

In most cases, there is no way to undo a loss. However, there are ways to cope with the grief a loss can bring. Here are a few **coping strategies**, or ways of dealing with the sense of loss people feel when someone close to them dies. Coping strategies can help you recover after a loss.

GRIEF CAN CAUSE FEELINGS OF SADNESS, LONELINESS, OR ANGER.

- Give yourself time. Grieving is a long, slow process, and it can't be rushed. Don't try to make any major decisions in your life while you are still grieving.
- Accept all your thoughts and feelings, both positive and negative.
- Don't be afraid to cry. Shedding tears serves as a useful release for strong emotions.

- Remember to take care of yourself. Take the time to eat right and get enough sleep and physical activity.
- Get back into your normal routine as soon as you can. Keeping up with your usual daily tasks can keep you from feeling too overwhelmed.
- Talk about your feelings. You can write in a journal, talk to a trusted friend, or join a support group. If you are having trouble coping, seeing a counselor may help.
- Accept help from others. Friends and family often want to know what they can do for someone who is grieving. Don't hesitate to let them know what you need.

Reading Check

Identify Name two strategies people may use for coping with grief.

Friends can be a source of comfort in times of grief. **Describe how you can help someone who is grieving.**

Helping Others

When someone close to you has suffered a loss, just being there for that person can be a great help. Remember, though, that each person grieves in a different way. Some people may need to cry, while others may want to talk about death. Respect the other person's feelings. Physical touch, such as holding hands or hugging, can be very comforting. It is the decision of the grieving person to decide if she or he is comfortable with it, so follow that person's lead.

You can show empathy by acknowledging the pain the person is experiencing. However, try to avoid sharing stories of your own as a way to show that you understand. The grieving person may feel as if you are dismissing his or her pain. Try to listen without judging or giving advice. However, don't force the other person to share feelings if she or he doesn't want to.

Allow the person to decide how much time he or she needs to recover from the loss. Some people recover quickly, while others may not.

> REMEMBER, THOUGH, THAT EACH PERSON GRIEVES IN A DIFFERENT WAY.

Developing Good Character

Caring It can be hard to know how to deal with someone who is grieving. You may want to help but not know how. Often, just being there is the best thing you can do. However, you can also help in more practical ways. You can bring a gift of food so that the person does not have to cook. You can also offer to help with household tasks, such as mowing the lawn or caring for pets.

Lesson 4 Review

What I Learned

1. **DEFINE** What is the *grief reaction?*

2. **LIST** What are the five stages of grief that many people experience?

3. **EXPLAIN** How can you show support for someone who is grieving?

Thinking Critically

4. **EVALUATE** How long should it take to grieve over a death?

5. **ANALYZE** Keiko's parents plan to get a divorce. She is upset and wants them to stay married. She has promised to get straight A's in school from now on if they will stay together. What stage of grief do Keiko's actions most reflect?

Applying Health Skills

6. **STRESS MANAGEMENT** The death of a loved one can cause a great deal of stress. Refer back to the strategies for managing stress. Write a paragraph explaining which stress reducing strategies you think might be useful for someone who is grieving.

Mental and Emotional Disorders

LESSONS

1 Mental and Emotional Disorders

2 Suicide Prevention

3 Help for Mental and Emotional Disorders

5

Mental and Emotional Disorders

Before You Read

Quick Write Think about a time when you were very sad. Tell what you did to overcome your sadness.

Vocabulary

depression
mental and emotional disorders
disorder
anxiety disorder
panic
phobia
mood disorder
major depression
personality disorder
schizophrenia

BIG IDEA Mental and emotional disorders are real illnesses, just like physical disorders.

What are Mental and Emotional Disorders?

MAIN IDEA Mental and emotional disorders are health disorders.

Mental and emotional problems are more common than many people realize. Each year, they affect about one out of four adults and one out of five children in the United States. As common as they are, however, these problems are often misunderstood. Some people think that problems such as **depression** are just part of someone's personality. Depression is an emotional problem marked by long periods of hopelessness and despair. Some people may even see these problems as a character weakness. The truth is that **mental and emotional disorders** are illnesses that affect a person's thoughts, feelings, and behavior. They are illnesses just like heart disease or cancer.

Mental and emotional disorders can have various causes. Some of them may be caused by genetics that passed from one family member to another. They can result from having too much or too little of certain chemicals in the brain. Physical injury, such as a blow to the head, can also affect mental and emotional health. In other cases, a **disorder**, which is a disturbance in the normal function of a part of the body, may arise from life experiences. They may result from violence, stress, or the loss of a loved one.

Reading Check

Explain What causes mental and emotional disorders?

Long-term feelings of sadness can be a sign of depression. **Explain why it is important to seek treatment for depression.**

Anxiety Disorders

MAIN IDEA Intense anxiety or fear keeps a person from functioning normally.

Everybody feels anxious sometimes. It's normal to feel nervous when you're trying out for a play or starting an important test. For some people, however, anxiety never goes away. It takes over their lives. These people have an **anxiety disorder**, which are extreme fears of real or imaginary situations that get in the way of normal activities. Doctors do not know exactly what causes these disorders. However, they can still treat them. Usually, treatment involves both drugs and counseling. There are several common types of anxiety disorders.

- **Generalized anxiety disorder** makes people worry all the time, often for reasons that aren't that important. It can also cause physical problems, such as headaches.
- **Panic disorder** causes periods of intense fear called **panic** attacks. Panic is a feeling of sudden, intense fear. They cause symptoms such as a pounding heart, sweating, or nausea. They make people feel so awful that they will go out of their way to avoid anything that could cause another attack.
- A **phobia** is an intense and exaggerated fear of a specific situation or object.
- **Social anxiety disorder,** also known as social phobia, is a fear of dealing with people. Some people are afraid of specific situations, such as public speaking. Others are afraid to be around other people at all. They may fear that others are watching or judging them.
- **Obsessive-compulsive disorder (OCD)** traps people in a pattern of repeated thoughts and actions. It may force them to do certain tasks over and over, such as washing hands or counting objects.
- **Post-traumatic stress disorder (PTSD)** occurs after a very stressful event. Physical or sexual violence can trigger this disorder. It also affects many war veterans. People with this disorder have flashbacks in which they relive the event. Some people recover quickly from this disorder, but others need more time.

Reading Check

Identify What are the symptoms of OCD?

People with OCD may feel the need to perform certain actions over and over, such as washing their hands. **Predict how this person might feel after turning off the faucet.**

Acrophobia	Fear of heights
Agoraphobia	Fear of crowded places
Astraphobia	Fear of thunder and lightning
Aviophobia	Fear of flying
Claustrophobia	Fear of enclosed spaces
Hemophobia	Fear of blood
Hydrophobia	Fear of water
Odontiatophobia	Fear of dentists
Trypanophobia	Fear of injections
Zoophobia	Fear of animals (usually spiders, snakes, or mice)

This table lists several of the most common phobias. **What is the name for fear of flying?**

Mood Disorders

MAIN IDEA Extreme or inappropriate changes in mood that last a long time may indicate a mood disorder.

Everyone feels happy on some days and sad on others. During your teen years, in fact, it's normal to go from feeling happy to feeling sad quite suddenly. However, emotions that are extreme and last a long time may be a sign of a **mood disorder**. Mood disorders are mental and emotional problems in which a person undergoes mood swings that seem extreme, inappropriate, or last a long time. The two main types of mood disorders are depression and bipolar disorder.

Depression

You might describe yourself as feeling "depressed" because you did badly on a test or your boyfriend or girlfriend broke up with you. This kind of sadness is normal and happens to everyone sometimes. What doctors call depression, on the other hand, is a serious problem. Clinical depression, or **major depression**, is a serious mood disorder in which people lose interest in life and can no longer find enjoyment in anything. It causes feelings of extreme sadness, hopelessness, worthlessness, and guilt. These feelings may last for weeks, months, or even years.

People who are depressed often lose interest in activities they once enjoyed. They may also feel angry or irritable for no reason. Depressed people may withdraw from family and friends. Some are unable to sleep, while others sleep all the time. Depression can also cause physical problems, such as tiredness and headaches. If depression is not treated, people may begin to think about suicide.

Reading Check

Define What is a mood disorder?

Reading Check

Identify Name two personality disorders.

PEOPLE WHO ARE DEPRESSED OFTEN LOSE INTEREST IN ACTIVITIES THEY ONCE ENJOYED.

Depression can cause some people to lose interest in or to withdraw from things they once enjoyed. **What other problems can depression cause?**

Bipolar Disorder

Bipolar disorder, also known as manic-depressive disorder, causes extreme mood swings. People switch between low periods and high periods. Their low periods cause all the symptoms of depression. During their high, or *manic*, periods, people are in a constant state of excitement. They may feel either happy or irritable. They sleep less and have huge amounts of energy. They may also behave in reckless ways. However, in between the cycles, people may go through calmer periods. At these times, their behavior is fairly normal.

The highs and lows of bipolar disorder can make a person feel torn in two. **What are some of the symptoms people show during the manic phase?**

Personality Disorders

MAIN IDEA Three types of personality disorders include paranoid, antisocial, and dependent personality disorders.

All people have a basic need to belong and be accepted by others. People with a **personality disorder**, however, find it very hard to relate to others. A personality disorder is a variety of psychological conditions that affect a person's ability to get along with others. They get stuck in harmful patterns of thinking and acting that cause major trouble in their social lives.

Personality disorders fall into three main groups. The first kind causes people to have a hard time trusting others. One example is *paranoid personality disorder*. People with this disorder often suspect others of wanting to harm them. They are suspicious and often hostile. Disorders in the second group cause extremes of emotion and behavior. *Antisocial personality disorder* is one example. People with this disorder have little respect for others. They may lie, steal, or harm people in other ways, including physically. They often get into trouble with the law.

Finally, some disorders make people feel or act afraid all the time. People with *dependent personality disorder*, for example, look to others to take care of them. They do not like to make decisions for themselves. Often, they would rather put up with unkind or abusive treatment than be alone.

Schizophrenia

MAIN IDEA Schizophrenia is a severe mental disorder that is treatable.

Schizophrenia (skit•zoh•FREE•nee•uh) is a severe mental disorder in which a person loses contact with reality. The causes of schizophrenia are not known. It is a mental health disorder that requires medical attention. People with this disorder often have *delusions*. These are beliefs that have no basis in reality. A person with this disorder may also *hallucinate*, hearing voices or seeing people who are not there. These symptoms are very frightening and can cause the person with schizophrenia to behave in strange and unpredictable ways. Their thinking may be disorganized and they may express themselves in unusual ways. People with this disorder often have trouble holding a job or caring for themselves.

Among teens, the symptoms of schizophrenia may be subtle. Some teens appear to be withdrawn and depressed, show little interest in schoolwork, have trouble sleeping, and lack motivation. Teens with schizophrenia are less likely to suffer some of the most severe symptoms of the disease. Many teens do not have delusions or hallucinations.

ALL PEOPLE HAVE A BASIC NEED TO BE LOVED AND BE ACCEPTED BY OTHERS.

Reading Check

Explain How is schizophrenia treated?

A PERSON WITH SCHIZOPHRENIA IS SUFFERING FROM A MEDICAL CONDITION AND NEEDS MEDICAL HELP.

A person with schizophrenia may be afraid of what they are feeling. The strange behavior associated with schizophrenia causes fear in others. A person with schizophrenia is suffering from a medical condition and needs medical help. The symptoms of schizophrenia can be treated with drug therapy. Many people with the disorder also need therapy and community support.

With therapy and community support, people with schizophrenia can rebuild the lives. **Describe the symptoms of schizophrenia.**

Lesson 1 Review

What I Learned

1. **DEFINE** What are *mental and emotional disorders?*

2. **IDENTIFY** Name three kinds of anxiety disorders.

3. **DESCRIBE** What are the symptoms of schizophrenia?

Thinking Critically

4. **INFER** Charley is afraid of heights. He can't stand to go above the second floor of any building. How could this phobia affect Charley's daily life?

5. **ANALYZE** How are the symptoms of depression and bipolar disorder similar? How do they differ?

Applying Health Skills

6. **ACCESSING INFORMATION** Work with a group to develop a fact sheet about one of the disorders discussed in this lesson. Discuss the symptoms of the disorder and the kinds of treatment available. Provide a list of the sources you use and be prepared to explain why they are reliable.

Suicide Prevention

Before You Read

Quick Write Write a paragraph describing what you could do to help a friend who is talking about suicide.

Vocabulary

suicide

BIG IDEA Knowing the warning signs of suicide could help you save a life.

Depression Among Teens

MAIN IDEA Everyone feels depressed at times. Severe or long-lasting depression may require treatment.

As people go through adolescence, lots of things are changing, including their bodies, cognitive abilities, and emotions. All of these changes are natural. In addition to these changes, the teen years are always stressful. Teens face pressure to succeed at school. They may have to deal with rejection by peers. Life-changing family events—such as moving, divorce, or the death of a loved one—can also feel overwhelming for young people. All these factors can lead to depression among teens.

A 2017 study found that about three million adolescents in the U.S. had at least one major episode of depression in the previous year. This kind of depression is different from ordinary sadness. Depressed teens feel worthless and empty inside. They lack energy, and they no longer feel pleasure in the activities they once enjoyed. Depression can also cause physical symptoms such as headaches or stomachaches, loss of appetite, or trouble sleeping. These problems can last for months or even years at a time.

Suicide

MAIN IDEA Suicide is not a solution to depression.

People suffering from major depression may come to feel that life is not worth living. Some of them begin to think about **suicide**, which is the act of killing oneself on purpose. Suicide is one of the leading causes of death among teens. This is one reason it is so important for people suffering from major depression to get help. Learning to recognize the warning signs of depression and suicide can help you help others—and maybe even save a life.

Causes of Teen Suicide

Risk factors for teen suicide vary. Mental illnesses, such as depression, account for 95 percent of suicides. Teens are at greater risk for thoughts of suicide if they:

- Feel stress due to changes in the family or new responsibilities.
- Experience bullying, abuse, violence, or substance abuse.
- Witness abuse, violence, or substance abuse within the family.
- Live in poverty.
- Use alcohol or drugs.
- Feel disconnected or rejected by peers.

Warning Signs of Suicide

Teens who are thinking about suicide often show warning signs ahead of time. They may even talk about their plans. They may not use the exact words, "I'm going to kill myself." However, they may drop hints, such as "It doesn't matter anymore" or "You won't have me around much longer." Remarks like this should always be taken seriously. Other possible warning signs include:

- Talking about suicide or death in general.
- Feeling hopeless, guilty, or worthless.
- Spending less time with family and friends.
- Losing interest in normal activities.
- Lacking attention to personal appearance.
- Engaging in self-destructive behaviors, such as violence, substance abuse, or running away.
- Feeling bored, having trouble concentrating, or experiencing a sudden drop in grades.
- Giving away favorite belongings.
- Becoming suddenly cheerful after a long period of depression (this may signal that the person has decided on suicide as a way to "solve" his or her problems).

All teens sometimes face stressful situation such as getting a poor grade in school. **Explain why some teens find it harder to cope with stress than others.**

Reading Check

Identify What are two warning signs of suicide?

Myth vs. Fact

Myth It is dangerous to ask people whether they are thinking about suicide. This might put the idea into their heads.

Fact Asking directly about suicidal thoughts gives the person a chance to talk about his or her problems. It also shows that there is someone who cares.

Providing Support

People who talk about or attempt suicide may be asking for help. Friends and classmates can provide some help. If anyone you know shows any warning signs of suicide, let the person know that you care for him or her. Encourage the person to talk to a parent or other adult to try to find solutions to the problems that are causing the thoughts of suicide. Remind the person that depression can be treated and urge him or her to seek professional help. Tell the person that help lines are available online or by telephone. These help sites can be accessed free-of-charge and keep all information confidential.

Never agree to keep someone else's suicide plans a secret. Tell a trusted adult as soon as possible, If you think the person might attempt suicide immediately, do not leave him or her alone. Call 911 or a suicide hotline for help.

Dealing with Depression

MAIN IDEA Help is available to help people deal with depression.

Depression can feel like a heavy weight. However, suicide is not the solution; medical help is. With counseling and therapy, people can escape from under the heavy weight of depression and feel like themselves again.

People who may be thinking about suicide need help right away. **Recall what you should do if someone you know shows warning signs of suicide.**

Group counseling is one way for depressed teens to receive emotional support. **List other available sources of help for troubled teens.**

If you are suffering from depression, it is important to remember that you are not alone. There are people who care about you and want to help you. Talk to someone as soon as possible. A trusted adult can help you get the medical attention you need. Don't wait and hope that the problem will go away on its own. The sooner you tell someone, the sooner you can get help.

Organizations exist that can help teens deal with suicidal thoughts. One is the National Suicide Prevention Lifeline. This free, 24-hour hotline offers help to anyone in crisis. The National Hopeline Network also has a hotline and a Web site with information about depression.

Reading Check

Name What are two organizations that can help teens thinking about suicide?

Lesson 2 Review

What I Learned

1. **VOCABULARY** Define the term *suicide* and use it in an original sentence.

2. **IDENTIFY** What are two reasons that increase the risk of suicide in teens?

3. **RECALL** What factors can make teens more likely to end their own lives?

Thinking Critically

4. **INFER** Suicide is fairly rare among children but much more common among adolescents. Why do you think this might be the case?

5. **APPLY** Karen's friend Clay broke up with his girlfriend. Ever since, he has been unhappy and withdrawn. Today, however, he suddenly seemed more cheerful. He gave Karen his new music player, saying he didn't need it anymore. How should Karen respond?

Applying Health Skills

6. **COMMUNICATION SKILLS** Write a dialogue between a teen who has been showing warning signs of suicide and a concerned friend. Show how the second teen uses communication skills to show empathy and concern for the first teen.

Help for Mental and Emotional Disorders

Before You Read

Quick Write Create an outline on treatment options for people with emotional problems. Use the terms and headings from the lesson as a guide.

Vocabulary

referral
therapy
family therapy
psychologist
psychiatrist
clinical social worker (CSW)

Reading Check

Analyze Why is it important to seek help for mental and emotional problems?

BIG IDEA Mental and emotional disorders can be treated.

When to Get Help

MAIN IDEA Mental disorders are treatable medical conditions.

Mental and emotional disorders are treatable illnesses. People with mental health disorders can be helped. However, nearly two-thirds of all people with mental and emotional disorders do not seek help. Many people who have these disorders feel ashamed, embarrassed, or afraid of how others will perceive the disorder. A person with a broken leg would seek help immediately. So should people with mental health disorders. Medical professionals work to reduce the stigma associated with mental health disorders.

Seeking help for a mental health disorder is an important first step. If feelings of sadness and depression persist for many weeks, seek help. Other signs of serious mental health disorders include:

- feeling sad or angry for two weeks or longer
- wanting to spend all time alone
- feelings begin to affect sleep
- feelings begin to affect eating habits
- feelings begin to affect schoolwork
- feelings begin to affect relationships
- feeling out of control
- feeling worried or nervous all of the time

A PERSON WITH A BROKEN LEG WOULD SEEK HELP IMMEDIATELY. SO SHOULD PEOPLE WITH MENTAL HEALTH DISORDERS.

A person with a mental health disorder may feel ashamed that they have a disorder. Some people with mental health disorders are treated badly by others. A person who teases or bullies a person with a mental health disorder is being hurtful. Mental health disorders are like any other medical disorder. A person with a mental health disorder should be treated like any other person with an ongoing illness.

A person who has a mental illness may be less likely to get treatment if the person sees others with mental health disorders teased or bullied. Like any illness, a person with a mental health illness cannot recover without treatment. Some mental problems are too serious to ignore and require treatment so that the condition can improve.

Where to Get Help

MAIN IDEA Trusted adults can help you find a health care professional.

If you think you need help, start by talking to a trusted adult. This could be a parent or guardian, a school nurse, a counselor, or a teacher. Often, just talking about the problem is an important step toward recovery.

A person with a mental health disorder cannot recover without treatment. **List three warning signs that a person should seek help for a mental health disorder.**

When you have a problem, talking to a trusted adult can help. **List some people you could talk to if had a problem.**

However, it is only the first step. A trusted adult can help steer you toward valid resources to find professional help. These may include doctors, counselors, and support groups. Many health care professionals can provide a **referral**, or a suggestion to seek help or information from another person or place.

Therapy for Mental Disorders

Treatment for mental and emotional problems may include some form of **therapy**, or professional counseling. Working with a professional counselor can help a person learn to manage their thoughts and feelings in a way that is not harmful to their health. Therapy can be very helpful to anyone who is managing stress or experiencing stressful times. An example is dealing with dealing with grief.

Therapy can take several different forms. In most cases, a person will meet with a therapist in the therapist's office. This is called individual or one-on-one therapy. All conversations with the therapist are confidential. Ask the therapist if your conversations are confidential. When a person under age 18 is being treated, a therapist is not required to offer confidentiality unless the patient requests it.

In some cases, a therapist may meet with a whole group of teens who are all going through the same kind of problems. This is known as group therapy. The group may be called a support group. Group therapy has a few advantages. It shows people that they are not the only one with a mental health disorder. Group members can share ideas about ways to manage their illness. They may also talk about the problems they share. Another form of group therapy is **family therapy**, or counseling that seeks to improve troubled family relationships. A therapist can help family members learn to communicate, strengthen their relationships, and solve problems as a group.

Treatment for mental illnesses may also include taking medications as well as therapy. Medication may be used to treat depression, as well as other mental health disorders. These drugs are used in combination with therapy. Together, therapy and drug therapy can help a person with a mental health disorder life a healthful, happy life.

Reading Check

Identify What are two types of therapy?

Developing Good Character

Respect One reason many people do not seek help for mental health problems is that they see mental illness as shameful or disgraceful. People often look down on those who are mentally ill in a way they do not do with people who have physical disabilities. Learning to treat mentally ill people with respect and dignity can be a key to helping them.

Identify some of the harmful labels that society attaches to people who are mentally ill. What can you do to fight these labels?

Mental Health Professionals

There are several different types of mental health providers. All of them have been trained to treat people with mental and emotional problems. However, they differ in the services they can provide. Which type of mental health professional to see depends on the specific needs of the person getting treatment.

- **Psychologists** (sy•KAH•luh•jists) are mental health professionals who are trained and licensed by the state to counsel. They have advanced degrees in the field of mental health. They can diagnose mental health problems and provide counseling.
- **Psychiatrists** (sy•KY•uh•trists) are medical doctors who treat mental health problems. They can provide medication as well as counseling. In most states, they are the only mental health professionals who can prescribe medicines.
- A **clinical social worker (CSW)** is a licensed, certified mental health professional with a master's degree in social work. This professional works to help people overcome both health problems and social problems. A CSW has a master's degree and is trained in therapy.

Reading Check

List Name three types of mental health professionals.

> THERE ARE SEVERAL DIFFERENT TYPES OF MENTAL HEALTH PROVIDERS. ALL OF THEM HAVE BEEN TRAINED TO TREAT PEOPLE WITH MENTAL AND EMOTIONAL PROBLEMS.

Lesson 3 Review

What I Learned

1. **VOCABULARY** Define *therapy* and use the word in an original sentence.
2. **IDENTIFY** Name two warning signs of a mental illness that require treatment.
3. **EXPLAIN** What is the first step in getting help for a mental or emotional problem?

Thinking Critically

4. **ANALYZE** Compare and contrast the advantages of individual therapy and group therapy.
5. **INFER** What factors might determine which type of mental health provider a person chooses to see?

Applying Health Skills

6. **ADVOCACY** Create a poster about the importance of seeking help for mental and emotional disorders. Include information about signs of mental disorders and ways to treat them.

Conflict Resolution

LESSONS

1 Conflicts in Your Life

2 The Nature of Conflicts

3 Conflict Resolution Skills

4 Peer Mediation

6

Conflicts in Your Life

Before You Read

Quick Write List all the words that come to mind when you think of the word *conflict*.

Vocabulary

conflict
violence
revenge

Reading Check

Explain How can a conflict be helpful?

BIG IDEA Learning to resolve conflicts in healthful ways can help your overall well-being.

What Is Conflict?

MAIN IDEA Conflicts are a part of life and can be dealt with in a positive way.

Xavier and Evan both wanted to play first base during a softball game. They both became angry while insisting on playing that one position. This is an example of a **conflict**, or a disagreement between people with opposing viewpoints, interests, or needs. Conflicts happen to everyone. They are a normal part of life.

Conflicts can involve two or more people. These are interpersonal conflicts. They can occur between family members, friends, acquaintances, and strangers. Conflicts can also be internal. An example of an internal conflict is not being able to decide on an action to take.

Sometimes conflicts can be helpful. They can reveal problems and give people a chance to raise issues that need to be worked out. They also allow people to see the consequences of their behavior.

> **A VALUABLE TOOL IN PREVENTING CONFLICTS IS TOLERANCE.**

A conflict may be an exchange of words. Conflicts can be ongoing and severe. They can damage or even destroy relationships. Conflicts can sometimes result in **violence**, which is an act of physical force resulting in injury or abuse.

Gangs are groups of young people that come together to take part in illegal activities. They sometimes become involved in confrontations with other gang members, members of the community, or the local authorities.

Conflicts occur for many reasons. Understanding why conflicts happen can help you prevent them. Often, conflicts occur because of an act or an event. Differences of opinion or jealousy can cause conflicts.

Causes of Conflict

MAIN IDEA Understanding what causes conflicts can help you learn to prevent them.

You will have many disagreements throughout your life. If those are settled in positive ways, you can avoid arguments and fights. Some of the causes of conflicts are:

- **Property or territory.** Sometimes teens do not respect others' property. Conflict may also happen when teens feel someone has trespassed on their territory. If you and your friends always sit in the same place in the cafeteria, you may feel like someone else who sits there is trespassing.
- **Hurt feelings.** Teens may feel hurt when a friend pays attention to someone else. They may also feel hurt when they are left out of activities. Insults, teasing, and gossip can hurt feelings.
- **Revenge.** When someone is insulted or hurt, that person may want to take **revenge**. Revenge is a punishment, injury, or insult to the person seen as the cause of the strong emotion.
- **Differing values.** Conflicts can arise when people have different values, cultures, religions, or political views.
- **Prejudice.** Sometimes conflicts begin because of prejudice. Conflicts can arise when people are not accepted because of their differences. Showing respect for all people, even those who are different, can help avoid conflicts.

A valuable tool in preventing conflicts is tolerance. Accepting people who are different from you can help you build and maintain positive relationships. It also helps you learn more about yourself and others as you learn more about the world around you.

Reading Check

List What are five causes of conflict?

Conflicts can occur for many reasons. **Identify the cause of the conflict between the teens in the photo.**

Common Conflicts for Teens

MAIN IDEA Conflicts can be minor or major, interpersonal, or internal.

Conflicts happen at various times and for various reasons. Some conflicts are minor. They may involve a simple exchange of words or a mild argument. Other conflicts can be major and turn violent. Gang confrontations often lead to violence. Violent conflicts that involve weapons can be dangerous or even life-threatening.

The term *conflict* usually refers to conflicts between two or more people, or interpersonal conflicts. Conflicts can also be internal. Imagine you decide to run for class president. Then you learn that your best friend is also running. Campaigning against your best friend might cause an internal conflict. You want to avoid competing with your friend.

Conflicts that involve teens can occur at home or at school. You and your sibling might disagree on which TV show to watch. A classmate might want to copy your homework, but you know this would be against school rules. You have a conflict with your classmate.

Conflicts at Home

Whenever you are with other people, it is possible that a conflict may arise. Many times, teens want to become more independent from their parents or guardians. This may cause conflicts at home.

Many conflicts that arise between teens and their family members occur over limits. Teens usually want fewer limits than their parents are willing to allow. For example, Juan was going out with some friends and promised his parents that he would be home by 10:00 p.m. During the evening, his friends asked him to stay out with them until midnight. When Juan arrived home, he argued with his mother and father about the curfew set for him.

While teens like Juan expect greater freedom, their parents expect them to accept more responsibility. Parents often ask that teens contribute more to the household. This can be a source of conflict. Parents and guardians set limits and expectations to help teens grow and develop. In many families, teens can gain greater freedom by showing their parents that they are willing to take on added responsibility within the family.

Reading Check

Identify What is a common source of conflict among teens and their parents or guardians?

CONFLICTS OFTEN ARISE FROM DIFFERENCES IN PERSONALITY, BELIEFS, AND OPINIONS.

Siblings can also be a source of conflict at home. Sometimes siblings do not respect one another's space or property. You might have a brother or sister who uses your property without asking permission. Respecting each other's property and space and discussing these issues can help keep conflicts to a minimum.

Conflicts at School

Many of the conflicts teens have outside of the home relate to school. Conflicts at school may involve teachers or administrators. Teens might also experience conflicts with friends, peers, or acquaintances at school.

Conflicts often arise from differences in personality, beliefs, and opinions. Conflicts can also be the result of one incident. For example, imagine you are standing in line to get into the basketball game. Someone bumps into you causing you to bump into the person in front of you. The person ahead of you turns around and is upset with you for bumping him. In this case, the conflict may begin with a simple exchange of words. However, it could escalate into pushing and shoving or another form of physical violence.

Conflicts often arise because of differences. **Name three differences that may lead to conflict.**

Conflicts between parents and teens are very common. **Characterize how you might respond when you are told something you do not want to hear.**

Conflicts may be one-sided and unprovoked. This can be caused by people not taking the time to understand one another. It can also be caused by people seeking power and attention by putting others down. This type of behavior is called bullying.

Bullying is not always physical. It can come in many forms. Bullying behaviors include:

- Teasing someone or saying hurtful things to a person.
- Labeling, or name-calling, based on prejudice.
- Leaving a person out of group activities or events.
- Sending untrue or hurtful e-mails to others.
- Becoming physically violent.

TRY TO REMEMBER THAT PARENTS AND GUARDIANS SET EXPECTATIONS TO HELP TEENS GROW AND DEVELOP.

Bullying is never okay. No one deserves to be bullied. Use the resources available to you to make your environment as safe as possible. If you are a target of bullying, here are some ways to stop it when it happens:

- **Tell the bully to stop.** Look at the person and speak in a firm, positive voice with your head up. Explain that if the bullying continues, you will report the behavior.
- **Try humor.** This works best if joking is easy for you. Respond to the bully by agreeing with him or her in a humorous way. It could catch the bully off guard.
- **Walk away and stay away.** Do this if speaking up seems too difficult or unsafe.
- **Avoid physical violence.** Try to walk away and get help if you feel physically threatened. If violence does occur, protect yourself, but avoid doing anything that might escalate it.
- **Find an adult.** If the bullying is taking place at school, tell a teacher, counselor, or other school official immediately.

> **BULLYING IS NEVER OKAY.**
>
> **NO ONE DESERVES TO BE BULLIED.**

Lesson 1 Review

What I Learned

1. **VOCABULARY** What is a *conflict*?

2. **LIST** What are three causes of conflicts among teens?

3. **IDENTIFY** What are the two main places teens have conflicts?

Thinking Critically

4. **ANALYZE** Ivan and his dad argued about the amount of time Ivan spends playing video games on his computer. Ivan is feeling angry with his dad. What can they do to maintain a positive relationship with each other?

5. **APPLY** Imagine you have a younger sister who gets into your belongings without asking permission. This makes you upset and angry with her. How can you resolve this issue?

Applying Health Skills

6. **ANALYZING INFLUENCES** In what ways do movies and television programs misinform teens about how to handle conflicts? Why do you think movies and television programs portray conflicts this way?

The Nature of Conflicts

Before You Read

Quick Write Write a short paragraph explaining how you resolve conflicts. How effective are the results?

Vocabulary
escalate
mob mentality

Reading Check

List What are three signs of conflict?

BIG IDEA Factors that make conflicts worse include anger, jealousy, group pressure, and the use of alcohol and other drugs.

How to Spot a Conflict

MAIN IDEA Conflicts can often be resolved if the signs of conflict are recognized early.

Conflicts generally occur because you and another person have different viewpoints. Recognizing the signs of conflict early is the key to resolving them. Early signs of conflict include emotional and physical responses. These include:

- **Disagreement.** If you have a disagreement with another person, be aware that this can lead to conflict.
- **Strong Emotions.** Anger, sadness, jealousy and other emotions can be an early sign of conflict. Not feeling in control of emotions, wanting to attack or lash out at others, and wanting to escape are also strong emotions that can be signs of conflict. Learn to control your emotions or to walk away to avoid conflict.
- **Body Language and Behavior.** Feelings of anger can also be reflected in body language. Some people may make fists, cross their arms, or tighten their lips. The conflict is also felt internally. A person's heart may beat faster, they may have an upset stomach, or they may feel a surge of energy.

Managing Your Anger

Sometimes, conflicts **escalate**, or become more serious. Identifying the emotions and other factors that cause conflicts to escalate can help prevent this. Some of these factors include anger, jealousy, group pressure, and the use of alcohol and other drugs.

Anger is a normal emotion that everyone feels. It is important, however, to express your anger in healthful ways. Holding in your anger, or trying not to be angry, can be harmful. Some healthful ways to manage anger follow:

- **Step away from the situation** to get your emotions under control. This allows you time to be quiet and to collect your thoughts. It can help to keep you from saying things you will regret later.
- **Share your feelings** with another person, such as a parent, guardian, or friend. Talking about your anger can help.
- **Engage in physical activity.** Running, biking, or other physical activity can often help diffuse anger.

Reading Check

Recall What should you do when you are angry?

When you feel that your emotions are under control, let the other person know how you feel. Speak calmly. Remember to use "I" messages, and discuss the problem, not the person.

Handling Feelings of Jealousy

Have you ever felt like someone got better treatment than you? Did you feel like that treatment was unfair? If so, you were probably experiencing jealousy. Like anger, we all feel jealous at one time or another. Perhaps another student seems good at everything. It may appear to you that he or she tries very little to succeed. Feeling jealous some of the time is normal. It can make you work harder to succeed.

Jealousy, however, can also be destructive. It can lead to feelings of anger and resentment, and can ruin friendships. Strong feelings of jealousy may result in wanting to get revenge. Seeking revenge can also turn a minor problem into a major conflict. One or both parties could get hurt—physically or emotionally. Managing negative feelings such as jealousy in a positive way can help you avoid these situations.

If you have feelings of jealousy, it is important to deal with them in a healthful way. Talking to a good friend or trusted adult can help. Writing in a journal can also help you better understand some of the feelings.

CONFLICTS GENERALLY OCCUR BECAUSE YOU AND ANOTHER PERSON HAVE DIFFERENT VIEWPOINTS.

Talking about feelings of jealousy with a trusted adult is better than seeking revenge. **Identify some situations that might cause a teen to feel jealous of someone else.**

THE WARNING SIGNS OF BUILDING CONFLICT

Physical Signs	Emotional Signs
A knot in the stomach	Feeling concerned
Faster heart rate	Getting defensive
A lump in the throat	Wanting to cry
Balled-up fists	Not feeling valued
Cold or sweaty palms	Wanting to lash out
A sudden surge of energy	Wanting to escape

Some warning signs that a conflict is building are physical. Other signs are emotional. **Identify some other signs of building conflict.**

Reading Check

Explain How can pressure from a group of peers cause a conflict to escalate?

Group Pressure

Have you ever noticed that during a disagreement in public, a crowd will often form? Sometimes people start gathering out of curiosity. As others join in, some people may begin encouraging the conflict to escalate to include violence. Others in the crowd may join in. This behavior may then become a **mob mentality**, which is acting or behaving in a certain and often negative manner because others are doing it.

When a mob mentality takes over, the people involved in the conflict might become more aggressive than they would normally. They might get caught up in the crowd's encouragements to do harm. Individuals might forget their own values and instead begin to do what the crowd wants. Each person may forget their values and copy the crowd's behavior. Mob mentality can lead to physical violence.

Conflicts and Substance Use

Using alcohol and other drugs can worsen a conflict. When a person uses alcohol or other drugs, it can affect their emotional state. People under the influence of alcohol and other drugs lose inhibitions and behave in ways that they normally would not. In many cases, alcohol and other drugs can cause a conflict to become violent. For teens, alcohol use is illegal. Using drugs, other than for their intended purpose, is illegal for everyone.

Controlling Conflicts

MAIN IDEA Conflicts can usually be prevented from building by handling the problem in an appropriate way.

Conflicts are a part of life. However, they can be handled in ways that help prevent them from escalating. Using these strategies can help you handle conflicts in healthful ways.

- **Understand your feelings.** Anger is one of the strongest emotions involved with conflicts. If you take the time to understand why you are angry, you may better understand the conflict. Some emotions last only a few minutes. Take time to let your emotions subside before dealing with a conflict.

Body language can be a sign of conflict. **Describe how this teen's body language may be signaling the start of a conflict.**

- **Show respect for others.** It may be difficult to show respect for people who are not kind to you. Sometimes, respect can overcome unkindness. It is also important to show respect for yourself. When you avoid letting others force you into doing or saying things that you are not comfortable with, you show respect for your values.

- **See the other person's point of view.** We all have different backgrounds and points of view. If you try to understand these differences, you may be more accepting and considerate of others. This shows you want to resolve conflicts in a positive way.

- **Keep your conflicts private.** Avoid sharing information about your conflicts. If you have a conflict, find a quiet spot to share your differences. Being private will give you a much better chance of resolving the conflict peacefully.

- **Avoid alcohol and other drugs.** Using alcohol or other drugs is illegal for teens. If you are under the influence of one of these substances, you may behave in ways that are not normal for you. You may allow your emotions to control you. When faced with a conflict, you need to think as clearly as possible.

- **Leave the scene if necessary.** At times, the best thing to do is to walk away from a conflict. This allows you and the other person to think about the conflict in a rational way. Then, you will have a much better chance of resolving the conflict peacefully.

Reaching an agreement that puts an end to a conflict is gratifying. **List some other ways to help prevent conflicts from building.**

Reading Check

Explain Why is it important to show respect for others?

Developing Good Character

Citizenship When a fight is developing, you can show good citizenship by encouraging those involved to find a positive way to resolve the conflict. However, if the conflict turns violent, do not get involved. Instead, get help from a trusted adult right away. Identify an appropriate person at your school to whom you could report a fight.

Lesson 2 Review

What I Learned

1. **VOCABULARY** What does the term *mob mentality* mean?

2. **RECALL** What factors cause a conflict to escalate?

3. **LIST** Name two ways to prevent conflicts from building.

Thinking Critically

4. **ANALYZE** Imagine that you have walked away from a conflict. The other person involved calls you a "chicken." What would you say?

Applying Health Skills

5. **PRACTICING HEALTHFUL BEHAVIORS** Compose a list of factors that can prevent conflicts from escalating. You might list *understanding your feelings*, *showing respect for others*, and so on. After you have made your list, indicate which items you feel you do well. Then identify which items you need to improve on. For each item that needs improvement, write out how you can work toward improving in this area.

Conflict Resolution Skills

Before You Read

Quick Write List two ways of communicating that could lead to conflict. Next to each, write how the same idea could be expressed in a more positive way.

Vocabulary

conflict resolution
negotiation
mediation
collaborate
compromise
win-win solution

BIG IDEA You can deal with conflict in constructive ways.

Finding Solutions to Conflicts

MAIN IDEA Conflict resolution involves solving a disagreement in a way that satisfies everyone involved in the conflict.

Sometimes the solution to a conflict is easy to find. Other times, it might be more difficult to find a solution. In cases such as this, you might be tempted to ignore the conflict and not to deal with it. Ignoring or not resolving a conflict can be damaging to a relationship. Avoiding a conflict could make it worse.

Conflict resolution skills are a life skill that involves solving a disagreement in a way that satisfies both sides. These skills allow both parties involved to work together for a positive and healthful resolution to the problem. Two conflict resolution strategies are **negotiation** and **mediation**. Negotiation is the process of talking directly to the other person to resolve a conflict. Mediation is resolving conflicts by using another person or persons to help reach a solution that is acceptable to both sides. Each can be used to resolve a conflict in a peaceful way.

Using Negotiation Skills

Negotiation is a powerful skill for addressing disagreements and other conflicts. During negotiation, the two parties of the conflict meet and share their feelings, expectations, wants, and reasons for their wants. The meeting is to arrive at a peaceful solution to the conflict.

The T.A.L.K. strategy is often used during negotiations. This strategy allows both parties in the conflict to **collaborate**, or work together, to arrive at a solution. You may find that you build a relationship with the other person when collaborating. The T.A.L.K. strategy can help you remember the steps of conflict resolution.

> **AVOID LETTING YOUR EMOTIONS KEEP YOU FROM TRYING DIFFERENT SOLUTIONS TO YOUR PROBLEM.**

- **T**ake a time-out. Thirty minutes is usually enough time for both sides to calm down and get their emotions under control. This will allow each side to think clearly before they talk.

Negotiation is an important tool in conflict resolution. **Describe the process of negotiation.**

- **A**llow each person to talk. It is important that each person be able to share his or her feelings calmly without being interrupted by the other person. It is also important that the speaker not use angry words or gestures.
- **L**et each person ask questions. Each person should be allowed to ask questions of the other person. Questions should be asked and answered in a calm manner. It is important for both sides to be polite and respectful.
- **K**eep brainstorming. Continue to think of creative solutions until one that satisfies both parties is reached.

It may be difficult to keep your emotions in check during negotiations. However, it is important to do so. Avoid letting your emotions keep you from trying different solutions to your problem. Negotiations can be difficult. It is important to negotiate in private, if possible. It is also important to remember some don'ts for negotiation:

- Don't touch the other person.
- Don't point a finger at the other person.
- Don't call names.
- Don't raise your voice.

There are also some do's to remember:

- Do take a break if either party begins to get angry.
- Do stop negotiations if you are threatened by the other person.
- Do leave and tell a trusted adult if you feel threatened.

Reading Check

Identify What are the four steps in the T.A.L.K. strategy?

These teens are working through a disagreement. **What is the listener doing to demonstrate that he is paying attention?**

Reading Check

Contrast What is the main difference between a compromise and a win-win solution?

Negotiations can be helpful for solving conflicts. However, remember that some issues are not open for negotiation. If something is illegal or potentially harmful to you or others, do not negotiate. Say no or use refusal skills, and if possible, leave the scene of the conflict.

Possible Outcomes of Negotiation

During conflict negotiation, it may be necessary for both sides to **compromise**. Compromise occurs when both sides in a conflict agree to give up something to reach a solution that will satisfy everyone. Compromise works well when negotiating such things as which television program to watch or what to have for dinner. However, for some issues, such as whether to accept a ride from someone who has used alcohol, compromise is not a solution. Laws, your values, school rules, and limits set by your parents, for example, should not be compromised.

Another possible outcome of negotiation is a **win-win solution**, or an agreement that gives each party something they want. Often people think that if you negotiate, someone will win, and someone will lose. With a win-win solution, both sides come away with something. There are no losers, so each side is satisfied with the outcome. A win-win solution can improve a relationship if both parties feel that the other allowed them to win something.

Compromise is one possible outcome of conflict. **Explain how it differs from a win-win solution.**

Lesson 3 Review

What I Learned

1. **VOCABULARY** What is *conflict resolution*?

2. **EXPLAIN** Why is it important to take a time-out before beginning a negotiation?

3. **IDENTIFY** List and explain two different outcomes of negotiation.

Thinking Critically

4. **EVALUATE** Think of a conflict you have had. Describe how you handled the situation. How might the outcome have been different if you had used the T.A.L.K. strategy to solve the problem?

5. **DETERMINE** Why is it important to brainstorm all the possible solutions to a problem?

Applying Health Skills

6. **COMMUNICATION SKILLS** Imagine that you are involved in a conflict with a classmate. What communication skills could you use to help reach a solution?

Peer Mediation

Before You Read

Quick Write Have you ever had a conflict with a friend that you were able to work out? Write two to three sentences describing what you did to help resolve the conflict successfully.

Vocabulary

peer mediation
neutrality

Reading Check

Explain When might you need to turn to a mediator?

BIG IDEA Mediation can provide a solution that is acceptable to both parties.

What Is Mediation?

MAIN IDEA Mediation can help the parties in a conflict work together to solve the problem.

At times, you are able to resolve a conflict so that both you and the other person are satisfied with the result. You achieve a win-win solution. Knowing that you worked together to arrive at a solution makes you feel good. Unfortunately, not all conflicts are easy to resolve.

Imagine that you and a classmate have teamed up to create a project for your health class. You and your partner disagree on the method you should take to complete the project. Because of the conflict, you have not made any progress on the project.

Sometimes you will not be able to resolve a conflict on your own. You may find you need help to resolve a conflict. A third person who is not involved in the conflict can help people move closer to a solution. This process is called mediation. Mediation is similar to negotiation except that a third party, a mediator, is involved. A mediator doesn't resolve the conflict. She or he helps the parties involved arrive at a solution.

Mediators can be guidance counselors or other trusted adults. They can also be students who have been trained in mediation strategies. Some schools have **peer mediation** programs to help resolve conflicts between teens. Peer mediation is a process in which a specially trained student listens to both sides of an argument to help the people reach a solution.

The Mediation Process

Before mediation can begin, both parties must agree to allow a mediator to try to help them. The process begins in a private location. It is important that the only people present are the two parties in conflict and the mediator. If others are present, it could be distracting to the mediation process. It will be counter-productive if others began to take sides in the conflict.

The mediator then asks each party to present its side. The mediator must listen carefully. The mediator does not allow one party to talk while the other party is sharing its side. The mediator may ask questions for clarification. This is to make sure the mediator understands the situation or the point of view of each party involved.

Once the mediator understands the situation, that person will try to help the parties come up with solutions. The mediator and the two parties will brainstorm possible solutions. Then they will evaluate the possible solutions. Some of the solutions may not be practical. Others may not be well received by both parties.

Then the mediator will encourage the parties to decide on a solution. The mediator does not participate in the selection process. This choice must be made by the parties involved. Both sides of the conflict must be in agreement about the solution. If both parties are not in agreement, the solution will not work.

When following these steps, a mediator can often help the parties in conflict arrive to a solution. Sometimes the parties will have to make compromises. The overall solution, however, must be acceptable to both parties. The figure in the lesson details the steps in the mediation process.

In order to be effective, mediators must be able to both talk and listen. A mediator must be able to help parties communicate without taking sides. Traits of a successful mediator include being:

- a good communicator.
- a good listener.
- fair.
- neutral.
- an effective problem solver.

Reading Check

Identify What are two skills shared by effective mediators?

Negotiation is an important tool in conflict resolution. **Describe the role of a peer mediator.**

STEPS IN THE MEDIATION PROCESS

1. The parties involved in the conflict agree to seek an independent mediator's help.
2. The mediator hears both sides of the dispute.
3. The mediator and the parties work to clarify the wants and needs of each party.
4. The parties and mediator brainstorm possible solutions.
5. The parties and mediator evaluate each possible outcome.
6. The parties choose a solution that works for each of them.

A key element to successful mediation is cooperation. Each party must be willing to work with the other and the mediator. **Explain how a mediator helps people in a conflict find a solution.**

Fitness Zone

Relieving Stress I tried to please everyone all of the time. I would take what people said about me too seriously. I got into conflicts with my classmates because of the stress. All the peer mediators at school knew me by name. Finally, one suggested physical activity to help relieve my stress. I joined the swim team, and guess what? It worked! I have less stress, I'm in better shape, and I haven't been to peer mediation for months.

Being a Peer Mediator

Many schools have peer mediation programs. In these programs, student mediators help other students solve their conflicts. What does it take to become a peer mediator?

A peer mediator must like to help others and be good problem solvers. Peer mediators must also be able to maintain **neutrality**, or a promise not to take sides, during the mediation process. Mediators also need to be good communicators.

If two people who are in a conflict recognize that they need help, they can consider going through the mediation process. This process includes six steps to help resolve the conflict. The steps are:

1. Both parties agree to seek an independent mediator's help.
2. The mediator hears both sides of the dispute.
3. The mediator and the parties work to clarify the wants and needs of each party.
4. The parties and the mediator brainstorm possible solutions.
5. The parties and mediator evaluate the different outcomes.
6. The parties choose a solution that works for each of them.

The mediation process should take place in a private location. The mediator should agree to keep both sides of the dispute confidential.

While both people in the conflict present their side, the mediator must listen carefully. The mediator will allow only one party to talk at a time. The mediator may ask questions for clarification. This is to make sure the mediator understands the situation or the point of view of each party involved.

The mediator then helps the two people come up with a solution that makes both people happy. The mediator and the two parties will brainstorm possible solutions. Then they will evaluate the possible solutions. Some of the solutions may not be practical. Others may not be agreeable to both people.

Mediators encourage people to decide on a solution. A mediator should not participate in the selection process. Both people must agree on the solution, or else the solution will not work. A mediator can often help the parties in conflict arrive to a solution. Sometimes the parties will have to make compromises. The overall solution, however, must be acceptable to everyone involved.

Does the idea of becoming a peer mediator sound interesting to you? If so, you might want to find out whether your school has a peer mediation program. Many schools like to use peer mediators because it allows teens to put problems into words that other teens understand. If you decide to become a peer mediator, you will need to go through training. Peer mediators volunteer their services, so training would likely happen on your own time. Find out whether your school has a peer mediation program. If your school has a program and you are interested, ask how you can get involved.

Problem Solver
Easy to Talk to
Enthusiastic
Responsible

Mature
Effective Listener
Decisive
Interested
Alert
Trustworthy
Open-minded
Reliable

One trait of an effective peer mediator is good listening skills. **Explain why you think it is important for a peer mediator to have these traits.**

Lesson 4 Review

What I Learned

1. **VOCABULARY** What is *neutrality*?

2. **LIST** What are five traits of a good peer mediator?

3. **EXPLAIN** When should the parties of a conflict turn to mediation?

Thinking Critically

4. **APPLY** Think about a conflict you experienced. Then explain how your skill of conflict resolution helped or could have helped you reach a solution.

Applying Health Skills

5. **CONFLICT RESOLUTION** Imagine you are involved in a conflict with a classmate. You go to the peer mediation program at your school. One of the other students acts as your mediator and your conflict is eventually resolved. Both you and your classmate are satisfied with the result. Why do you think you were able to work out the issue with the peer mediator?

Violence Prevention

LESSONS
. . . .

7

Understanding Violence

Before You Read

Quick Write What steps do you take to keep yourself safe? Write a paragraph about these strategies.

Vocabulary

homicide
assault
rape
gang

BIG IDEA Even the threat of violence can affect a person's well-being.

Violence in Society

MAIN IDEA Many factors contribute to the violence in society.

Have you noticed how many news stories are about acts of violence? Popular TV shows, movies, music, and video games often feature violent content.

Violence is a major national health concern. A Centers for Disease Control and Prevention study found that violence sends more than 750,000 young people to the hospital each year.

Types of Violence

The most common violent crimes in the U.S. are **homicide**, **assault**, **rape**, and robbery. Homicide is a violent crime that results in the death of another person. Assault is an attack on another person in order to hurt him or her. Rape is forced sexual intercourse. Robbery involves taking property by force. In addition to robbery, other property crimes include burglary and vandalism.

Violence can take many forms. It can be verbal, physical, or sexual. Examples of each type are:

- **Verbal** Name-calling and teasing.
- **Physical** Harming someone physically, such as hitting, slapping, or kicking.
- **Sexual** Unwanted touching if consent is not given. It can include hugging or kissing if the person has said no to being touched. Hugging or kissing might seem like friendly behaviors. Both are abusive if the other person says no. Before touching another person, always ask the person if the touching is okay.

Another form of violence is bullying. When one person intimidates another for a long period of time, it is bullying. Another form of bullying is spreading rumors about someone.

All types of violent behaviors have causes. One person may want to control or have power over another person or group of people. Some people may become violent because they are frustrated or angry. They don't know how to express their frustration or anger in a healthful way. Sometimes a person who is violent has been a victim of violence themselves. Drug use is another factor that can contribute to violent behavior.

Reading Check

Identify What are the four most common violent crimes?

Factors in Teen Violence

MAIN IDEA A variety of factors influence teens toward violence.

Some use violence to get respect from their peers. Others use it to show their independence. Some use it when they feel they are being controlled and think it is their only choice.

Most teens do not use violence to solve problems. Teens who commit violence may not have learned to deal with feelings such as anger in healthful ways. Using violence to solve problems is not a real solution. It can affect personal goals if a teen is suspended or expelled from school. Teens who use violence to solve problems may also engage in other risk behaviors.

Teens who become violent are influenced by similar factors. Some of these factors are external or societal. They are prejudice, peer pressure, media influence, drug use, gang activity, or weapon use. Many times, teens who use violence have seen violence at home.

- **Prejudice.** Having an unfair negative opinion of a group of people is prejudice. Prejudice is a form of discrimination. When a group of people share the same prejudice, they may begin to feel that the prejudice is accepted. This may escalate their behavior to include violence. Prejudice can lead to hate crimes. A hate crime is an illegal act that targets a member of a particular group of people.
- **Peer pressure.** Pressure from others may cause a teen to go against his or her personal values. In some cases, a group might pressure a teen into committing a violent act.

Reading Check

Evaluate How do external factors, such as prejudice or violence in the media, influence attitudes about violence?

More than 750,000 young people go to the emergency room each year due to violence. **List the three types of violence.**

- **Media influence.** Studies show that by age 18, teens will see thousands of violent acts on TV. Violence is also a common theme in movies, music, and video games. Research shows a link between media influences and teen violence.

- **Drugs.** Studies show that teens who use drugs are more likely to engage in acts of violence. Drugs and alcohol affect your judgment and ability to make healthful decisions.

- **Gangs**. A gang is a group of young people that come together to take part in illegal activities. Some teens join a gang to gain a sense of belonging, for protection, or because a family member or friend has joined. However, gang members may carry weapons, sell drugs, and commit violent acts. The risks of being in a gang include an increased risk of committing a crime, an increased risk of using drugs, increased risk of being sentenced to a youth detention center or jail, and a decreased chance of graduating from high school.

- **Weapons.** In a recent survey, 17 percent of teens admitted to carrying a gun or other weapon at some point. Statistics show that people who carry guns are more likely to be injured as a result of gun violence. It is important to immediately report a weapon that is found or that a peer has possession of.

One way to defeat violence is through education. **Cite an example of a way to help others learn to avoid homicide, rape, and assault.**

Playing violent video games can affect the way teens think about violence. **Explain how playing video games might affect a teen's thinking.**

Effects of Violence

MAIN IDEA Victims of violence can suffer physical as well as mental and emotional effects.

A teen is twice as likely as an adult to be a victim of violent crime. Violent crimes, including sexual assault and rape, always have an effect on the survivors. They may have physical injuries. They may also have emotional injuries.

At times, the emotional injuries that result from being the victim of a crime may cause difficult emotions. For some victims, the emotional injuries may cause more difficulty than physical injuries. Survivors of violent crimes, including rape, need help to emotionally recover. Victims of sexual assault and rape may feel depressed or develop post-traumatic stress disorder. A trusted adult or mental health professional can help. With the right help, victims of violent crime can recover.

IF YOU ARE THE VICTIM OF A VIOLENT CRIME, TAKE ACTION TO HELP YOURSELF RECOVER.

Experiencing violence can cause emotional injuries. **Explain how survivors of violence can get help for those injuries.**

A TEEN IS TWICE AS LIKELY AS AN ADULT TO BE VICTIM TO A VIOLENT CRIME.

Reading Check

Identify What should you do if you were to experience a violent crime?

One of the effects of violent crime is physical injuries. Another effect is injury to mental/emotional health. Survivors of sexual assault or rape say that the assault has caused depression. Some survivors can develop post-traumatic stress disorder (PTSD) as a result of the crime. Sometimes, a witness to a crime can develop PTSD too. These survivors need help from a person trained in recovery from violence. A trained counselor can help survivors recover.

Victims of violent crimes need help to recover. They also need to report the crime in order to make sure that the attacker is caught and punished. If you are the victim of a violent crime, take action to help yourself recover.

- **Get medical attention.** You may have injuries you are not aware of. You may be in a state of shock, which can temporarily block out pain. Rape victims need to be tested for STDs that may have been transmitted during the rape.
- **Report the incident to the police.** It is important to file a report. This will help bring the responsible person to justice and help prevent that person from harming someone else.
- **Get treatment for the emotional effects of the crime.** A violent crime is a traumatic experience. Professionals can help you work through the emotional and psychological pain caused by the crime. Counseling can help survivors of a crime recover from the experience. They can then move on with their lives.

The person who commits a violent crime will also be affected by it. The person may become violent because they need help dealing with problems in their life. The person may feel angry or frustrated about the things happening in their life. Those problems cannot be solved through violence. In fact, becoming violent will only make them worse.

Becoming violent can lead to consequences that affect a person for life. They may be charged with a crime by police. That person may spend time in jail or a facility that houses juvenile offenders. Goals for the future might also be affected. They may harm their chances to get some types of jobs or even the college of their choice.

Myth vs Fact

Myth Victims of violence are partly to blame for the attack.

Fact People who experience violence are not at fault for the acts committed against them. Violence directed toward a person, except in self-defense, is a crime.

Lesson 1 Review

What I Learned

1. **IDENTIFY** What are three actions that the subject of a violent crime should take?

2. **DESCRIBE** What are the risks associated with being in a gang?

3. **CONTRAST** What are the differences between verbal, physical, and sexual violence?

4. **EXPLAIN** What are some causes that different types of violent behavior have in common?

Thinking Critically

5. **EXPLAIN** Why should a student report to school authorities if another student brings a weapon to school?

6. **DESCRIBE** What are three risks that teens face if they become involved in gangs?

7. **EXPLAIN** How can prejudice and/or discrimination lead to violence?

Applying Health Skills

8. **REFUSAL SKILLS** Your friend's brother has recently joined a local gang. You know that the gang sometimes commits violent crimes. They also use drugs and carry weapons. You are approached about joining the gang. How would you respond?

Violence Prevention

. . . .

Before You Read

Quick Write Find a news story about someone who has experienced crime or violence. Write a one-paragraph summary of the story.

Vocabulary

youth court

. . . .

BIG IDEA You can help prevent violence.

Protecting Yourself from Violence

MAIN IDEA You can take a number of steps to help protect yourself from violence.

Acts of violence can happen anywhere, including at home, at school, or in your neighborhood. However, you can take a number of steps to help protect yourself. You can learn to avoid unsafe situations that may lead to violence.

- Remember that guns can cause serious injury or death. Avoid picking up a gun, even if you think it is unloaded.
- If a stranger stops his or her car to ask you for help or directions, walk or run in the opposite direction. Do not get close to the car.
- Avoid carrying your backpack, wallet, or purse in a way that is easy for someone to grab.
- If another person threatens you with violence, give that person the money or valuable. Your safety is more important than your possessions.
- Make sure your family always knows where you are, where you are going, and when they can expect you home.
- Walk with a friend. After dark, walk only in well-lit areas. Avoid dark alleyways and other places where few people are nearby.
- When you are at home, lock your doors and windows. Open the door only for people you know. If your family has a rule against answering the door, do not open it at all.
- If someone is using a weapon at school, follow your school's plan to protect yourself or to escape.
- Avoid giving out personal information when you answer the phone or respond to a text message or e-mail. Do not tell anyone you are home alone.
- Tell a parent or other trusted adult if you know of a situation that could cause injury to yourself or another person.

You can also take steps to protect yourself from gangs. If you live in an area that has gang activity, avoid wearing gang-related colors or clothing. When you are walking on the streets, avoid wearing expensive jewelry. Avoid carrying expensive electronics that might be a target of theft. Another way to help stay safe from gang violence is to choose friends who are not members of a gang.

A positive way to promote good overall health is to avoid gang involvement. To avoid gangs, practice positive alternatives. These activities can also promote good health. Some teens join an afterschool program. A sports team or club are two examples of afterschool programs that can help a teen resist gang involvement. Afterschool activities may also allow you to develop a new hobby or explore new interests. Another way is to volunteer in your community. Getting involved in the community can help make it safer for everyone. It's also a positive alternative that prevents you from being affected by gangs.

Practicing safe habits online is another way to protect yourself. If you are like many teens, you may communicate with many people online. However, avoid meeting an online friend without first learning more about that person.

- Is your online friend really another teen? Sometimes, an adult who could harm you may pretend to be a teen online.
- If you do agree to meet an online friend in person, share your plans with a parent or another trusted adult.
- Make sure you meet your online friend in a public place. First-time meetings are always safer in an area where other people are around in case you need help.

During the teen years, many people form new friendships and start to date. Make sure your friendships are built on respect. Healthy relationships also include caring and honesty. If a friend or dating partner becomes violent or abusive, avoid the person and tell a parent or other trusted adult.

Avoiding unsafe situations can help prevent violence. **List two ways to avoid unsafe situations.**

Dating Violence

Dating violence can affect anyone. It is not limited to people of a certain economic status, gender, race, or ethnic group. A 2019 fact sheet by the CDC notes that among high school students, about 10 percent of females and about 4 percent of males have experienced physical and/or sexual dating violence in the past year.

It's difficult to think that a person you care for can become mean or violent. Sometimes, the violence may begin when a dating partner says mean things to you. The violence can then escalate to assault and even rape. Unhealthy relationships that may lead to dating violence include those in which one or both partners:

- believe that dating violence is acceptable
- are depressed or have symptoms of trauma
- are aggressive to peers and others
- use drugs or other illegal substances
- engage in early sexual activity and have multiple sexual partners
- know someone who is involved in dating violence
- have serious conflicts with a dating partner
- live in a home where violence occurs

> **HEALTHY RELATIONSHIPS INCLUDE CARING AND HONESTY.**

Dating violence is an issue that even adults have a difficult time handling. To avoid the likelihood of dating violence, build your friendships on respect. Healthy relationships include caring and honesty. If a friend or dating partner becomes violent or abusive, avoid the person and tell a parent or other trusted adult.

Avoiding dangerous situations may not always be possible. Having good communications skills can help you get help. Using good communications skills can help others understand the nature of the threat you are facing. If you feel that you are in a dangerous situation, some information you should tell is:

- Were you threatened by another person? What did the person say? What did the person threaten to do?
- Does the threatening person have a weapon? If so, what type of weapon?
- Is the threatening person using drugs or under the influence of alcohol?
- Is the threatening person a member of your family? Is this a situation involving domestic violence?

Ways to Reduce Violence

MAIN IDEA Everyone can make an effort to help reduce the spread of violence.

Violence affects victims, their families, and our society. A CDC study shows that the cost of violence among youth costs the U.S. more than $21 billion each year. Some of these injuries were so severe that they person injured will need support for the remainder of his or her life.

- Develop your own personal zero tolerance policy regarding violence. This might include making a commitment never to fight with or threaten others.
- Encourage others to resolve conflicts peacefully. You can be a role model for nonviolence.
- Encourage your family to become a member of a Neighborhood Watch program. These programs include volunteers who work closely with law enforcement.
- Report any acts of violence you witness. Talk to your parents, guardians, or a trusted adult and ask what you should do next. The adult may help you contact law enforcement officials.

You can also become an advocate, or supporter, of safety and victims' rights. Some organizations you might become involved with are the National Center for Victims of Crime or the National Crime Prevention Council. These organizations can tell young people where they can go for help if they experience violent crime.

Reading Check

Explain Why should you tell a trusted adult when you witness an act of violence?

Many schools today are stepping up measures to prevent acts of violence. **What steps have officials in your school taken to make you feel safer?**

Reading Check

List What are four ways schools are working to reduce violence?

Reducing Violence in Schools

You spend a lot of time at school, so it is important to know what your school does to help keep you safe. Most schools are safe places. However, each year more than 3 million students have experienced crime at school. Two million of these crimes are violent. Many schools have taken the following steps to help reduce violence on campus.

- **Anti-violence training.** Many schools hold training for students, teachers, and other school personnel. These trainings teach everyone how to respond if violence erupts while in the school. This type of training can be aimed at reacting to a shooting incident at the school.
- **School uniforms.** Some schools use dress codes or uniforms to help keep students safe. When all students wear similar clothing, it's difficult to tell if one student is wealthy and another is not. As a result, there may be less violence. Dress codes and uniforms also affect gang members. Dress codes make it difficult for gang members to dress in clothing that shows what gang they are in.
- **Security systems.** Many schools have security systems. They may limit entry to just one door. All of the other doors are kept locked. They may have metal detectors, security cameras, or security guards. These help keep weapons out of schools. Schools may also have a school resource officer from the local police department. Resource officers are usually on campus during school hours. They get to know the students, which helps to prevent problems.
- **Locker searches.** Some schools periodically search students' lockers. Schools may also search students' backpacks. Sometimes trained dogs are brought in by the police. The dogs are used to sniff out drugs and weapons.
- **Conflict resolution programs.** More and more schools are educating students in conflict resolution. Peer mediation and **youth court** programs are popular. Youth court is a special school program where teens decide punishments for other teens for bullying and other problem behaviors. These programs involve teens working to help other teens resolve problems in nonviolent ways.

Reducing Violence in Communities

Students can often influence their peers to make healthful choices. In some schools, students have started their own programs to reduce violence. They put their ideas into practice in their school and in their community.

Many communities work together with law enforcement to make their neighborhoods safer. **What is being done in your community?**

Communities are also taking steps to reduce violence. Some communities use their resources to create after-school programs. These programs may be academic, recreational, or cultural. They offer a safe place for teens to spend their afternoons. Teens can stay until their parents or guardians get home from work.

Improved lighting in parks and at playgrounds is another way communities work to reduce violence. Crimes are less likely to happen in well-lighted areas. People who commit crimes are more likely to be seen and recognized if the area has lots of lights.

> IN SOME SCHOOLS, STUDENTS HAVE STARTED THEIR OWN PROGRAMS TO REDUCE VIOLENCE.

Neighborhood Watch programs are also popular in many communities. Members of the program watch their neighborhoods for signs of trouble. They report to authorities if they see suspicious activities.

Some communities put their patrol officers on foot, on bicycles, or on horses. This allows the officers to be closer to people. They get to know the people of the community. This helps them prevent criminal activity.

Protect yourself by avoiding dangerous situations. Walk directly to and from your home. Travel with another person or in a group. Avoid taking shortcuts through unfamiliar or unsafe areas of your community.

Developing Good Character

Speaking Out About Violence Good citizenship includes helping to keep your school safe. If you know of violent activities at school, or of violence that is going to take place there, tell school officials right away. Find out if your school has a hotline or a special way to report rumors of violent activities. If it doesn't, think about how you can report a potentially dangerous situation and still feel safe yourself. Write your ideas in a brief paragraph.

Lesson 2 Review

What I Learned

1. **EXPLAIN** What are communities doing to prevent the spread of violence?

2. **RECALL** What can you do to help reduce the spread of violence?

3. **IDENTIFY** What are schools doing to help eliminate violent behavior?

4. **DESCRIBE** How does the presence of a gun increase the risk of serious injury or death?

Thinking Critically

5. **ANALYZE** Some people argue that school searches violate a person's privacy. How would you respond to this argument?

6. **APPLY** What strategies could you use to keep yourself safe walking home from a movie theater?

7. **SYNTHESIZE** List ways you can escape if you are in a situation where someone is using a weapon.

Applying Health Skills

8. **GOAL SETTING** Choose one way you can help to reduce the spread of violence at school or in your community. Use the goal-setting steps you have learned to make a plan. Show your plan to your teacher. Follow your plan for a week, and then write a paragraph about your experience.

Abuse

Before You Read

Quick Write Write a paragraph that describes problems that might affect a relationship. Identify healthy ways of dealing with the problems.

Vocabulary

battery
domestic violence
human trafficking

BIG IDEA Abuse affects the physical, mental/emotional, and social health of the person who is abused.

What Is Abuse?

MAIN IDEA Abuse can happen in various ways and take many forms.

Relationships, even close ones, have their good days and bad days. In a healthy relationship, people respect and care for each other. Sometimes relationships become unbalanced and difficulties arise. When this happens, abuse, or the physical, emotional or mental mistreatment of another person, can occur.

Abuse is never the fault of the person who is being abused. It can affect people of all ages, races, and economic groups. All forms of abuse are wrong and harmful. A person who is being abused needs to tell someone about the abuse. Teachers, counselors, nurses, and physicians are required to report suspected abuse to the police. Police investigate reports of abuse to determine whether a crime has been committed. The chart below gives some statistics about abuse.

Abuse can happen to people of all ages. **Why do you think people in certain age groups may be more likely to be abused?**

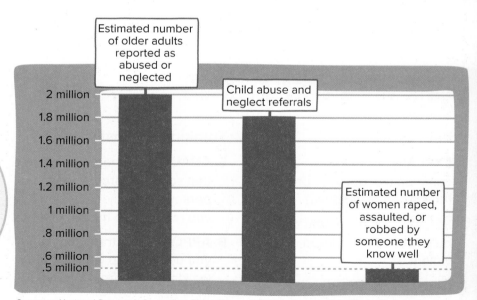

Estimated number of older adults reported as abused or neglected

Child abuse and neglect referrals

Estimated number of women raped, assaulted, or robbed by someone they know well

2 million
1.8 million
1.6 million
1.4 million
1.2 million
1 million
.8 million
.6 million
.5 million

Sources: National Research Council on Elder Abuse, U.S. Department of Justice, U.S. Department of Health and Human Services.

Types of Abuse

An abuser might use several different methods to hurt others. They are known as the six types of abuse. They include:

- **Physical** abuse which occurs when a person is physically hurt.
- **Emotional** abuse is harming another person's self-esteem by saying hurtful, unkind things to that person.
- **Sexual** abuse can include rape and incest. It can also include sexual activity when one person is below the age of consent. Human trafficking is a form of sexual abuse.
- **Spiritual** abuse is the use of a toxic culture or shaming within a church or other congregation to control others.
- **Financial** abuse is the theft or misuse of another person's money or valuables.
- **Technical** abuse or the use of technology to hurt another person. This can include spreading untruthful rumors about another person through social media, online gaming sites, or posting embarrassing photos of another person. A technical abuser might use technology to identify possible victims of abuse.

When children are abused within a family, this type of abuse is referred to as neglect. It is a type of physical and emotional abuse.

Abuse can occur in all kinds of relationships. Within families, the most frequent forms of abuse are physical and emotional. It should never be tolerated. Abuse can occur within families. Children may be abused by parents or guardians. One spouse may abuse the other spouse. Siblings may abuse siblings. A younger family member may abuse an older relative.

Abuse can also occur between dating partners. Healthy dating relationships are based on each partner respecting the other. When one partner doesn't respect the other, abuse may occur. The abuse may be physical, emotional, or psychological.

Often, abusers will try to make the people they abuse believe that they deserve to be treated harshly. This is never true. Abuse is not the same as discipline. A parent may use discipline, such as a time-out or grounding, to correct behavior or help shape a person's character. On the other hand, abuse causes severe harm to a person—physically, mentally/emotionally, or socially. No one ever deserves to be abused.

Emotional abuse can include saying hurtful things to another person. **Explain what a child who is being abused can do to stop the abuse.**

Reading Check

Identify Name some ways in which a person can be affected by abuse.

ABUSE IS NEVER THE FAULT OF THE PERSON WHO IS BEING ABUSED.

Physical Abuse

Physical abuse causes physical harm. Another form of physical abuse is **battery**, or the beating, hitting, or kicking of another person. Battery is common in cases of **domestic violence**, which is the physical abuse that occurs within a family. Domestic violence takes place in a family setting. This is where most abuse happens. Domestic violence is about power and control. The abuser tries to keep strict control over one or more family members. Any type of physical abuse is assault. Assault is a crime and should be reported to the police.

Domestic violence includes pushing, slapping, punching, and choking. Sometimes household items are used as weapons. Often, when a young victim of abuse is taken for medical attention, the abuser makes excuses for the injuries. The abuser might claim that the child is clumsy and fell down. Similar excuses are often used when the person who has been abused is an older adult. Domestic violence can also include financial abuse. This type of abuse occurs when the abuser controls the finances in the relationship. This will make it more difficult for victim to leave.

Emotional Abuse

Physical abuse often leaves physical signs. It may not be as obvious that a person is being emotionally abused. However, the effects of emotional abuse can be harmful too. The effects of emotional abuse may last even longer than the effects of physical abuse.

Emotional abuse involves words and gestures. These are used to make a person feel worthless, stupid, or helpless. Bullying, yelling, and teasing are all forms of emotional abuse. Insults, harsh criticism, and threats of violence are also forms of emotional abuse. People who have been emotionally abused often feel bad about themselves.

Emotional abuse can occur in the home, at school, or with friends. Sometimes emotional abuse occurs in a dating situation. For example, your partner may be very jealous and want to know where you are all of the time. Your partner may keep you from spending time with family and friends. These are forms of emotional abuse. Spiritual abuse can occur when a faith leader uses their position to shame or control members of the congregation. If you or someone you know has been emotionally abused, get help from a trusted adult.

Emotional abuse can make a person feel isolated or unwanted. **Describe ways you could reach out to a person who needs help.**

Neglect

Everyone has needs. We all need food, clothing, and shelter. We also need medical care and education. Children need supervision. Children are dependent on someone else to provide for their needs. Sometimes, through neglect, children's needs are not met. This type of abuse sometimes also affects adults who cannot care for themselves.

People also have emotional needs. People need to feel loved and nurtured. If caregivers don't provide for the needs of the people who depend on them, they may be guilty of neglect. Neglect is against the law and should be reported to the police.

About 1 out of every 10 children in the United States is a victim of neglect or child abuse. This means that each year about ten percent of the children are not having their needs met. Neglect can have long-lasting physical and emotional effects on children.

Sexual Abuse

Every two minutes someone in the U.S. is sexually abused. It is any type of unwanted touching, kissing, or other sexual activity. Sexual abuse is always a crime. It should be reported to police.

Abusers may use complex techniques to identify potential victims. The abuser will use manipulation to control the victim. One way is to make the victim think he or she is to blame for the abuse. A victim is never at fault for the abuse.

Some abusers target victims using technology, a form of technical abuse. Chat rooms or other social media sites allow you to meet new people. However, some people use these sites to find others to exploit. The abuser may lie about his or her age. The abuser will try to develop a friendship with the victim.

Computers and other technology can be used to identify others for exploitation. Photography and video can also be used to exploit a victim. It is illegal to take photographs or videos of a person under the age of 18 who is undressed or engaging in sexual activity. Using technology to distribute photos and videos is also illegal. It is also illegal to force a person under the age of 18 to look at sexually explicit materials.

> **SEXUAL ABUSE IS A SERIOUS CRIME.**

Several resources are available to help teens who are being sexually abused. Sexual abuse is a crime. One site that offers information to children and teens include KidsHealth.org.

Sexual abuse is a serious crime. In 2012, Erin's Law was introduced to raise awareness of sexual abuse and to stop it from happening.

Reading Check

Explain What is the purpose of Erin's Law?

It can be difficult for a person who has been abused to talk about it. **If you thought someone was being abused, who might you suggest that person turn to?**

Erin's Law provides tools to help adults recognize the signs of child sexual abuse. The law requires that states provide students with information telling them what they can do if they are being abused or know someone who is being abused. As of 2019, Erin's Law was the law in 37 U.S. states. Supporters of the law hope to have it passed in all 50 U.S. states soon.

Sexual harassment is another form of sexual abuse. Sexual harassment is the use of words, touching, jokes, and gestures that have a sexually explicit manner or meaning. Sexual harassment is illegal and should be reported to a trusted adult immediately.

You can learn the laws in your state related to sexual abuse and sexual harassment. Go online and search for your state's laws. Each state has a website listing legislation passed by the Governor and state representatives.

Human trafficking is a form of sexual abuse. It is the recruitment, transportation, transfer, harboring, or receipt of persons by improper means for the purpose of using those people as forced labor or for sexual exploitation. Human trafficking is a form of slavery. It is a crime. Criminals transport victims to other countries so that they can control them. The victims of human trafficking can be any age or gender. Teens who are at risk for becoming human trafficking victims include:

- Homeless or runaway teens
- Teens living in poverty
- Victims of physical or sexual abuse
- Teens in the foster care system
- LGBTQ teens
- Teens who are substance abusers
- Teens who have experienced disruptions in normal development

To combat human trafficking, the U.S. passed the Trafficking Victims Protection Act (TVPA) in 2000. The law was renewed in 2013. The law defines a human trafficking victim, enhances laws that prosecute the criminals who traffic other people, and helps the victims of trafficking. The TVPA also supports awareness campaigns that help people recognize human trafficking victims. It also raises awareness of the toll that human trafficking takes and to reduce its demand.

Effects of Abuse

MAIN IDEA People who are abused can experience long-term effects.

Physical scars from abuse may go away after time. That is not always the case with the emotional scars. People who experience abuse may blame themselves. They may be afraid or ashamed to get help. Children who suffer abuse or neglect often have a number of problems.

Physical health concerns include:

- Impaired brain development.
- Impaired physical, mental, and emotional development.
- A hyper-arousal response by certain areas of the brain. This may result in hyperactivity and sleep disturbances.
- Poor physical health, including various illnesses.

Mental/emotional health consequences include:

- Low self-esteem.
- Increased risk for emotional problems such as depression, panic disorder, and post-traumatic stress disorder.
- Alcohol and drug abuse.
- Difficulty with language development and academic achievement.
- Suicide.
- Eating disorders.

Social health consequences include:

- Difficulty forming secure relationships.
- Difficulties during adolescence.
- Criminal and/or violent behavior.
- Abusive behavior.

It is important for people who have been abused in any way to report the abuse. Reporting will help prevent the abuse from occurring again. It can also help the person begin to heal. Help is available. If you or someone you know experiences abuse, seek help. Talk to your parent or guardian, a school official, the police, or another trusted adult.

Abuse can affect the physical, social, and mental/emotional health of a person who has been abused. **Name some specific effects of abuse.**

Lesson 3 Review

What I Learned

1. VOCABULARY What is *abuse*?

2. DESCRIBE Name and describe the six types of abuse.

3. IDENTIFY What are four social health concerns for survivors of abuse?

Thinking Critically

4. ANALYZE Why are children and older adults often the subjects of abuse?

5. EVALUATE Do you think it is important for survivors of abuse to get help? Why or why not?

Applying Health Skills

6. ADVOCACY Locate and research local agencies and organizations that provide help for people who have experienced abuse. For each agency or organization, write down the name, contact information, and whether they specialize in a particular type of abuse. Make copies of the list for your school counselor to give to students.

Preventing and Coping with Abuse

Before You Read

Quick Write Think about adults you trust. Write one or two sentences that describe what makes them trustworthy.

Vocabulary

cycle of abuse
crisis hot line

BIG IDEA They cycle of abuse can be stopped, but it often requires outside help.

Warning Signs of Abuse

MAIN IDEA Abuse has warning signs and risk factors.

You may notice injuries, such as bruises and burns, on a person who has been abused. The person may have no explanation for the injuries. Sometimes a person who is being abused might appear to be withdrawn or depressed. A person who is being abused may also become aggressive toward others. If you can spot the signs of abuse, you may be able to help.

All families have problems now and then. Usually the problems are dealt with in healthful ways. Good communication skills can help families solve problems in healthful ways. However, some families may not know how to handle problems when they arise. This can lead to extra stress. The stress of family problems can also increase the risk of abusive behavior. The figure in this lesson shows additional risk factors.

The Cycle of Abuse

Often, abuse within a family may have begun long ago. Health experts have found that patterns of abuse may go back many generations. A child who has been abused or who witnessed abuse may grow into an adult who abuses others. The **cycle of abuse** is a pattern of repeating abuse from one generation to the next. It does not have to continue forever in a family. However, breaking the cycle of abuse often requires outside help.

Breaking the Cycle of Abuse

Each person has the power to help end the cycle of abuse. People who have been abused—especially children—often feel ashamed. This is especially true if the abuse was sexual. People who have experienced domestic violence may have similar feelings. A person might be afraid that the family will be separated if news of the abuse gets out. A child might also be afraid of getting the abuser in trouble.

However, the cycle of abuse will not end until it is reported. Again, a person who is abused needs help, and so does the abuser. In many cases, the abuser will be required to seek treatment. If the abuser can learn to manage his or her behavior in more healthful ways, it can help end the cycle of abuse.

Reading Check

Define What is a cycle of abuse?

Abusers will often threaten their victims or make them promise not to tell anyone. If you or someone you know is being abused, remember that keeping it secret is not healthful. The only way anyone will get help is if the abuse is reported. Remember that many people have experienced abuse. People who have been abused have many sources of help. The National Domestic Violence Hotline (https://www.thehotline.org/), helps people who are being abused. The hotline is free and open 24 hours a day, every day of the year.

Reading Check

Identify What are three reasons that people who are abused often stay silent?

Help for Survivors of Abuse

MAIN IDEA Many resources are available to help people who have been abused.

Abuse in a family affects the whole family. It affects the abuser, the person who has been abused, and other relatives. The effects can be long-lasting. Even if a child only witnesses abuse, he or she may grow up thinking abuse is acceptable behavior. In such cases, every family member needs help. They need to learn that abuse is not acceptable.

Why Survivors Stay Silent

Many people who have been abused do not tell anyone about it. This prevents them from getting the help they need. Survivors of abuse might choose to stay silent because:

Warning Signs of Abuse	
• Illness	• Alcohol or other drug use
• Divorce	• Unemployment and poverty
• Lack of communication and coping skills	• Feelings of worthlessness
• History of having been abused as a child	• Emotional immaturity
	• Lack of parenting skills
	• Inability to deal with anger

This table presents some factors that can increase the risk of a person becoming abusive. **Explain how identifying and working to improve these factors can help prevent abuse.**

- Some people think no one will believe them.
- A child may fear that adults will think he or she is lying.
- A person may think abuse is a private matter.
- Some people believe they deserved the abuse because of something they did. No one ever deserves to be abused.
- A person may think their abuser will seek revenge.
- Some people are ashamed that abuse occurs within their family.

Men and boys often think that because they are males, they should be able to protect themselves from abuse. This is not true. Abusers usually have an advantage over their victims. They may be stronger and older. An abuser may be in a position of trust and authority. Abusers may make a person feel afraid. They may threaten to harm the victim and that person's family if the victim tells.

Developing Good Character

Helping a Victim of Abuse If someone you know has been a target of violence, you can show concern and compassion for that person. You can listen if the person wants to talk. You can help him or her know when to seek help from a parent or other trusted adult. Identify some communication skills that would be helpful in this type of situation. Describe some other ways in which you could show the person that you care.

Help is available within communities for people who have been abused and their families. **List two resources for students in your school to contact if they need to report abuse.**

Sometimes, adults who experience domestic violence make excuses for their injuries or for their abuser. This behavior is called enabling. Enabling creates an atmosphere in which a person can continue unacceptable behavior. Enabling can establish a pattern of abuse. People who experience abuse should instead seek help for themselves and for their abusers. Remember, people who are abused are never to blame for the abuse.

Sources of Help for Victims of Abuse

If you have a friend you think is being abused, try talking to that person about it. Encourage your friend to seek help. If your friend is too afraid to seek help, go talk to a trusted adult yourself. Let the adult know that you are concerned and worried that your friend is being abused. Remember that child abuse is a crime, and it should be reported to the police.

Talking with a parent or other trusted adult can be one way to help make abuse stop. A **crisis hot line** is a toll-free telephone service where abuse victims can get help and information. It offers many types of assistance to people who have been abused.

Services provided by crisis hot lines may include help for the abused person, family members, and others who may be affected. People who answer the phones at hotlines have been trained to deal with abuse problems. They know how to help people who have been abused. The caller's identity is kept anonymous.

Crisis hot line conversations are always kept confidential. A chart listing the names of organizations offering help, with contact information, is included on the next page. Additional hotline numbers can be found online.

Families experiencing extreme abuse can also escape abuse by moving into a shelter. A shelter is run by the community. It is a place where people can feel safe. The location of shelters is private. An abuser can't find family members and continue the abuse.

A shelter is a safe place to live. Family members receive help to break the cycle of abuse. Other services provide help finding jobs.

In the United States, contact the National Domestic Violence Hotline. The telephone number is 1-800-799-7233. The web site, https://www.thehotline.org/, also has a chat feature.

Recovering from Abuse

Many times, people who have been abused need help in order to recover. Professional counselors are trained to help people overcome emotional trauma. A counselor can help a survivor of abuse by helping the person understand his or her feelings. The counselor can also help a survivor learn ways to manage feelings. Group counseling is another option. Teens meet and talk with other teens who have also experienced abuse.

Organization	Whom They Help
Childhelp USA 1-800-4-A-CHILD (1-800-422-4453) www.childhelphotline.org	Child abuse victims, parents, concerned individuals
YouthLine 1-877-968-8491 Text: teen2teen to 839863 oregonyouthline.org	Teens helping teens. YouthLine is a national teen help and crisis hotline based on Oregon.
Love is Respect 1-866-331-9474 Text: LOVEIS to 22522 www.loveisrespect.org	Child sexual abuse victims, parents, offenders, concerned individuals
National Domestic Violence Hotline 1-800-799-7233 www.thehotline.org	Children, parents, friends, offenders
Girls and Boys Town 1-800-448-3000 www.yourlifeyourvoice.org	Abused, abandoned, and neglected girls and boys, parents, and family members

NAIC, U.S. Department of Health and Human Services.

These organizations help people who have been abused by providing hot lines where people can talk about abuse and get advice. **Identify some others who may be helped by these organizations.**

Children and families may need professional counseling to recover from the abuse. This may take the form of family counseling, where everyone in the family attends counseling together. Children and teens often find group counseling comforting and helpful. In group counseling, children and teens meet with their peers who have also experienced abuse in the family. During counseling sessions, they support each other and learn new strategies for recovering from the abuse.

If you have experienced abuse, use what you have learned to get help. If you suspect a friend has been abused, share your knowledge. Encourage that person to get help. Remember that with help, recovery is possible for both the victim and the abuser.

Reading Check

Identify What are some sources of help for people who have been abused?

Lesson 4 Review

What I Learned

1. **IDENTIFY** What are three warning signs of abuse?

2. **EXPLAIN** What is the cycle of abuse?

3. **LIST** Why do some people not report abuse?

Thinking Critically

4. **ANALYZE** Why are some abusers older than their victims?

5. **EVALUATE** Why do you think crisis hot lines keep their callers' identities anonymous?

Applying Health Skills

6. **COMMUNICATION SKILLS** Write a dialogue between yourself and a person who has been abused. That person wants to report the problem, but doesn't know where to begin.

Nutrition

LESSONS

1 Nutrients Your Body Needs

2 Creating a Healthful Eating Plan

3 Managing Your Weight

4 Body Image and Eating Disorders

8

Nutrients Your Body Needs

Before You Read

Quick Write Make a list of foods that you think are high in nutrients. After reading the lesson, check to see if you were right.

Vocabulary

nutrients
nutrient deficiency
nutrition
proteins
amino acids
carbohydrates
fiber
fats
saturated fats
trans fatty acids
unsaturated fats
cholesterol
vitamins
minerals

> EATING A VARIETY OF FOODS CAN PROVIDE ALL THE NUTRIENTS THAT A PERSON NEEDS.

Reading Check

Identify What is the basic substance that makes up proteins?

BIG IDEA Each nutrient plays a specific role in keeping your body healthy.

Understanding Nutrients

MAIN IDEA Nutrients are important to keeping you healthy.

Imagine that you're riding in a car, and it suddenly runs out of gas. The car will stop, because it can't run without fuel. Your body is the same way, but it gets its fuel from food. Your body is always using energy, even while you sleep. **Nutrients** are substances in food that your body needs to grow, have energy, and stay healthy. Nutrients come in two types. One type gives you the energy you need for work and play. The other type of nutrient provides the building blocks needed to grow and to repair your body. Since different foods contain different nutrients, you need to eat a variety of foods each day.

Eating a variety of foods can provide all of the nutrients that a person needs. Some people may prefer to eat only certain foods. These people are at risk for a **nutrient deficiency** or a shortage of a nutrient. This condition is rare in industrialized countries such as the U.S. It can cause numerous health problems.

Learning **nutrition** is the process of taking in food and using it for energy, growth, and good health. It can help you make more healthful choices about the foods you eat. Scientists have identified more than 40 nutrients. However, they all can be grouped into six main classes: proteins, carbohydrates, fats, vitamins, minerals, and water. Each of these nutrients plays an important role in keeping you healthy.

Proteins

Proteins are a nutrient group used to build and repair cells. Proteins help your body heal, grow, and develop. Foods containing protein also aid in the repair of tissues. Proteins are made of chemical building blocks called **amino acids** which are small units that make up a protein. Different foods contain different amino acids. Foods such as meat, fish, eggs, dairy products, and soybeans provide all the amino acids your body needs. They are called complete proteins. These foods are good sources of protein. Most plant foods, however, are missing at least one amino acid. A person who follows a vegetarian diet may need to combine certain plant foods, such as beans and rice to gain all the needed amino acids. Eating a wide variety of plant foods should give you all the amino acids you need.

Carbohydrates

What do a steamy plate of spaghetti and a ripe peach have in common? Both foods contain large amounts of **carbohydrates**, or starches and sugars found in foods. Your body needs carbohydrates to use as energy. Your body uses carbohydrates throughout the day and even during the night. Even as you sleep, your body continues to use energy. About 45 to 65 percent of your daily energy needs come from carbohydrates.

Two types of carbohydrates are found in foods: simple and complex. *Simple* carbohydrates are sugars. They occur naturally in foods like fruit, milk, and honey. Sugars may also be added when foods are processed. Sugars are used by the body for energy. Any sugars that are not used are stored as body fat. *Complex* carbohydrates are starches, which are made up of long chains of sugar linked together. Complex carbohydrates are found in foods such as potatoes, seeds, and whole-grain cereals. Your body cannot use these nutrients directly. First, it must break them down through the process of *digestion*.

WHAT DOES A STEAMY PLATE OF SPAGHETTI HAVE IN COMMON WITH A RIPE PEACH?

Carbohydrates are sugars that provide energy to the body. **Describe what happens to any sugars that are not used by the body for energy.**

The sugars that occur naturally in foods, such as fruit, are a healthier choice than those found in processed foods, such as cookies. As well as containing sugars, fruit also contains high amounts of fiber. In general, foods that are high in sugar tend to be low in other nutrients.

Another type of complex carbohydrate, **fiber**, cannot be digested. Fiber is a complex carbohydrate that the body cannot break down or use for energy. Fiber carries wastes out of your body. Eating high-fiber foods can lower your risk of heart disease and some types of cancer.

Fats

You may hear that you should avoid foods containing **fats**, which are nutrients that promote normal growth, give you energy, and keep your skin healthy. While it is important to reduce the amounts of foods you eat that are high in fat, your body needs some fat to function properly. Your body uses fats to build and maintain your cell membranes. These nutrients also carry certain vitamins in your bloodstream. Additionally, the fats in the foods you eat help you feel full after a meal.

A great way to include fiber in your diet is by eating whole grain cereals. **Explain why it is important to get enough fiber.**

Not all fats are the same, however. Foods like butter, cheese, and many meats are high in **saturated fats**. These are fats that are usually solid at room temperature. Over time, eating too much of these fats can increase your risk of health problems such as heart disease. The same is true of **trans fatty acids** or trans fats, which are a kind of fat formed when hydrogen is added to vegetable oil during processing. They start off as liquid oils and are made solid through processing. Stick margarine is one example. However, **unsaturated fats**, which are fats that usually remain liquid at room temperature, can actually lower the risk of heart disease. These fats come mainly from plant foods, such as olive oil, nuts, and avocados.

You probably have heard of **cholesterol**. It's a waxy, fat-like substance that the body uses to build cells and make other substances. Your body makes two kinds of cholesterol. HDL cholesterol protects against heart disease. LDL cholesterol sticks to the walls of blood vessels, which can lead to heart disease. Eating foods high in saturated fats can raise blood levels of LDL cholesterol. Dietary sources of cholesterol include meat, eggs, some seafood, whole or reduced-fat milk, many cheeses, and butter.

Reading Check

Compare and Contrast What is the difference between saturated and unsaturated fats?

FATS PROMOTE NORMAL GROWTH, GIVE YOU ENERGY, AND KEEP YOUR SKIN HEALTHY.

Choose unsaturated fats, or healthier fats, to cook foods. **Explain why unsaturated fats are a good choice to use when cooking.**

Most teens don't get enough of the nutrients listed in the chart. **Identify which of the foods listed in the chart that are part of your regular diet.**

VITAMINS AND MINERALS

Vitamins	Functions	Food Sources
Vitamin A	Important for good vision and healthy skin	Dark green leafy vegetables (such as spinach), milk and other dairy products, carrots, apricots, eggs, liver
B Vitamins	Helps produce energy; keeps nervous system healthy	Eggs, meat, poultry, fish, whole-grain breads and cereals
Vitamin C	Helps keep teeth, gums, and bones healthy; helps heal wounds and fight infection	Oranges, grapefruits, cantaloupe, strawberries, tomatoes, cabbage, broccoli, potatoes
Minerals	**Functions**	**Food Sources**
Calcium	Helps build strong bones and teeth	Milk and other dairy products, fortified breakfast cereals, oatmeal, dark green leafy vegetables, canned salmon
Iron	Contributes to healthy blood, which helps your body fight disease	Red meat, poultry, dry beans, fortified breakfast cereals, nuts, eggs, dried fruits, dark green leafy vegetables
Potassium	Helps maintain your body's fluid balance	Baked potatoes, peaches, bananas, oranges, dry beans, fish

Vitamins and Minerals

Two other kinds of nutrients that the body needs are **vitamins** and **minerals**. Vitamins are compounds that help to regulate body processes. They also help your body fight disease. Others help your body produce energy. Minerals are substances the body uses to form healthy bone and teeth, keep blood healthy, and keep the heart and other organs working properly. Your body uses only small amounts of minerals, but they are essential to your health. Minerals help to regulate body functions. The table below shows how your body uses several important vitamins and minerals.

Some vitamins, such as vitamins C and B, are *water-soluble*. This means that they dissolve in water. Your body can't store these vitamins, so it needs a fresh supply of them each day. *Fat-soluble* vitamins, such as vitamins A and D, dissolve in fat. Your body can store these vitamins until they needed.

How can you make sure you are getting enough vitamins and minerals? The best way is to eat a variety of foods. The table shows which foods are good sources of some important vitamins and minerals. Taking vitamin supplements can also help, but they are not a replacement for healthy eating.

Water

Water is essential to every body function you have. In fact, a person can live for only about a week without it. Your body uses water to carry other nutrients to your cells. Water also helps with digestion, removes waste, and cools you off. Teen girls should drink about nine cups of water per day and teen boys should drink about 13 cups of water each day.

To make sure your body is getting enough water, drink water whenever you are thirsty as well as with your meals. You can also get water from many foods, such as fruits and vegetables, and from beverages such as milk. Beverages containing caffeine cause the body to lose water. These beverages, such as coffee, some teas, and cola drinks should be used only in moderation.

If you are very active, you may need to drink more water. Your body loses water when you sweat during hot weather or when exercising. At these times, you will need to drink extra water to meet your body's needs.

Reading Check

Explain Why is water important for your body?

Myth vs. Fact

Myth Eating any food with fat in it is unhealthful.

Fact Some dietary fat is needed to keep you healthy. Those fats should be mostly from the more healthful fats, such as a small amount of unsaturated fats. To learn more, review the Dietary Guidelines for Americans at www.usda.gov.

Active people may need to drink more water than inactive people. **Explain why active people need to drink more water.**

Lesson 1 Review

What I Learned

1. **VOCABULARY** Define the term *nutrition*. Use the word in an original sentence.

2. **NAME** What are the six nutrients?

3. **EXPLAIN** How do vitamins help your body?

4. **LIST** What foods contain calcium and iron? How do these minerals function in the body?

Thinking Critically

5. **HYPOTHESIZE** How can the foods you choose to eat today affect your health in the future?

6. **ANALYZE** Research foods that are high in fiber. Identify at least three foods that are high in fiber. Then, find all of the high-fiber foods in your house. How many foods did you find? Is your family eating enough high-fiber foods?

Applying Health Skills

7. **DECISION MAKING** Keep track of the foods you eat daily for three days. Do you feel that you are meeting nutritional requirements based on your list? Use your decision-making skills to write a plan to meet your nutritional needs with the foods you eat daily.

Creating a Healthful Eating Plan

· · · · · · · ·

Before You Read

Quick Write Describe your current eating habits in a short paragraph. Include a list of the foods you eat the most and the kinds of snacks you enjoy. After you read the lesson, review your response. Would you make any changes to your current eating habits or food choices?

Vocabulary

appetite
hunger
U.S. Department of Agriculture (USDA)
MyPlate
empty-calorie foods
calorie
sodium
nutrient dense
foodborne illness

· · · · · · · · · ·

BIG IDEA Learning to make healthful food choices will help you maintain good health throughout your life.

What Influences Your Food Choices?

MAIN IDEA A variety of factors influence food choices.

Think about the last meal you ate. Why did you choose the foods you ate? Perhaps you were grabbing a quick breakfast food while rushing to school, and you needed something that was easy to carry. Maybe it was lunchtime, and you had to choose a meal in the school cafeteria. Or maybe you just had a craving for a favorite snack, like popcorn.

All kinds of factors can affect your food choices. One of the biggest factors is your own personal preferences. You may simply like the appearance, flavor, or texture of certain foods. These foods appeal to your **appetite**, which is a psychological desire for food. Appetite is different from **hunger**. Hunger is the physical need for food. When you are hungry, your body needs fuel in the form of food. When you are hungry, you may feel weak or tired. Your body is telling you that it needs fuel.

Certain foods may trigger your appetite because of their connections to your memories and feelings. For some people, the smell of baking bread triggers appetite. A person who has fond memories of times with his or her family may feel like eating bread even if his or her body does not need fuel. Sometimes, seeing one of these foods in the media, or smelling a food as it is being cooked can make you want it.

Other Influences on Food Choices

When you make food choices based on appetite, you may eat more food than you need. When you are making food choices, try to determine whether you are choosing certain foods based on your appetite or hunger. Other factors that can affect your food choices include:

- **Your family and friends.** At home, you might eat more vegetables because your parents encourage you to eat them. With your friends, on the other hand, you might choose fast food because that's what the group likes.

- **Your cultural background.** Your culture influences your food choices. You may learn about these cultural influences from your family members. Among some cultural groups, people eat more fish than meat because fish is easily available while meat is not. Some cultures prefer eating large amounts of vegetables because these are easy foods to acquire. Different cultural groups may also choose different types of foods to eat during celebrations.
- **Convenience.** You might grab a snack from a vending machine because it is handy. If your family grows vegetables or you live near a farmers' market, you may be more likely to eat fresh vegetables and fruits.
- **Media.** Advertisers use many techniques to make you want to buy their foods. For instance, you might want to try a new pizza place if the ad shows teens having a great time eating there.
- **Economics.** The amount of money that a family has to spend on food can influence what they eat. Families that have smaller food budget might need to choose foods that are less expensive. Also, foods that are shipped from other countries usually cost more. Fruits and vegetables that are not in their growing season might be shipped from other countries where they are in-season. Other foods, such as organically grown produce, also cost more because of the methods used to produce it.
- **Geography.** Geographic location also influences food choices. Living close to a grocery store and owning a car makes food easy to acquire and transport home. In neighborhoods without a grocery store, people are forced to travel longer distances to buy food. If the family does not have a car, they can only buy the amount of food that can be carried home easily. The USDA defines a food desert as an area in which residents do not live near supermarkets or other food stores that offer affordable and nutritious food. Food deserts are usually found in low-income areas. Finally, due to geographic location, foods that are not in their growing season will not be available.

Reading Check

Explain What is the benefit of buying foods that are currently in-season and grown locally?

CALORIES ARE A UNIT OF HEAT THAT MEASURES THE ENERGY AVAILABLE IN FOOD.

Farmer's markets help bring fresh fruits and vegetables to urban areas. **Explain how a farmer's market can reduce the impact of food deserts.**

Using MyPlate

MAIN IDEA MyPlate provides a visual guide to help consumers make more healthful food choices.

When choosing foods, there are many different factors determining what foods you eat. A group of leading health and nutrition experts have worked with the **U.S. Department of Agriculture (USDA)** to develop **MyPlate**, which is a visual reminder to help consumers make healthier food choices. MyPlate is based on research into healthy nutrition. The USDA is a department of the Federal government providing leadership on food, agriculture, nutrition, and other topics.

The idea behind MyPlate is simple: Foods are placed into one of five food groups. Each group provides a different set of nutrients. For each meal, you should aim to choose foods from all five groups. This will provide your body with all the nutrients you need to stay healthy. The USDA web site also offers tools to help you design a healthy eating plan.

When creating a meal plan, keep in mind that you may not eat something from every food group at every meal. For example, if you go out for a pizza lunch with friends, the pizza will provide grains, dairy, and fat. Pizza will probably offer very few vegetables and no fruit. However, your next meal may be a salad with plenty of vegetables and an apple as a snack. A meal plan will be easier to follow if you include some flexibility. It is also okay to have a treat now and then as long as you follow the MyPlate guidelines.

> **WHEN CHOOSING FOODS, THERE ARE MANY DIFFERENT FACTORS DETERMINING WHAT YOU EAT.**

MyPlate shows how much of the food you eat should come from each of the five food groups. **Identify the two groups that should make up the largest share of your daily food choices.**

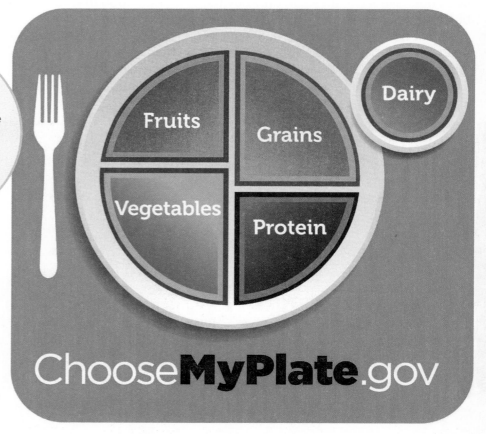

The Five Food Groups

MyPlate is a visual guide to help you make more healthful food choices. The MyPlate guide shows five food groups that should be part of each meal. The list below describes the types of food to select from each group.

- **Grains.** This group includes all grains such as rice, corn, wheat, oats, and barley. It also includes bread and pasta. About half of the grains should be whole grains. These are foods that contain the entire grain kernel, such as whole-wheat bread, oatmeal, brown rice, and popcorn.
- **Vegetables.** Choose from many vegetables to get a variety of nutrients. They can be fresh, frozen, or canned. Vegetables include leafy greens, such as spinach and lettuce. It also includes starchy vegetables such as potatoes.
- **Fruits.** Like vegetables, fruits can be fresh, frozen, or canned.
- **Dairy.** This group includes milk and many other products such as cheese and yogurt. Keep in mind that cream and butter contain little calcium and are not part of this group. Low-fat or fat-free dairy products are a healthy choice.
- **Protein.** This group includes meat, fish, beans and peas, eggs, and nuts. One ounce might be one egg or a tablespoon of peanut butter.

The grain group includes foods that are high in fiber. **Identify the percentage of the grains you eat that should be whole-grains.**

Reading Check

List What are the five basic food groups in MyPlate?

The MyPlate plan tells you how much to eat from each food group. **Identify how much protein should 12-year-old girls and boys each day.**

Daily Food Plan ChooseMyPlate.gov		Grains	Vegetables	Fruits	Dairy	Protein
12-year-old girl 4'10" tall, 90 lbs.	Active 30 to 60 min.	6 oz.	2.5 cups	1.5 cups	3 cups	5 oz.
12-year-old boy 4'10" tall, 90 lbs.	Active 30 to 60 min.	7 oz.	3 cups	2 cups	3 cups	6 oz.
13-year-old girl 5'1" tall, 100 lbs.	Active 30 to 60 min.	6 oz.	2.5 cups	2 cups	3 cups	5.5 oz.
13-year-old boy 5'2" tall, 105 lbs.	Active 30 to 60 min.	8 oz.	3 cups	2 cups	3 cups	6.5 oz.
14-year-old girl 5'4" tall, 110 lbs.	Active 30 to 60 min.	6 oz.	2.5 cups	2 cups	3 cups	5.5 oz.
14-year-old boy 5'5" tall, 115 lbs.	Active 30 to 60 min.	8 oz.	3 cups	2 cups	3 cups	6.5 oz.

MyPlate suggests eating a variety of vegetables each day. **Tell why a variety of vegetables should be eaten each day.**

Planning Healthful Meals

MAIN IDEA Healthful eating is a positive behavior that can promote health and prevent disease.

To put the five food groups together into healthy meals and snacks, remember these three words: *variety, moderation,* and *balance.* Choosing a *variety* of foods from the different food groups will provide all the nutrients you need. It will also keep your meals and snacks interesting. *Moderation* means keeping your portions to a reasonable size. It also means limiting **empty-calorie foods**, which are foods that offer few, if any, nutrients but do supply calories. You can do this by eating a piece of fruit or drinking water instead of sugary drinks. Finally, find the right *balance* between the amount of food you eat and your level of physical activity. This will help you maintain a healthy weight.

Preparing Healthful Meals

Most teens prepare some of their meals at home. Meal preparation may be simple, such as putting a frozen pizza in the oven. Meal preparation may also be more complex. It can include chopping vegetables, deciding which spices or herbs to use in a recipe, or thinking about how a food will be cooked.

When preparing foods at home, you control the ingredients that are used to prepare the food. When reading the recipe, decide whether you can reduce the amounts of sugar, salt, and fats in it. In many recipes these ingredients can be reduced without affecting the taste of the dish. Another thing to think about is how the food is prepared. Frying a food adds fat to the dish. If the food is breaded, it will have even more fat.

Some tips for preparing healthful meals at home are:

- Bake or broil meats and other foods rather than frying or deep-frying them.
- Use low-fat alternatives to high-fat items such as cream-based soups, cheese, mayonnaise, butter, and salad dressings.
- Add beans or vegetables to some prepared foods to make them more healthful.
- Choose cuts of meat that are low-fat. Remove skin from poultry.

Balancing Calories and Physical Activity

Eat food containing only as many **calories** as your body needs. A calorie is a unit of heat that measures the energy available in foods. Some calories are burned to perform daily body functions. Some calories are burned through physical activity. The more active a person is, the more calories that person can eat.

Teens who are on sports or dance teams will likely need to consume more calories than teens who are not on a team. These teens are more active and will burn more calories. Even though these teens are more active and will burn more calories, they should still choose foods that are high in nutrients.

Eating Right at Every Meal

The MyPlate advice can help you select foods that will fit into a healthy eating plan. You have probably heard people say that breakfast is the most important meal of the day. After sleeping for eight to ten hours, your body needs fuel. Eating a healthful breakfast will boost your energy and increase your attention span, help you concentrate, and retain information at school. One study found that eating a healthy breakfast every day had a positive effect on math grades and academic performance. Eating a healthy breakfast can also increase athletic performance.

Good breakfast choices include foods with complex carbohydrates and protein, such as oatmeal with milk or eggs with whole-grain toast. If you are in a hurry, you can still choose a healthful breakfast. Foods such as fresh fruit, whole-grain bread, and string cheese are all easy to grab and go.

At lunch and dinner, fill half your plate with vegetables and fruits. If you are packing your own lunch, consider a salad to go with—or instead of—your sandwich. Speaking of sandwiches, using whole-grain bread is a good way to get your daily servings of whole grains. Fresh fruits, such as apples and bananas, are easy to carry and make a healthful dessert.

For growing teens, snacks are also an important part of a day's meal plan. They help meet your nutrient needs and keep you going through the day. However, many popular snack foods, such as potato chips and cookies, contain a lot of fat, sugar, or **sodium**, a nutrient that helps control the amount of fluid in your body. More healthful choices are **nutrient dense** foods, which are foods that have a high amount of nutrients relative to the number of calories. These include fresh or dried fruit, air-popped popcorn, whole-grain crackers with cheese, or nuts.

BREAKFAST IS THE MOST IMPORTANT MEAL OF THE DAY.

Reading Check

Identify What are some healthful breakfast foods?

Eating a variety of fruits and veggies is the best way to make sure you get all the nutrients you need. **Select which of these healthful foods you enjoy.**

Nutrition Facts

Serving Size 1 cup (226g)
Servings Per Container 2

Amount Per Serving

Calories 250   Calories from Fat 110

	% **Daily Value***
Total Fat 12g	**18**%
Saturated Fat 3g	**15**%
Trans Fat 3g	
Cholesterol 30g	**10**%
Sodium 470mg	**20**%
Potassium 700mg	**20**%
Total Carbohydrate 31g	**10**%
Dietary Fiber 0g	**0**%
Sugar 10g	
Protein 5g	

Vitamin A 4%	•	Vitamin C 2%
Calcium 20%	•	Iron 4%

*Percent Daily Values are based on a 2,000 calorie diet. Your Daily Values may be higher or lower depending on your calorie needs.

	Calories	2,000	2,500
Total Fat	Less than	65g	80g
Saturated Fat	Less than	20g	25g
Cholesterol	Less than	300mg	300mg
Sodium	Less than	2,400mg	2,400mg
Total Carbohydrate		300g	375g
Dietary Fiber		25g	30g

The Nutrition Facts label on a food package gives you important information about a food's nutritional value. **Determine how many calories a serving of this food contains.**

What Teens Want to Know

Is a vegetarian diet right for me?

People choose vegetarian diets for various reasons. Meat can be expensive in some parts of the world. Health issues or religious beliefs may restrict certain kinds of meat. Some people prefer not to eat meat, or they like other foods better. Some vegetarian diets limit all animal products. Others may include dairy products, eggs, or fish. Vegetarian teens must be sure to get enough key nutrients. Everyone should eat a variety of vegetables, fruits, and grains every day. Many cereals and breads also contain added vitamins and minerals. See the Dietary Guidelines for Americans for more information.

Getting the Nutrition Facts

Some foods are easy to identify as healthful or less healthful choices. However, with foods that come in a package, it is not always easy to tell. Is a granola bar really better for you than a candy bar? Are pretzels a better choice than potato chips?

The Nutrition Facts Label on packaged foods can tell you if a food is healthful. It shows the number of servings in the package. It also shows how many calories per serving the food contains. That means if a package has two servings, eating the entire package delivers twice the amount of calories listed on the label.

Keep Foods Safe to Eat

When food is not handled, stored, or prepared properly, bacteria or other organisms can grow rapidly in the food. For example, food left out at a picnic can be the source of a **foodborne illness** after an hour or so. Foodborne illness is a sickness that results from eating food that is not safe to eat. To keep food safe, follow these steps:

1. **Keep your hands, utensils, and surfaces clean.** Always wash hands and items used to prepare foods in warm, soapy water.

2. **Separate raw, cooked, and ready-to-eat foods.** Keep raw, cooked, and ready-to-eat products separate from raw foods to avoid contamination. Use different cutting boards for raw meats and vegetables. Use separate cutting boards for vegetables and meats and for cooked and uncooked foods, or wash the cutting board with hot soapy water in between.

3. **Cook foods thoroughly.** Use a food thermometer to make sure that foods are cooked to the proper temperature:

   - Fish and meats such as beef and lamb should be cooked to 145 degrees.
   - Ground beef should be cooked to 160 degrees.
   - Poultry should be cooked to between 170 and 180 degrees.
   - Leftovers should be reheated to at least 165 degrees.

4. **Chill when necessary.** Refrigerate foods that spoil easily such as meat, fish, chicken, and eggs. Frozen foods should be thawed out in the refrigerator rather than on a countertop. Put leftovers in the refrigerator right after a meal.

5. **Serve safely.** Keep hot foods hot and cold foods cold.

6. **Follow directions.** Read food labels and follow suggestions for cooking or refrigeration.

7. **When in doubt, throw it out.** Do not eat any food that you suspect has not been cooked, handled, or stored properly.

Reading Check

Describe What is the proper way to clean up after preparing raw meat for cooking?

Lesson 2 Review

What I Learned

1. **VOCABULARY** Define *nutrient dense*. Give an example of a nutrient dense snack.

2. **NAME** What are three factors that can influence your food choices?

3. **EXPLAIN** How many cups of milk or dairy products do teens need each day?

4. **DESCRIBE** What are the benefits of eating breakfast every morning?

Thinking Critically

5. **ANALYZE** Find an ad for a food or food product. What does the ad tell you about the food? What methods does it use to encourage you to buy the food? Does the ad make you want to try the food? Discuss your findings with your classmates.

6. **SYNTHESIZE** Research more information about food deserts. Write an essay summarizing what you found. Include possible solutions for eliminating food deserts.

Applying Health Skills

7. **COMMUNICATION** Faye wants to start eating healthier foods and using healthy cooking methods. How can she use her communication skills to talk with her parents about making these changes?

Managing Your Weight

Before You Read

Quick Write Write a paragraph describing what actions you currently take to try and maintain a healthy body weight. After you read the lesson, review your response. What changes might you make to your action plan?

Vocabulary

overweight
underweight
appropriate weight
body mass index (BMI)
energy equation

BIG IDEA You can maintain a healthy weight by balancing the food you eat with physical activity.

Your Weight and Your Health

MAIN IDEA Maintaining a healthy weight can help prevent serious health problems during all stages of life.

One advantage of eating right is that it helps you maintain a healthy weight. Keeping your weight in a range that is right for your body is important for your overall health. Teens who are overweight are at a higher risk for a variety of health problems. Being **overweight**, or weighing more than the appropriate weight for your gender, height, age, body frame, and growth pattern, puts extra strain on your heart and lungs. Teens who are overweight are at an increased risk of type 2 diabetes. They are also more likely to develop high blood pressure, heart disease, and cancer later in life.

Being **underweight**, or weighing less than the appropriate weight for gender, height, age, body frame, and growth pattern, can also lead to health problems. It may be a sign that you are not getting the nutrients your body needs to grow and develop. Teens who are underweight are at risk for anemia, a blood condition caused by a lack of iron in the diet. Anemia can make you feel tired and rundown. They may also have a harder time fighting off illness. Finally, teens who are too thin may not have enough stored body fat to keep them warm and provide an energy reserve.

During the teen years, maintaining a healthy weight may be more difficult. These years are a time when teens develop an adult body. Teens will experience times when they are somewhat overweight. They will also experience times when they appear to be underweight. These changes in weight are due to growth spurts.

A teen who is close to experiencing a growth spurt may appear to be overweight. Then a growth spurt will occur. After the growth spurt, a teen may appear to be underweight.

> KEEPING YOUR WEIGHT IN A RANGE THAT IS RIGHT FOR YOUR BODY IS IMPORTANT FOR YOUR HEALTH.

Finding Your Healthy Weight Range

MAIN IDEA Several factors help to determine your healthy weight range.

Of course, a weight that is healthy for one person will not be the same for another. A six-foot-tall, male football player would not be healthy at the same weight as a five-foot-tall, female ballet dancer. Your **appropriate weight**, or the weight that is best for your body, is not a single number but a range that depends on your age, gender, height, and build.

So how can you tell if your weight is healthy? One way to check is to calculate your **Body Mass Index (BMI)**, which is a method for assessing your body size by taking your height and weight into account. If your BMI falls into the appropriate weight range on the chart, it means that you are most likely at a healthy weight for your height. However, it is important to understand that the BMI measurement is not perfect. For one, it does not take a person's body build into account. This means that people who are heavily muscled or have a stockier build may be labeled as overweight, even if they do not have a lot of body fat.

It is also important to remember that during your teen years, your body is changing rapidly. Growth patterns may cause you to become overweight or underweight for a time. This is usually normal. If you are unsure whether your weight is within a healthy range, see a health professional. Do not try to lose or gain weight unless the provider recommends it and suggests a specific eating and physical activity plan.

Reading Check

Describe What are the limitations of the Body Mass Index as a measure of body size?

First calculate your BMI using an online tool. Then find your age on the graph. Trace an imaginary line from your age to your BMI to see what range it falls into. **Explain why age and height are important factors in determining BMI.**

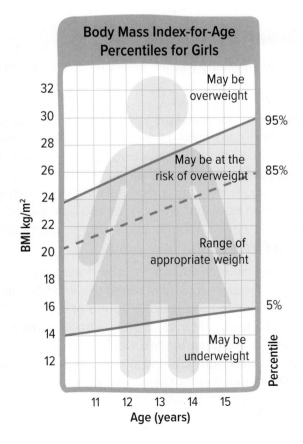

Body Mass Index-for-Age Percentiles for Girls

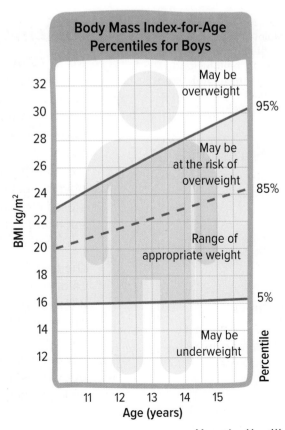

Body Mass Index-for-Age Percentiles for Boys

Maintaining a healthy weight may mean saying no to unhealthful foods. **Explain how to burn extra calories to avoid gaining weight.**

Reading Check

Explain What happens if you take in more calories than your body uses?

Reaching a Healthy Weight

MAIN IDEA To maintain a healthy weight, balance the calories you eat with the calories your body uses.

So, if you are overweight or underweight, what can you do to reach a healthy weight? The key is determining how many calories you need during the day. When you digest food, your body converts food calories into energy. It uses this energy to power your daily body processes and physical activity. When you take in *more* calories than your body uses, it stores the extra calories as body fat. This makes you gain weight. If you eat *fewer* calories than you need, though, your body converts stored body fat into energy. As a result, you lose weight.

The balance between the calories you take in from food and the calories your body uses through activity is called the **energy equation**. It takes about 3,500 calories to produce one pound of body fat. This means that if you eat 250 fewer calories than your body burns each day, you can lose one pound of weight after two weeks. The same thing happens if, by exercising, you burn 250 more calories per day than your body uses. While 250 calories may not seem like very much, it can make a big difference in your weight over time.

If you are trying to lose weight, choosing nutrient-dense foods will help. A turkey breast sandwich on whole wheat bread, for instance, contains about 250 calories. A chocolate bar has roughly the same number of calories, but it has far fewer nutrients. It will not do as much to fill you up or nourish your body. However, this does not mean that if you need to gain weight, you should eat more candy bars. Instead, choose nutrient-rich foods that are high in calories, such as nuts. You can also try drinking milk or juice with your meals instead of water.

Physical activity is another key to managing your weight. If you need to lose weight, increasing your level of activity will help. However, teens who want to gain weight still need a certain amount of physical activity to stay healthy. The MyPlate food guidance system recommends that all teens get approximately 60 minutes of physical activity on most days. If you want to gain weight, do not decrease your activity below this level. Instead, increase the number of calories you consume.

KEEPING YOUR WEIGHT IN A RANGE THAT IS RIGHT FOR YOUR BODY IS IMPORTANT FOR YOUR HEALTH.

To maintain a healthy weight, eat regular meals, choose snacks wisely, and be more physically active. **Explain how planning ahead could help with weight management.**

Avoiding Unhealthy Diets

Most teens should not diet to lose weight. Many diet plans are ineffective or even harmful. Some unhealthy diets restrict the types of foods eaten. Others require a lot of certain specific foods, such as grapefruit. Still others avoid some types of food completely, such as high-carbohydrate foods. Some weight-loss plans rely on pills, supplements, or extreme workout programs that promise fast results. Diet pills, or weight-loss pills, may be ineffective at helping a person loss weight. They may also interfere with the body's ability to absorb nutrients. Some contain caffeine or other stimulants that can increase heart rate and blood pressure.

To lose weight in a healthy way, it is often better to eat a selection of healthy foods in the amounts that are appropriate for a person's age and activity level. The USDA's web site offers a tool, the MyPlate Plan, to help determine the correct number of servings from each food group. In addition to eating healthy foods, you can prepare foods using healthy cooking methods. This includes baking, broiling, or roasting fish, poultry, or lean meat instead of frying them in butter or lard. Steaming vegetables is a great way to prepare them without adding any fat. You can avoid adding salt to food while cooking by using spices and herbs instead. These add flavoring without adding salt or fat.

The truth is, when it comes to managing weight, there really are no shortcuts. Although people who use unhealthy diets may lose weight, many gain the weight back as soon as the diet ends. In the meantime, they miss out on important nutrients. Instead, try these tips for managing your weight in a healthy way:

- Eat a variety of healthful foods, but reduce the portion sizes.
- Drink plenty of water, and avoid high-calorie beverages.
- Make time for regular meals. Rather than snacking a lot, try to eat only when you are hungry. It is easier to keep track of calories at set times.
- Boost your level of physical activity.
- Take your time when you eat. Chew your food thoroughly. This gives your stomach time to signal your brain when you have eaten enough.
- Talk to your health care provider. He or she can recommend a safe, healthy approach that will help you reach your goal in a reasonable amount of time.

Reading Check

List What are three ways that a person can assume personal responsibility for maintaining a healthy weight or addressing weight issues.

> THE TRUTH ABOUT MANAGING WEIGHT, THERE REALLY ARE NO SHORTCUTS

Remember that physical activity can help maintain a healthy weight. Teens are advised to get about 60 minutes of physical activity most days of the week. Choosing healthful foods most of the time does not mean that you cannot enjoy foods that are less healthful at times. The goal is to choose foods most of the time that provide nutrients and are low in fat and sodium.

Some people do have health problems that cause them to gain weight. Other health problems can make it difficult to lose weight. People with health issues that cause weight gain should seek medical help. Some people may need to take medication to correct a health issue that causes weight gain. It is that person's responsibility to make sure that any medications are taken according to the doctor's advice.

> TEENS ARE ADVISED TO GET ABOUT 60 MINUTES OF PHYSICAL ACTIVITY MOST DAYS OF THE WEEK.

What Teens Want to Know

How do I know if I need to lose weight? Teen bodies come in many sizes and shapes. You may not know whether you are underweight or overweight. Calculate your Body Mass Index (BMI) using an interactive online tool such as the one found at https://www.cdc.gov/healthyweight/bmi/calculator.html. Is your BMI close to the recommended value for your age? Follow the Dietary Guidelines for Americans to help make healthful food choices. If you eat a balanced diet and get at least 60 minutes of physical activity each day, you should reach your ideal weight.

Lesson 3 Review

What I Learned

1. **VOCABULARY** Define *Body Mass Index*.

2. **IDENTIFY** What is one health problem associated with being overweight and one health problem associated with being underweight?

3. **LIST** What are two factors that help determine your healthy weight?

Thinking Critically

4. **APPLY** Suppose that a teen takes in 2,000 calories each day and burns 2,300 calories. Over time, what will happen to the teen's weight? Explain your answer.

5. **EVALUATE** A friend of yours wants to lose some weight. She says she plans to try weight-loss pills that contain caffeine that claim that users will lose ten pounds in two weeks. What could you say to convince her that this is not a good idea?

6. **APPLY** Jenny is planning to cover fish in breading and then fry it in butter. How can she prepare this dish using healthy cooking methods?

Applying Health Skills

7. **COMMUNICATION SKILLS** Write a short dialogue between you and a friend who has just been teased for being overweight. How would you help the friend manage his or her feelings and make healthful choices?

Body Image and Eating Disorders

Before You Read

Quick Write Write a paragraph explaining why you think some teens develop an eating disorder.

Vocabulary

body image
eating disorder
anorexia nervosa
bulimia nervosa
binge eating
obese

BIG IDEA A teen's body goes through many changes during puberty. Many teens gain or lose weight.

Body Image

MAIN IDEA Your body image, or the way you view your body, is closely related to your weight.

Many people wonder if they need to gain or lose some weight. A person's weight affect **body image**. It's the way you see your body. People who have a poor body image may believe they are too thin, overweight, or not muscular enough. People who are unhappy with their bodies may try to change their weight in extreme ways. This can damage their health and may even be life threatening.

Developing a Positive Body Image

You can develop a positive body image. Think about the internal and external influences that affect you. An internal influence is how you feel about the way you look. External influences may include the things other people say to you about your weight. Media images are another external influence. It shows people with unusual body types. These include professional athletes and models. Comparing yourself to them can lower self-esteem.

Very few people have bodies similar to those you see in the media. The shape and size of your body depends mostly on your gender and traits you inherited from your parents. You cannot control these things. Your growth stage also affects your body shape. During puberty, you will experience growth spurts. Before a growth spurt, most teens appear to be overweight. Once the growth spurt occurs, the extra weight seems to disappear. To develop a positive body image, try think about these tips:

- Accept yourself as you are.
- Focus on the things you like about yourself.
- Avoid comparing yourself to an unrealistic body type.
- Maintain a healthy body by eating an appropriate amount of food and getting enough exercise.
- Spend time with people who like you and appreciate you for yourself.
- Show that you value yourself by taking good care of your body and mind.
- Eat well, get plenty of rest and exercise, and make time for the activities you enjoy.

Eating Disorders

MAIN IDEA An unhealthy body image can lead to an eating disorder.

A person with an unhealthy body image may be at risk of developing an eating disorder. **Eating disorders** are extreme eating behaviors that can lead to serious illness or even death. Their cause is unknown. However, people who feel bad about themselves or are depressed are more likely to develop an eating disorder. Eating disorders can affect anyone, but are more common to appear first during the teen years. Males and females can develop eating disorders.

A person with an eating disorder is not getting enough nutrients to help his or her body grow. Eating disorders are very dangerous, and can lead to death. Often people with eating disorders may deny that they have a problem. If you think that someone you know has an eating disorder, discuss the problem with an adult whom you trust. You can also help by encouraging your friend to ask for help.

Reading Check

Describe What is the key to having a positive body image?

People with eating disorders feel that they are overweight even if they have a healthy BMI. **Tell how eating disorders can affect a person's health.**

• • • • • • • • • •

Reading Check

Analyze When should a friend seek help for someone who may have an eating disorder?

• • • • • • • • • •

Anorexia Nervosa

Anorexia nervosa is an eating disorder in which a person strongly fears gaining weight and starves herself or himself. People with anorexia nervosa may feel a need to be very successful or be accepted by all people. They eat far fewer calories than they need to stay healthy. They may also exercise excessively. Even after they have become extremely thin, they still see themselves as overweight.

People with this disorder can become dangerously underweight. Also, because they eat so little, their bodies do not get the nutrients they need to grow and repair themselves. Their bones may become thin and brittle from lack of calcium. Their blood pressure and body temperature may drop. If they do not get treatment, people with this disorder can literally starve to death. They may also die from heart failure, kidney failure, or other medical complications. In addition, the depression that often comes with anorexia may lead to thoughts of suicide.

People with anorexia may see themselves as overweight even if they are very slim. **Describe the symptoms of anorexia.**

Bulimia Nervosa

Bulimia nervosa is an eating disorder in which a person repeatedly eats large amounts of food and then purges. One way to purge is to vomit, or throw up. Another way is to use laxatives. People with this illness may have a normal weight, but still feel the need to go on an extreme diet. When they cannot stay on the diet, they suddenly eat large amounts of food. Then, after eating, they purge. They may also try to burn the calories with constant exercise.

Unlike anorexia, bulimia usually does not lead to extreme weight loss. Its signs are usually subtler. For instance, people with this disorder may often go to the bathroom after eating a large meal. While in the bathroom, they may run the water to cover the sound of vomiting. Another sign is swollen cheeks caused by vomiting.

Bulimia robs the body of nutrients. Several body organs can be damaged. The coating on the teeth may wear off, damaging the teeth. The person may also be dehydrated.

Binge Eating

Binge eating is an eating disorder in which a person repeatedly eats too much food at one time. It is also called compulsive overeating. A *compulsion* is something that you feel you cannot control. They eat enough to make them feel physically uncomfortable. They may eat alone so that others do not see how much food they eat. They may hide food for the same reason.

As you might guess, binge eating can lead a person to become overweight or obese. Being **obese** means that a person is more than 20 percent higher than what is appropriate for their height, age, and body frame. Binge eaters can develop all the health problems related to being overweight. These include heart disease, diabetes, and some types of cancer. In addition, the guilt that compulsive eaters feel about their problem can lead to depression and low self-esteem.

Reading Check

Identify What are two signs of bulimia nervosa?

> PEOPLE WHO FEEL BAD ABOUT THEMSELVES OR ARE DEPRESSED ARE MORE LIKELY TO DEVELOP AN EATING DISORDER.

Males and Eating Disorders

As you have learned, eating disorders are linked to body image. People who do not like the way they look may try extreme measures to change their weight. When you think of anorexia and bulimia, extreme weight loss often comes to mind. Binge eating often leads to extreme weight gain. These disorders affect both males and females. Males and females experience similar signs and symptoms.

A related disorder involves an obsession with adding muscle. Muscle dysmorphia can also affect anyone, but it is more common in males. During the teen years, a girl may mistakenly think she needs to lose weight. A boy the same age may see himself as being too small or not weighing enough. As a result, he may want to gain weight and add muscle. In addition to compulsive eating, this disorder often involves the use of steroids or other dangerous drugs in an effort to bulk up. Eating disorders may be less noticeable in males and more difficult to diagnose. However, they are just as harmful and require similar treatment.

Treatment for Eating Disorders

Eating disorders are a mental health problem. They can also cause serious physical harm. People with eating disorders need medical help to recover.

Many types of health care providers work together to treat eating disorders. These include doctors, counselors, and nutritionists. These health care providers can help the person rebuild his or her physical and mental/emotional health. In extreme cases, a hospital stay may be needed to treat serious physical problems or severe depression.

Eating disorders can affect both males and females. **Explain why extreme weight loss or extreme weight gain can be harmful to a teen's physical, mental, emotional, and social health.**

Eating disorders are mental health problems. **Infer how you think that depression may be linked to eating disorders in some people.**

Getting help with an eating disorder is the first step toward recovery. **Discuss ways that health care providers work to treat eating disorders.**

Cultural Literacy

Compulsive Exercising

Compulsive exercising is closely related to eating disorders. It is possible to become dependent on or obsessed with exercise. Compulsive exercisers may add extra workouts. They may also force themselves to exercise when feeling ill or get very upset about missing a workout. The effects of compulsive exercising are similar to those of other eating disorders. Use online resources to learn about the warning signs of compulsive exercising. Find out what to do if you suspect that you or a friend may have this disorder.

People with eating disorders often cannot admit that they have a problem. Family members and friends can help them to recognize the problem and seek treatment. If you are concerned that you or someone you know may have an eating disorder, talk to a trusted adult. A person with an eating disorder needs to get help right away. The sooner a person gets treatment, the better his or her chances of recovering.

Lesson 4 Review

What I Learned

1. **VOCABULARY** What is another name for *binge eating disorder*?

2. **LIST** What are two health risks associated with anorexia nervosa?

3. **EXPLAIN** How can friends and family members help a person with an eating disorder?

Thinking Critically

4. **ANALYZE** How are the symptoms of anorexia nervosa similar to the symptoms of bulimia nervosa? How are they different?

5. **ADVOCACY** With a group, write and perform a short dramatic sketch about the body image and eating disorders. In it, show how negative body image and eating disorders are linked. Present your sketch for the class.

Applying Health Skills

6. **COMMUNICATION SKILLS** Suppose you think a friend might have an eating disorder. Write that friend a letter giving her or him facts about the problem and suggesting that he or she get professional help.

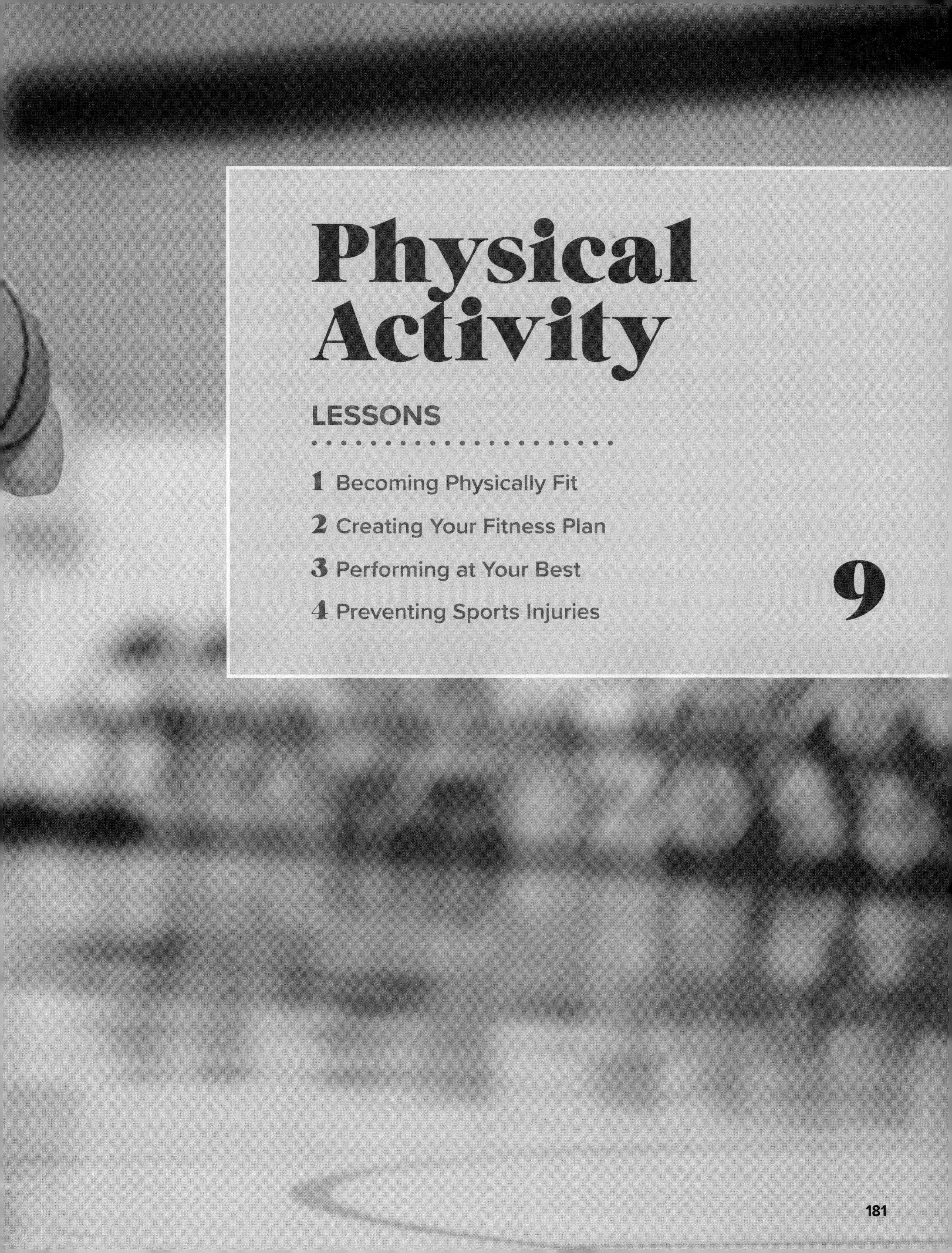

Physical Activity

LESSONS

1 Becoming Physically Fit

2 Creating Your Fitness Plan

3 Performing at Your Best

4 Preventing Sports Injuries

9

Becoming Physically Fit

Before You Read

Quick Write Write a short paragraph about the kinds of physical activity you do in a typical day.

Vocabulary

physical activity
balance
coordination
physical fitness
fitness
exercise
strength
endurance
flexibility
body composition
heart and lung endurance
muscle endurance
stamina
muscle strength
joints

BIG IDEA Being physically active benefits your total health in a variety of ways.

Choosing an Active Lifestyle

MAIN IDEA An active lifestyle will help to keep you healthy throughout your lifetime.

Anna and Sonja are identical twins. Although they look alike, their habits are different. Anna often plays basketball with her friends. Sonja prefers to stay inside and play video games. Anna would rather take the stairs than wait for the elevator. Can you guess which choice Sonja makes?

Anna is getting plenty of **physical activity**, which is any movement that makes the body use extra energy. Physical activity helps to balance all sides of your health triangle. On the physical side, being active helps you build strong bones and muscles. It also helps you manage your weight. It keeps your blood pressure at a healthy level and strengthens your heart and lungs. Physical activity can also improve your **balance**, or the feeling of stability and control, and **coordination**. Coordination is the smooth and effective working together of your muscles and bones. When you are active, you have more energy. You are also in less danger of developing certain diseases, both now and later in life.

Physical activity is good for your mental and emotional health, too. It helps you sleep better and concentrate better in school. It can also improve your self-confidence and relieve stress. Becoming physically active can also help your social health. You have more chances to make new friends by trying new activities or sports. As you continue to make new friends, you will continue to get opportunities to try new activities and meet new people. Being willing to try new things will give you even more chances to expand your social network. This expands your social network. Also, while playing sports, disagreements may occur. Physical activity helps you practice resolving conflicts in a healthful way.

Teens need at least 60 minutes of physical activity every day, according to guidelines released by the CDC. However, this doesn't have to mean an hour of activity all at once. For instance, suppose you walk to school several days a week. If it takes half an hour each way, your walking time will add up to 60 minutes. Experts also say teens should get some vigorous activity at least three days a week. Examples include jumping rope, swimming, or playing soccer.

Exercising for Physical Fitness

Leading an active life is the key to **physical fitness**, or the ability to handle the physical demands of everyday life without becoming overly tired. When you are fit, you have the energy for everything you want to do.

You can work to build **fitness** through **exercise**. Fitness is being able to handle physical work and play each day without getting overly tired. Exercise is planned physical activity done regularly to build or maintain one's fitness.

Some people think physical activity means working out or playing sports, but that isn't really true. In fact, there are lots of ways to be active. For example, you may walk to and from school every day. You probably have various chores to do at home. Vacuuming, doing laundry, and scrubbing the bathtub clean are all ways you can be more physically active. You might do yard work in the summer or rake leaves in the fall.

You might like to garden. If you have a bike, you may enjoy riding it for fun. Walking in the park is another great way to be physically active. You can look for other ways to increase your physical activity, such as taking the stairs instead of the elevator, walking your or a neighbor's dog, or dancing to your favorite music. All these daily activities can help you stay fit. However, if you are trying to improve your physical fitness, you will need to exercise as well.

Reading Check

Organize What strategies can you use to access information and tools to help you lead a healthy, active lifestyle?

Playing soccer is one way to get 60 minutes of physical activity each day. **Name two other ways that a teen can get 60 minutes of physical activity each day.**

Elements of Physical Fitness

MAIN IDEA Elements of fitness include strength, endurance, flexibility, and body composition.

Think about someone who is very strong but cannot walk a mile without becoming tired. Would you consider that person physically fit? Probably you would not. True fitness includes **strength**, which is the ability of your muscles to use force. It also includes **endurance**, which is the ability to perform difficult physical activity without getting overly tired, and **flexibility**, or the ability to move joints fully and easily through a full range of motion. It also means having a healthy **body composition**, which includes the proportions of fat, bone, muscle and fluid that make up body weight.

Reading Check

Identify What are the four elements of fitness?

Different sports require different types of endurance. **Name the types of endurance needed to play tennis?**

PHYSICAL ACTIVITY HELPS TO BALANCE ALL SIDES OF YOUR HEALTH TRIANGLE.

Endurance

The word *endure* means "to last," and endurance is a measure of how long you can last during physical activity. If you can climb several flights of stairs and not feel out of breath, that means you have good **heart and lung endurance**, which is also known as cardiovascular endurance. It is the measure of how efficiently your heart and lungs work when you exercise and how quickly they return to normal when you stop. This kind of endurance is important for many physical activities, such as running, swimming, and playing team sports. The ability of a muscle to repeatedly use force over a period of time is called **muscle endurance**. Activities that build muscle endurance include jumping rope, dancing, and riding a bike. Both kinds of endurance help build **stamina**. Stamina refers to your ability to stick with a task or activity for a long period of time.

Strength

The stronger you are, the more work your muscles can do. Having strong arms and a strong back means that you can move heavy objects. Strong legs will help you with running and jumping. You build **muscle strength** every time you make your muscles work against a force, such as gravity. Muscle strength is defined by the most weight you can lift or the most force you can exert at one time. Building muscle strength will shape and tone your body and help you with activities such as sports.

Many of the activities you do every day can contribute to fitness. **Identify which elements of fitness you could improve by raking leaves.**

Flexibility

Flexibility allows you to turn, bend, and stretch. Being flexible is especially important for gymnasts and dancers. However, it plays a role in other sports as well. Being flexible makes it easier to change directions quickly. This can help you with everything from stopping and turning to throwing a ball. Also, when your muscles can stretch easily, they are less likely to become injured during sports and other activities. You improve your flexibility every time you stretch your muscles and **joints**, or the places where two or more bones meet. Activities that are good for building flexibility include yoga, swimming, and karate.

As you become more familiar with fitness terms, you may hear the term *strong core muscles* mentioned. Your core muscles are those in your abdomen. These muscles provide strength for the rest of your body. Coaches and other fitness trainers will focus on developing core strength by doing sit ups and other exercises that build strength in this area of the body.

This graph shows how many calories a 100-pound person uses during 60 minutes of different activities. **Identify the activity that burns the most calories. Identify the activity that burns the fewest calories.**

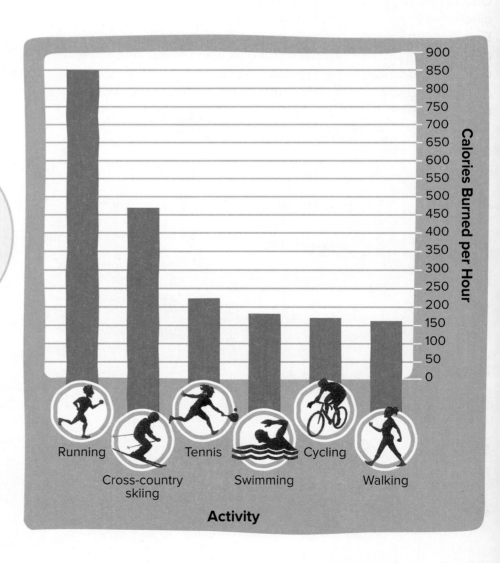

Body Composition

The last element of fitness is body composition. A healthy body generally has more bone, muscle, and fluid than fat. Body composition is not the same as how much you weigh. It is a measure of how much of that weight is lean tissue instead of body fat. Too much body fat can increase your risk of health problems.

Your body composition depends partly on the genes you inherit from your parents. It also depends on how much you eat and how active you are. When you are active, your body is burning up calories from the foods you eat. Burning extra calories means that your body won't store them as fat. The Activity and Calories Burned graph shows an average number of calories burned during different types of physical activity.

Being active also improves your body composition by helping you build up your muscles. Physical activity increases the amount of muscle on your body and reduces the amount of fat. Being active can help you develop a lean, fit, healthy body. In the long run, it will reduce your risk of many health problems, including heart disease and cancer.

One measure of body composition is a skinfold test, which involves pinching a fold of skin at two or three sites on the right side of the body. Each fold is measured with an instrument called a skinfold caliper. To ensure accuracy, a trained person should administer the skinfold test using standardized testing procedures and a high-quality caliper. Another way to measure your body fat is to calculate your body mass index (BMI).

Myth vs Fact

Myth The best measure of healthy body composition is your weight on a scale.

Fact Muscle tissue is actually denser than fat. This means that a pound of muscle takes up less space than a pound of fat. If you lose fat and gain muscle, your weight will actually go up, but you will look thinner. So, to tell if your body composition is improving, focus on how you look and how firm your body feels—not just on what the scale says.

Lesson 1 Review

What I Learned

1. **VOCABULARY** What is the difference between *physical activity* and *exercise*?

2. **IDENTIFY** Name one activity that builds muscle endurance and one activity that builds flexibility.

3. **EXPLAIN** How can you increase your physical activity on a daily basis?

4. **DESCRIBE** What is one way to improve your body composition?

Thinking Critically

5. **EVALUATE** In a short paragraph, describe your own thoughts and feelings about the benefits of being physically fit.

6. **ANALYZE** Write a one paragraph essay analyzing the influence of technology and the media on physical activity.

Applying Health Skills

7. **ANALYZING INFLUENCES** What is your favorite physical activity and why? Did friends or family spark your interest in this activity? Was it something you saw on television or in a magazine? Did social customs play a part? Explain your answer.

Creating Your Fitness Plan

· · · · · · · · · · ·

Before You Read

Quick Write List at least three ways you can fit more physical activity into your daily life.

Vocabulary

lifestyle activities
cross-training
aerobic exercise
anaerobic exercise
F.I.T.T. principle
frequency
intensity

· · · · · · · · · · ·

BIG IDEA You can reach your fitness goals by making a plan to include a different exercise in your life.

Measuring Fitness

MAIN IDEA Measuring your fitness level will indicate what areas of fitness you need to improve.

You wouldn't try to build a house without a blueprint, or plan. The same is true for building your physical fitness. The first step in making a physical fitness plan is to figure out how fit you are now. Learning your level of fitness is the first step in creating a fitness plan. It helps you to create a plan that will improve each element of fitness. You can use several tests to measure the different elements of fitness.

Measuring Flexibility

You can measure your flexibility with the V-sit reach, or sit-and-reach, test. Mark a line two feet long on the floor, marked in inches. Sit with your feet right behind the starting point of this line, with your heels 8 to 12 inches apart. Put one hand on top of the other, palms down. Have a partner hold your legs down while you reach your hands forward as far as you can. After three practice tries, reach forward and hold the stretch for three seconds. Have your partner note how many inches you could reach past the starting point of the line.

> **ALSO KNOWN AS RESISTANCE TRAINING, WEIGHT LIFTING CAN BE USED TO STRENGTHEN EVERY MUSCLE GROUP.**

Measuring Heart and Lung Endurance

To test your heart and lung endurance, time how long it takes you to run one mile. If you get too tired while running, you can switch to walking for a while. However, try to cover the distance as fast as you can.

Measuring Muscle Strength and Endurance

You can test the strength and endurance of your abdominal muscles by doing curl-ups. Lie on your back with your knees bent and your arms crossed on your chest. Your feet should be about 12 inches from your buttocks. Have a partner hold your feet down while you raise your body to touch your elbows to your thighs. Then lower your body back down until your shoulder blades touch the floor. Count the number of curl-ups you can do in one minute.

To measure upper body strength and endurance, do right-angle push-ups. Start out in push-up position: face down with your arms straight and your hands right under your shoulders. Your legs should be straight and slightly apart, resting on your toes. Keeping your back and knees straight, lower your body until your elbows are bent at a 90-degree angle. Your upper arms will be parallel to the floor. Then straighten your arms to raise yourself back up again. Do one complete push-up every three seconds and count how many you can do in a row.

Another approach to building muscle strength is weight lifting. Also known as resistance training, weight lifting can be used to strengthen every muscle group. Just make sure that a fitness instructor or other expert supervises your workout. If resistance training is part of your workout plan, keep in mind that choosing weights that are too heavy will not help you build muscles faster. You should choose weights that tire your muscles after 10 to 12 repetitions.

Checking Your Results

To see how you did on the fitness tests, look at the table with the lesson. It shows healthy results for teens of different ages. After you start your fitness plan, you can use these same tests as guides to see how your fitness is improving.

· · · · · · · · · · ·

Reading Check

Identify What activity measures heart and lung endurance?

· · · · · · · · · · ·

A physical fitness plan is a blueprint for becoming fit. **List the three measurements that should be taken before creating a physical fitness plan.**

The V-sit reach test measures flexibility. **Explain why flexibility is an important part of physical fitness.**

The curl-up test measures the strength and endurance of your abdominal muscles. **Describe a way to measure upper body strength and endurance.**

The push-up test measures upper-body strength and endurance. **Explain why upper-body strength and endurance is an important part of physical fitness.**

Setting and Reaching Fitness Goals

MAIN IDEA Fitness test results help to determine fitness goals.

Once you know how fit you are right now, you can start to figure out your fitness goals. Your scores on fitness tests might point you toward a fitness goal. For instance, if you did very well on the curl-ups and the one-mile run but not very well on the V-sit reach, you might set a goal to improve your flexibility. You can also set goals based on specific activities you would like to do. If you plan to try out for the track team, for example, then your goals might focus on building better heart-lung endurance.

This table shows the score you need on each test to do as well as half of all teens of your age and gender. **Identify any areas of fitness you may need to work on.**

	Age	Curl-UPs (# one minute)	OR	Partial* Curl-Ups (#)	Sit and Reach (centimeters)	One-Mile Run (min-sed)	Pull-Ups (#)
BOYS	11	47		43	31	7:32	6
	12	50		64	31	7:11	7
	13	53		59	33	6:50	7
	14	56		62	36	6:26	10
GIRLS	11	42		43	34	9:02	3
	12	45		50	36	8:23	2
	13	46		59	38	8:13	2
	14	47		48	40	7:59	2

Activity

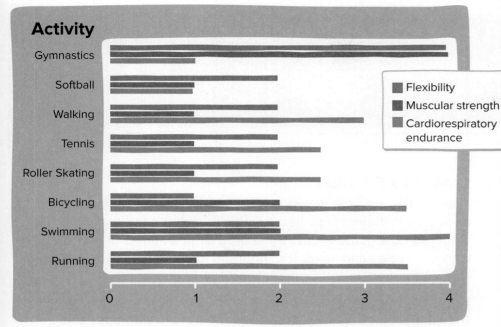

Legend:
- Flexibility
- Muscular strength
- Cardiorespiratory endurance

Rating scale: 1 = Low   2 = Moderate   3 = High   4 = Very High

Different activities promote different areas of fitness. **List your fitness goals. Identify the activities that can best help you reach those goals.**

Choosing Activities

Once you know what your goals are, you can start choosing activities to help you reach them. Different kinds of exercise are good for building up different elements of fitness. The graph with the lesson shows the fitness benefits of some popular **lifestyle activities**. These are physical activities that are part of your day-to-day routine or recreation. Consider these other points when choosing exercises.

- **Personal tastes.** If you choose activities you enjoy, you are more likely to stick to your plan. If you prefer group activities, you could try a team sport. If you would rather work out alone, you might try jogging or bicycling. Another way to keep your interest level high is **cross-training**, which means switching between different forms of physical exercise.

- **Requirements.** There's no point in choosing swimming as an activity if you don't have a pool you can use. Think about the requirements of different activities. Will you need special equipment? Do you need a partner or teammates? Are there special skills you need to learn first?

- **Time and place.** Figure out when you will have time to exercise. If you have to catch a school bus at 7:30, for instance, you may not be able to work out in the morning. Also, think about where you will exercise. If you want to start jogging, for example, you need to make sure the streets in your neighborhood are safe enough. Also, if you live someplace where it gets very cold in the winter, then you may need to run indoors on a track or treadmill.

> **DIFFERENT KINDS OF EXERCISE ARE GOOD FOR BUILDING DIFFERENT ELEMENTS OF FITNESS.**

Sun	Mon	Tues	Wed	Thurs	Fri	Sat
1 Ride bike (1 hr.)	**2** Gym class (40 min.) Walk briskly home from school (20 minutes)	**3** Soccer practice (1 hr.)	**4** Gym class (40 min.) Walk briskly home from school (20 minutes)	**5** Soccer practice (1 hr.)	**6** Karate class (1 hr.)	**7** Soccer game (90 minutes)

A written plan, like the one shown here, will help you stick to your fitness goals. **Identify the activities that are part of this teens schedule. Determine the number of hours per week that this teen is active.**

Creating a Schedule

In order to reach your fitness goals, you will most likely need to choose a variety of different activities. **Aerobic exercise** is the rhythmic, moderate-to-vigorous activity that uses large amounts of oxygen and works the heart and lungs. It's important for improving your heart and lung endurance.

Experts say that most of your 60 minutes of daily physical activity should be aerobic activity, done in a moderate to vigorous pace. Moderate activities include hiking, bicycling, and in-line skating. Some vigorous activities are swimming, jumping rope, and team sports. It is important to include vigorous activities like these in your schedule at least three times a week.

Some kinds of aerobic exercise, such as swimming, are great for your heart and lungs, but they do not do much for your muscles and bones. It is recommended you choose activities to strengthen muscles and bones at least three days a week. Many aerobic exercises, such as running or jumping rope, are also good for bone strength. To build up your muscles, however, you'll need to do some kind of **anaerobic exercise**. It's intense physical activity that builds muscle but does not use large amounts of oxygen. Weight lifting, sit-ups, push-ups, and climbing exercises are some good ways to help make your muscles stronger.

How can you fit all this into your schedule? Start by writing down all the physical activities that are part of your daily routine. Most teens have gym class at school on certain days. If you play a sport, include all your practices and games. Also include any outside activities that get your body moving, such as dance lessons. Once you have a written plan, you can see what you are doing now and what other activities you might need to add. Then you can find the gaps in your schedule where these new activities will go. To remain physically active throughout your life, you can continually check your schedule and the activities you can do. Your personal tastes may change as you age. You may have more resources available if you move to a new location. The important thing is that you remain physically active even as your life and circumstances change.

The F.I.T.T. Formula

Whenever you start a new activity, it's best to start small. If you've never run before, you are probably not ready to run a 5-kilometer race. Instead, start by running short distances. Then over time, you can gradually build up your distance.

A method for safely increasing aspects of your workout without injuring yourself is the **F.I.T.T. principle**. It includes four steps. They are listed below.

- **Frequency**, or the number of days you work out each week. When you first start a new activity, you might do it two to three times a week. As your fitness improves, you can work out more often.
- **Intensity**, or how much energy you use when you work out. This refers to how hard you should work different muscle groups. Take it easy at first and build up the intensity over time. Work each muscle group until it tires and then move to another muscle group. Over time, each muscle group will not tire from the weight being used. That's when it's time to increase the intensity or change the exercise. If you lift weights, increase the weight used. If you run, you can go a little faster each time.
- **Time.** Teens need 60 minutes of activity each day. With a new activity, you will not be able to do this all at once. Start with a shorter workout and build up to a full hour.
- **Type.** A complete fitness plan will include a variety of activities. Try switching among activities so that you can work different muscles on different days. This will also help keep you from getting bored.

Reading Check

Describe Cite two types of physical activity that all teens should include in a fitness plan.

Lesson 2 Review

What I Learned

1. **VOCABULARY** Define *aerobic exercise* and give an example.

2. **NAME** What are two factors you should think about when choosing physical activities?

3. **IDENTIFY** What are the four parts of the F.I.T.T. principle?

Thinking Critically

4. **EVALUATE** Alicia is an inactive teen who wants to get into shape. She has made a fitness plan with three hours of exercise and other physical activity per day. How likely do you think she is to succeed with this plan?

Applying Health Skills

5. **GOAL SETTING** Identify a personal fitness goal. Then make a fitness plan to meet that goal. Write down which activities you will do and when. Also, note how you can track your progress toward your goal.

Performing at Your Best

Before You Read

Quick Write Have you ever felt sore the day after you tried a new physical activity? Explain in a few sentences why you think this happened.

Vocabulary

warm-up
cool-down
resting heart rate
target heart rate
recovery heart rate
conditioning
dehydration

BIG IDEA Following some basic guidelines will help you get the most out of your workouts.

Keys to a Good Workout

MAIN IDEA Healthy workouts include a warm-up and cool-down.

Jay has decided to start jogging each day before school. He figures if he gets up just 15 minutes early, he can run a mile or so before breakfast. On the first day, he rolls out of bed, puts on a pair of jeans and a t-shirt, running shoes, and heads out the door. He runs for 15 minutes and then goes home. He feels a bit tired and sore for the rest of the day, but he figures that is normal. The next morning, he is so stiff and sore he cannot even think about getting up to run.

What did Jay do wrong? First of all, he forgot to warm up before his run and cool down after it. He did not drink any extra water after his run. He did not choose to wear athletic clothing that would have enabled him to move freely and easily. Finally, he tried to do too much too soon. He forgot that with a new activity, it is important to start slowly and work your way up to full speed. If you want to avoid Jay's problem, you need to plan ahead to get the most out of your workouts.

Another mistake Jay made was to wear jeans during his run. While jeans may be a good choice for many activities, they may not be a good choice to wear during a workout. Wearing loose clothing during exercise helps you move more freely and be more flexible. Exercise clothing may also offer protection from moisture or weather.

> STRETCHING SHOULD ALWAYS COME *AFTER* A WARM-UP AND SHOULD NOT BE USED *AS* A WARM-UP.

Warming Up and Cooling Down

Every workout should start with a **warm-up**. A warm-up is gentle exercise that get heart muscles ready for moderate-to-vigorous activity. Warming up increases blood flow and loosens up your muscles so that you are less likely to strain or tear them. Any light activity, such as walking or jogging in place, can make a good warm-up. Keep it up for about five to ten minutes. You will know you are warmed up when you begin to sweat and breathe more heavily.

After warming up, it's a good idea to stretch your muscles. This further reduces your risk of injury. Make sure to stretch the muscles that you will be using the most as you exercise. Stretching should always come *after* a warm-up and should not be used *as* a warm-up. Stretching your muscles while they are cold will actually increase the chances of injury.

At the end of your workout, you need to take some time to cool down. Like a warm-up, a **cool-down** should last five or ten minutes. It includes gentle exercises that let the body adjust to ending a workout. This will help your muscles relax so they will not feel stiff or sore afterwards.

Learning the right way to stretch can help prevent injuries. **Try the two stretches shown here. Describe where you feel the "pull."**

Monitoring Your Heart Rate

How can you tell if you are exercising hard enough? One way is to check your heart rate before, during, and after your workout. Your **resting heart rate** is the number of times your heart beats per minute when you are relaxing. It is a good measure of overall heart health. In general, a lower rate means a healthier heart.

You should fully warm up before stretching prior to beginning your workout. **What are some ways of knowing when your body is warmed up?**

When you work out, however, you want to get your heart pumping faster. Your goal should be to reach your **target heart rate** during exercise. It is the number of heart beats per minute that you should aim for during moderate-to-vigorous aerobic activity to help your circulatory system the most. To figure out your target heart rate, use the equations below.

(220 – your age) × 0.5 = bottom of target heart range
(220 – your age) × 0.85 = top of target heart range

Keeping your heart rate in this range as you exercise will help you work hard enough, but not too hard.

One way to check your heart rate during a workout is to stop for a minute and take your pulse. To do this, place two fingers against the base of your neck. (Do not use your thumb, which has its own pulse.) You should feel a throbbing sensation. This is the blood pumping through the blood vessels in your neck. Using a clock or watch with a second hand, count the number of pulses you feel in ten seconds. Then multiply this number by six to get your heart rate.

Measuring your pulse is one way to make sure your heart rate is in the right range for exercise. **Name another way to check your heart rate.**

An easier way to check your heart rate during moderate activity is the "talk test." If you don't have enough breath to talk at all, you're probably working too hard. However, if you have enough breath to sing, you're not working hard enough.

A third heart rate measurement is your **recovery heart rate**, or how quickly your heart rate returns to normal right after exercise is stopped. An aerobic fitness goal should be to achieve a lower recovery heart rate.

Staying in Shape for Sports

MAIN IDEA Sports conditioning will help you play to the best of your abilities.

Playing a sport can be a lot of fun. It is also a great way to stay active. However, playing a sport involves a lot more than just showing up for games. Sports put extra demands on your body, and meeting those demands involves proper training. You need to develop the skills and build up the muscles that are necessary for your sport. You also need to eat right so that your body can keep up with everything you are asking it to do.

Conditioning

Athletes devote many hours to **conditioning**, or training to get into shape for physical activity or a sport. They must exercise to get their bodies in shape and also practice the skills they need for their particular sport. This training takes place both on and off the field. Baseball players, for instance, practice batting and fielding before a game. Basketball players do passing drills and take practice shots from different parts of the court. Off the field, athletes may shape up their bodies with weight training or other exercises.

As the saying goes, "Practice makes perfect." **Think of a sport you enjoy. Name one skill you would need to practice for that sport.**

Drinking water before, during, and after a game is one good sports nutrition habit. **What is another good sports nutrition habit?**

Conditioning is not just important for team sports. Many kinds of physical activity can require training. For example, if you are a dancer, you need to practice regularly to keep your muscles and your skills in shape.

Sports Nutrition

Like other teens, teen athletes have special nutritional needs. They need to eat a variety of foods from all the major food groups. They also need to consume extra calories to replace all the energy they use during their chosen sport. Good food choices will include plenty of complex carbohydrates and healthful fats. Sugary snacks, such as candy bars, are not good choices. They can give you a burst of energy, but it will wear off quickly.

When you eat is as important as what you eat. It is best not to eat anything in the hour right before a game or a practice. Digesting food takes energy that your body needs to perform at its best. Also, having a full stomach during a workout can make you feel sick.

On the day of a game, it's best to eat a meal with some protein and carbohydrates about two to four hours before you compete. This will give you the energy you need to get through the game, but it will not slow you down by making you feel too full. If you don't have time for a meal, you can choose a light snack, such as fruit or crackers, one to two hours before a game or practice.

You should also make sure to get plenty of water—before, during, and after a game—to avoid **dehydration**. It is the excessive loss of water from the body. During sports and other kinds of intense activity, your body loses a lot of water through perspiration, or sweat. You need to drink more to replace this lost fluid. How much you need depends on your age, your size, the weather, and how hard you are working out. In general, though, you should drink some water every 15 to 20 minutes during activity. Don't wait until you feel thirsty—that's a sign that your body is already running low on fluids. It's okay to choose sports drinks if you like them better than water, but they aren't any better for you in most cases. If the weather is extremely hot or the practice lasts longer than an hour, the extra nutrients in sports drinks may be helpful.

> **WHEN YOU EAT IS AS IMPORTANT AS WHAT YOU EAT.**

Developing Good Character

Fairness When playing a sport, you show good character by displaying good sportsmanship. Play by the rules, take turns, and share the credit when your team wins. Learn to be a good loser too, and congratulate the other team if they win. When people play fair and show good sportsmanship, everyone can enjoy the game. Think of three ways you can show good sportsmanship while *watching* a game.

Reading Check

Give Examples List some of the foods an athlete might choose for a snack before a game.

Lesson 3 Review

What I Learned

1. **VOCABULARY** Explain what *warm-up* and *cool-down* mean.

2. **DESCRIBE** How can you figure out your target heart rate for exercise?

3. **EXPLAIN** Why do you need to drink extra water during physical activity?

Thinking Critically

4. **INFER** Why do you think teen athletes should eat a meal 3 to 4 hours before a game?

Applying Health Skills

5. **APPLY** Kerry is planning to start swimming on a regular basis. She says that because you stretch your muscles during swimming, there is no need to cool down. How would you respond?

Preventing Sports Injuries

Before You Read

Quick Write In a short paragraph, briefly describe how you or someone you know recently got hurt as a result of physical activity.

Vocabulary

sprain
fracture
dislocation
concussion
traumatic brain injury
stress fracture
tendonitis
sports gear
overworking
heat exhaustion
frostbite
P.R.I.C.E. formula

Reading Check

Define What is a stress fracture?

BIG IDEA You can prevent many sports injuries by taking precautions.

Common Sports Injuries

MAIN IDEA Some sports injuries can be treated at home, while others require medical care.

Each year, about 3.5 million children and teens are injured while playing a sport. A **sprain** and a strain are the most common sports injuries. A sprain is an injury to a ligament connecting bones to a joint. A strain is a stretch or tear in a muscle or a tendon (the tissue that connects muscles to bones and other muscles). Both sprains and strains can cause pain and swelling. However, they can usually be treated at home.

One serious sports is a **fracture**, which is a break in a bone. Another is **dislocation**. It's a major injury that happens when a bone is forced from its normal position within a joint. Another is **concussion**, which is a jarring injury to the brain that can cause unconsciousness. Fracture and dislocation can be very painful injuries. A concussion can result in a loss of consciousness and **traumatic brain injury**. It's a condition caused by the brain being jarred and striking the inside of the skull. These injuries need medical attention.

Some kinds of sports injuries don't happen all of a sudden. Instead, they start out small and get worse over time. These are known as *overuse injuries*. For example, a **stress fracture**, which is a small fracture caused by repeated strain on a bone, can happen as a result of jumping up and down on a basketball court over and over again. **Tendonitis** is painful swelling of a tendon caused by overuse. It can develop when a muscle is stretched too far, too often. Warming up and stretching before exercise can help prevent this problem.

Ways to Avoid Injury

MAIN IDEA Staying in shape, using the correct sports equipment, and avoiding bad weather conditions can help you avoid a sports injury.

As you can see, sports have their dangers. However, it is possible to make them a lot safer. You have already learned how warming up and cooling down can help you avoid fitness-related injuries.

Teen athletes need to take a few additional precautions as well. They need to have—and use—the right **sports gear**, which is sports clothing and safety equipment, for their chosen activity. They should also know their limits and avoid overtraining. Finally, teen athletes need to plan for the weather.

Using Proper Gear

Different sports call for different types of safety gear. The right shoes are important for just about any sport. For some sports, such as baseball and soccer, you will need cleats for better traction. Other sports, like basketball or track, have their own types of shoes. A coach or fitness trainer can tell you what kind of shoes will be best for your sport. Make sure your shoes fit well, and replace them when they wear out.

The type of clothing you choose for sports is also important. Choose clothing that is best for the weather. In winter climates, that means dressing in layers and wearing gloves and a hat while outside playing sports. Dressing in layers allows athletes to add or remove layers as their bodies heat up from activity. Athletic clothing that dries quickly helps to prevent skin irritation. Clothing should be loose fitting. This allows your body to move freely.

Many sports, from football to cycling, require a helmet. Make sure to choose one that's designed for the sport you're playing. Your helmet should fit snugly but comfortably, without tilting backward or forward. Check to make sure the helmet meets the standards of the Snell Memorial Foundation. The group is dedicated to protection from head injuries. It tests and certifies helmets for sports, bicycle, motorcycle, horse riding, and other uses.

Other types of protective gear include pads, mouth guards, athletic supporters, and eye protection. Mouth guards help protect against broken teeth or accidental tooth loss. Talk to your coach about what is needed for your sport.

Using the correct equipment for the sport you choose will help prevent injury. **Tell why sports that require running on grass or dirt require shoes with cleats.**

Mouth guards. These soft plastic shields protect your mouth, teeth, and tongue. Wear one for any sport where your mouth could be hit. Examples include baseball, football, and hockey. If you wear a retainer, take it out before you play.

Chest Protectors. A padded chest protector keeps the torso from being injured in sports such as baseball, softball, or hockey.

Face and throat protection. A face mask with a throat guard protects the face and throat from being hit by a ball or puck.

Pads. Pads are used to protect bones and joints from fractures and bruises.

Helmets. Always choose a helmet made for the sport you're playing. It should fit snugly but comfortably on your head. Be sure it doesn't tilt backward or forward. Never wear a cap under a helmet.

Elbow, knee, wrist, and shin guards. Elbow and wrist guards can prevent arm and wrist fractures. Knee and shin guards can protect these areas during falls.

Different kinds of protective gear are used for different sports. Think about a sport or activity you enjoy. **Write a goal statement describing why it is important to use safety equipment when playing the sport.**

Knowing Your Limits

As Aloysius dove to stop the ball from entering the net, he came down hard on his right hand. In his excitement, though, he barely noticed it. He did not mention it as he resumed his position as goalkeeper. By the end of the game, he was in real pain. That is when he discovered he had sprained his wrist and would have to sit out the next game.

Aloysius's soccer experience shows how important it is to listen to your body. If you feel pain during a game, you should stop right away and take a rest. If the pain doesn't stop, seek medical help. Avoid going back to playing too soon after an injury. Your body needs time to heal. How much time will depend on where and how badly you were hurt. A mild ankle sprain may heal in a few weeks, while a broken leg may sideline you for months.

It's also important to avoid **overworking**, which is conditioning too hard or too often without enough rest between sessions. You need to train to get in shape for a sport—but not too hard, and not too fast. Remember the F.I.T.T. principle. Sudden increases in the frequency, intensity, or duration of exercise can lead to injury. Talk to your coach or doctor about a training program to build your skills safely.

Dressing in layers can protect you from the cold during winter sports. **Describe some precautions you should take while exercising during hot weather.**

Watching the Weather

Both hot weather and cold weather can pose health risks. In hot weather, you sweat more and lose more fluid. This can lead to dehydration. When dehydration occurs, muscles may begin to cramp. Muscle cramps may be relieved by massaging the muscle. Dehydration and muscle cramps can be avoided by drinking water before and during physical activity. If dehydration occurs, a more serious condition that can occur is **heat exhaustion**. It's an overheating of the body that can result from dehydration. The symptoms of heat exhaustion include headache, dizziness, and nausea. Anyone who shows these symptoms should immediately lie down in a cool spot and drink cool water. If the person faints or seems confused, seek medical help right away.

Cold weather also has its dangers. Skin exposed to extreme cold can develop **frostbite**, which is freezing of the skin. This is most likely to happen to small, exposed body parts like hands, feet, ears, and noses. The skin will turn pale and feel numb. To avoid frostbite, make sure to dress properly for winter sports. Wear layers of clothing, plus a hat, gloves, and boots.

Reading Check

Describe What are the symptoms of heat exhaustion?

Sunburn is a problem that can occur in both hot and cold weather. The skin becomes red and sore, and blisters may form. You probably know that you need to protect your skin with sunscreen when you go to the beach.

However, you may not realize that sunscreen is important for cold-weather sports as well. The ultraviolet (UV) rays that can burn your skin reflect off snow and ice, so sunburn can develop quickly. To avoid sunburn, wear a sunscreen with an SPF of at least 15, and use lip balm as well. Be sure to use sunscreen even on cloudy days—harmful UV rays from the sun can penetrate through cloud cover.

IF YOU FEEL PAIN DURING A GAME, YOU SHOULD STOP RIGHT AWAY AND TAKE A REST.

Many minor sports injuries can be treated using the P.R.I.C.E. formula. **Identify the step in the P.R.I.C.E. formula that this teen is demonstrating.**

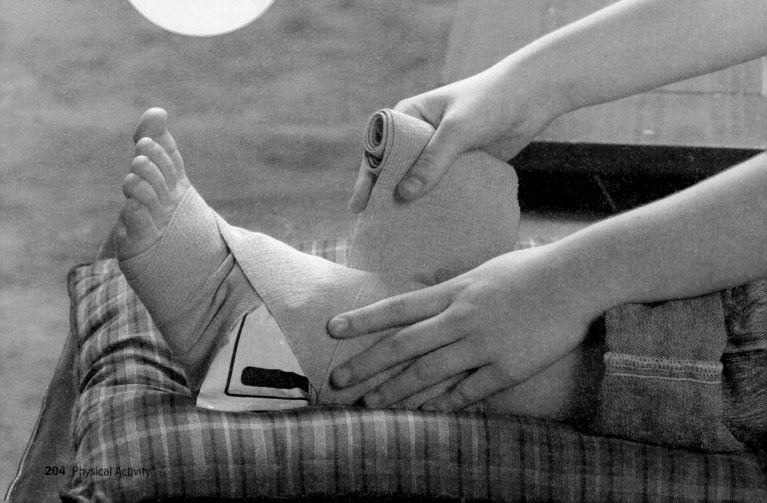

Treating Sports Injuries

MAIN IDEA The P.R.I.C.E. procedure is used to treat strains and sprains.

Sometimes, no matter how careful you are, injuries still happen. Minor injuries, such as sprains and strains, can be treated using the **P.R.I.C.E. formula**. P.R.I.C.E. means protect, rest, ice, compress, and elevate. This procedure works best if you apply it as soon as possible after an injury. It can also help if your muscles feel stiff or sore after a workout. The P.R.I.C.E. formula has five steps.

- **Protect** the injured body part from further injury. You can tape it up or splint it to take the strain off it.
- **Rest** the injured part. Give it at least a couple of days before you try to use it.
- **Ice** the body part. Apply an ice pack for 20 minutes at a time, four to eight times a day, until the swelling goes down (or for two days after the injury). Do not put ice directly on the skin.
- **Compress**, or apply pressure to, the injured part. Wrap it in an elastic wrap or bandage to reduce swelling.
- **Elevate** the injured part above the level of your heart, if possible. This will reduce swelling and pain.

> SOMETIMES, NO MATTER HOW CAREFUL YOU ARE, INJURIES STILL HAPPEN.

Reading Check

List What are the steps of the P.R.I.C.E. formula?

Lesson 4 Review

What I Learned

1. **VOCABULARY** Define the term *sprain*.

2. **DESCRIBE** What are two precautions that can help you avoid sports-related injures?

3. **NAME** What is one health risk associated with hot weather and one associated with cold weather?

Thinking Critically

4. **ANALYZE** Francis is shopping for a new pair of basketball shoes. He really likes the look of one pair, but when he tries them on, they feel a bit tight. The store does not have them in a larger size. He is thinking of buying them anyway, because they might feel better once he breaks them in. What advice would you give Francis?

Applying Health Skills

5. **INFER** How could being in good physical condition before starting a sport reduce your chances of injury?

The Life Cycle

LESSONS

10

Changes During Puberty

. . . .

Before You Read

Quick Write Make a list of five ways that you are different now than you were five years ago.

Vocabulary
adolescence
puberty
hormones
community service

. . . .

BIG IDEA You will go through many physical, mental/emotional, and social changes during your teen years.

Changes During Your Teen Years

MAIN IDEA Adolescence is a time of rapid change and growth.

When you were a baby, you grew and developed very quickly. Now, as a teen, you are going through **adolescence**. It's the stage of life between childhood and adulthood, usually beginning somewhere between the ages of 11 and 15. During this period, you will grow faster than you have at any time since infancy.

Along with changes in your body, you will experience changes in how you think and feel as well as changes in your relationships. During your teen years, you will be setting out on a journey of discovery. The object of that discovery is you. You will meet new people, explore new ideas, and take on new responsibilities. These years are a chance to practice for life on your own when you become an adult.

Physical Changes

The changes you undergo during adolescence are caused by **puberty**. These changes are caused by internal factors. They regulate your growth. During puberty, you will develop the physical characteristics of adults of your gender. Changes during puberty are caused by **hormones**. They are chemical substances made in the glands of your body. They regulate the way the body functions. They are released by the endocrine system. Hormones send signals to different parts of your body to prepare for adulthood.

> **AS YOU GET OLDER,
> YOU MAY WANT TO BEGIN
> TO HAVE NEW EXPERIENCES
> AND EXPAND YOUR WORLD.**

One of the first signs of puberty is often a growth spurt—a sudden, rapid increase in height. As you grow taller, your body shape will also begin to change. Boys' shoulders become wider and their bodies grow more muscular. Girls may notice that their hips are wider, and they're developing breasts. Both boys and girls may gain or lose weight as their bodies change during puberty, but this weight change is normal and no cause for concern.

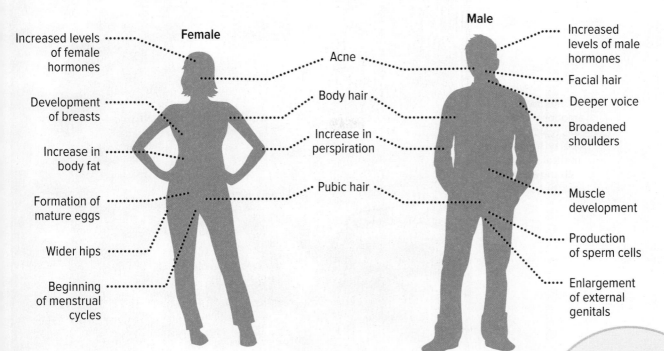

Female

- Increased levels of female hormones
- Development of breasts
- Increase in body fat
- Formation of mature eggs
- Wider hips
- Beginning of menstrual cycles

Male

- Increased levels of male hormones
- Facial hair
- Deeper voice
- Broadened shoulders
- Muscle development
- Production of sperm cells
- Enlargement of external genitals

- Acne
- Body hair
- Increase in perspiration
- Pubic hair

Puberty is the time during which you develop the physical characteristics of an adult. **Identify one change that happens to both males and females during puberty.**

Puberty begins for most girls between the ages of 8 to 13. For boys, it starts a little later. Many boys start puberty between the ages of 10 and 15. It starts at different times for everyone. Some teens begin puberty early. Some may experience changes much later. Everyone experiences puberty at their own time and pace.

Mental/Emotional Changes

Adolescence also causes changes to the way you think and feel. Studies have found that teens' brains expand right before puberty. Most of this growth takes place in the part of the brain related to planning and reasoning. During the teen years, your mental skills will improve. You will grow better at thinking critically and expressing yourself. You will also begin to develop your personal ideals and values.

The hormones that cause changes in your body also affect your emotions. Mood swings may cause sudden shifts from happiness to sadness or anger. You may feel that you are having more disagreements with family and friends. This can be confusing, but it is a normal part of adolescence. If you feel that the changes you're experiencing are troubling, talk to a trusted adult. Talking about your feelings can help you gain perspective. You may also learn new ways to manage your feelings.

During your teen years, you may also begin to develop feelings of attraction toward others. These feelings, combined with the changes in your body, may lead you to become interested in dating. However, not everyone is ready for dating at the same age.

Social Changes

During your teen years, you are likely to go through many changes in the way you relate to others. At home, your relationship with your parents may change as you become more independent. You may also want to spend less time with your parents. Your friendships will become more important to you.

Mental and emotional changes are typical during puberty. **Identify two mental skills that change during puberty.**

You will also become more aware of the world around you. As a child, your world probably centered around your home, your family, and your school. As you get older, you may want to begin to have new experiences and expand your world. One way to do this is by becoming involved in community service. Teens who volunteer learn about the impact that they can make on their community. Volunteering also offers the opportunity to meet new people.

CHANGES IN FAMILY RELATIONSHIPS As a child, you were very dependent on your parents or guardians. They made most of your decisions for you, like what you would eat and when you would go to bed. Now, as a teen, you are learning to act independently and make more decisions for yourself. Your increasing independence can sometimes lead to differences with family members. For instance, you might think that your parents should let you stay out later at night or shop for clothes on your own.

Many times, conflicts can arise when teens and parents or guardians disagree on how much independence a teen should receive. You may have a curfew to be home by 9 p.m. on school nights. Your parents or guardians set these rules to keep you safe. Other rules are set in the form of laws. If you continue to follow the rules your parents set and avoid breaking any laws, you show them that you respect the rules. Your parents will recognize that you are trustworthy. They may allow you to make more of your own decisions.

Listening to music can be a good way to deal with mood swings. **List two other good ways to manage your emotions.**

CHANGES IN PEER RELATIONSHIPS During your teen years, your peers are likely to play a growing role in your life. Many teens begin spending more time with friends and less with family members. Often, your peers will influence your ideas and your behavior. Your peers are an example of an external influence that affect choices you make or changes in relationships.

Some influences by peers are normal. Many of these make you feel as if you are part of a group. They include changing the way you dress or the music you prefer in order to be more like your friends. Peers can also influence you to try new activities. You may be encouraged to try a new sport, to act in a school play, or to study art. These are all positive influences.

> COMMUNITY SERVICE GIVES YOU A CHANCE TO HELP PEOPLE AND DEVELOP YOUR INDEPENDENCE AT THE SAME TIME.

If peers influence you to try risk behaviors, those behaviors can have negative consequences. Using tobacco, alcohol, and other drugs can lead to addiction. Becoming sexually active can result in an unintended pregnancy.

YOUR RELATIONSHIP WITH THE COMMUNITY You are part of several communities. They include your neighborhood, your school, and the city or town where you live. Your teen years can be a good time to make a positive contribution. **Community service** involves volunteer programs aimed at improving the community and the life of its residents. You will be able to make your own decisions about when and where to work. You can also learn new skills and develop a sense of responsibility.

Reading Check

Explain In what ways can peers have a positive influence on teens?

Developing Good Character

Respect Teens can be very sensitive about the changes their bodies are going through, especially if they start to go through puberty earlier than their friends. It's important not to tease others about the way they look. Remember, when your turn comes, you will want to be treated with respect. Explain what you could do to help a friend who is being teased about his or her appearance.

Lesson 1 Review

What I Learned

1. **VOCABULARY** Define the term *hormones?*

2. **NAME** What are three ways in which teens' bodies change during puberty.

3. **DESCRIBE** How are relationships with parents likely to change during the teen years?

Thinking Critically

4. **INFER** Why do you think teens go through puberty at different rates?

5. **EVALUATE** How do you think the changes you experience during adolescence help you prepare for adulthood?

6. **INFER** Is community an internal or external influence? How could your community influence your social growth?

Applying Health Skills

7. **COMMUNICATION SKILLS** Write a dialogue between a teen and a parent. In it, the teen should explain his or her need for more independence in a way that shows consideration for the parent's feelings.

Male Reproductive Health

Before You Read

Quick Write Do you plan to have children when you become an adult? How do you picture your family life when you are older?

Vocabulary

reproduction
reproductive system
sperm
egg cell
testes
semen
hernia

Reading Check

Define What is the reproductive system?

BIG IDEA The male reproductive system produces cells that can join with cells from a female's body to produce a child.

Reproduction

MAIN IDEA The reproductive system makes it possible for people to produce children.

Romi's older sister has recently come home with a new baby. Romi enjoys holding his niece and feeling her tiny hand squeezing his finger. Romi even helps change and bathe the baby. Playing with his niece and helping his sister and her husband take care of their new arrival has made him start to think about whether he would like to have children of his own someday.

Reproduction is the process by which living organisms produce others of their own kind. It is one of the most important functions of all living things. Individuals will eventually grow old and die, but through the process of reproduction, new ones are produced to keep the species going.

In human beings, the process of reproduction involves the joining of two cells that come from a male and a female **reproductive system**. This body system includes the body organs and structures that make it possible to produce children. Every human being on Earth begins life in this way. Puberty and adolescence is the time during which your body goes through a series of physical, mental/emotional, and social changes. These changes prepare your body to be able to produce children during adulthood. In this lesson and the next, you will learn more about how the male and female reproductive systems work and how to take care of them.

The Male Reproductive System

MAIN IDEA The main purpose of the male reproductive system is to produce and release sperm.

The male reproductive system has two main jobs. The first is to produce sperm. The second is to deliver **sperm**, or male reproductive cells, to the body of a female, where a sperm cell can join with a female **egg cell**. An egg cell is a reproductive cell from the female that joins with a sperm cell to make a new life.

The **testes** are a pair of glands that produce sperm. They also produce male hormones. The testes, or testicles, are located in a pouch called the scrotum. The scrotum helps keep the testes at the right temperature needed to produce sperm. Sperm is stored in a collection of tubes called the epididymis. The epididymis is located next to the testes.

Reading Check

Identify What are the two primary functions of the male reproductive system?

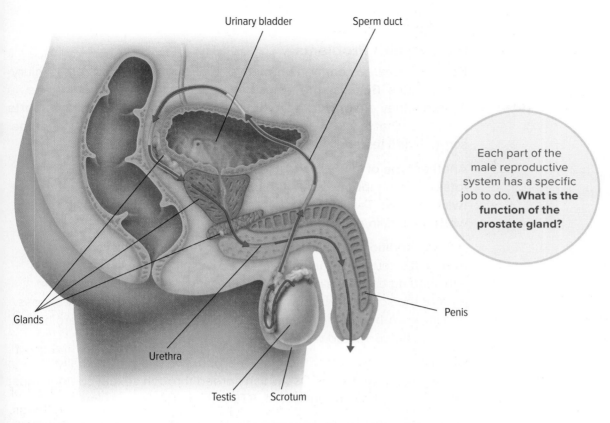

Urinary bladder

Sperm duct

Glands

Urethra

Testis

Scrotum

Penis

Each part of the male reproductive system has a specific job to do. **What is the function of the prostate gland?**

Before sperm leave the body, it mixes with fluids from both the prostate gland, the seminal vesicles, and the vas deferens. The resulting mixture, called **semen**, protects sperm and carries it through the tubes of the male reproductive system. The semen travels to the urethra and is released from the body through the penis. The penis is a spongy tissue made up of many blood vessels. This happens in a series of muscle contractions known as ejaculation. An ejaculation can contain as many as 500 million sperm.

The male and female reproductive systems make it possible for people to have children. **Name the process that allows living things to produce others of their own kind.**

Male Health Problems

For teen males, a common problem of the reproductive system is injury to the testes. Because the testes hang outside the body, they can be injured if they are hit, kicked, or crushed. This can happen during sports activities. That is why many sports require a protective cup to cover the scrotum and testes.

Another type of injury to the male reproductive system is testicular torsion. This happens when the cord that holds the testes together twists around. Testicular torsion cuts off the flow of blood to a testicle, which results in pain and swelling.

A third condition that can affect the male reproductive system is a **hernia**. It's caused by an internal organ pushing against or through a surrounding cavity wall. Part of the intestine can push into the groin area or the scrotum. Hernias are typically caused by muscle weakness and strain. A hernia may appear as a bulge or swelling in the groin area. This can be corrected with surgery.

Diseases can also affect the male reproductive system. Sexually transmitted diseases (STDs), which spread through sexual contact, must be treated. If they are untreated, STDs can result in infertility.

One of the most serious diseases that can affect the male reproductive system is cancer of the testes. This is the most common form of cancer in males aged 15 to 35. If it is not treated, the cancer may spread to other parts of the body. If it is caught early, it can almost always be cured. If the testicles are removed, however, the man will become sterile, or unable to produce offspring. Men may also become sterile as a result of drug use or being exposed to radiation or certain types of chemicals. Also, some STDs can cause sterility if they are not treated.

Reading Check

Identify Name two problems that can affect the male reproductive system.

Injuries to the testicles can occur during sports activities. **Name what the boy in the picture can do to protect himself from injury.**

IT IS IMPORTANT FOR MALES TO DO A TESTICULAR SELF-EXAM EVERY MONTH.

Caring for the Male Reproductive System

Like any other body system, the male reproductive system needs care. Taking good care of your reproductive system during your teen years will increase your chances of becoming a parent as an adult. Here are a few steps male teens can take to keep their reproductive systems healthy:

- **Practice good hygiene.** Take a bath or shower daily and be sure to clean the penis and scrotum carefully.

- **Wear protective gear during sports.** For contact sports such as baseball and football, a cup is needed. Other sports that involve running require an athletic supporter.

- **Conduct self-exams.** Once a month, check the testes for any lumps or swellings that could be a sign of testicular cancer. See the Health Skills Activity for more information.

- **Get regular checkups.** Your doctor or other health care provider can screen for reproductive health problems and catch them before they become serious.

- **Avoid wearing tight clothing.** Clothing that is too tight can harm the production of healthy sperm.

The American Cancer Society recommends that all males perform a testicular self-exam once a month. **Explain why it is a good idea for teen males to start performing testicular self-exams.**

Lesson 2 Review

What I Learned

1. **VOCABULARY** Define the term *hernia*.

2. **IDENTIFY** What are the two main functions of the male reproductive system?

3. **NAME** What are two types of gear that protect the reproductive system of a male athlete?

Thinking Critically

4. **EVALUATE** Why is it important to take care of the male reproductive system?

5. **APPLY** Anthony has noticed that one of his testicles is swollen. He did not feel comfortable talking about the problem, but it is not going away. What steps would you suggest that Anthony take?

Applying Health Skills

6. **ACCESSING INFORMATION** Some teens and adults use anabolic steroids to change their bodies. Use online resources to learn about the ways steroids affect the male reproductive system. Summarize your findings in a fact sheet about the dangers of illegal steroid use.

Female Reproductive Health

Before You Read

Quick Write What do you think is the most important thing teens can do to care for their reproductive systems?

Vocabulary

fertilization
ovaries
uterus
menstrual cycle
ovulation
menstruation
cervix
gynecologist

BIG IDEA The female reproductive system contains the organs that nurture a baby before birth.

How the Female Reproductive System Works

MAIN IDEA The female reproductive system has three main functions: to produce and store eggs, to create and nurture offspring, and to deliver a baby.

The first function of the female reproductive system is to produce, store, and release egg cells. Eggs, or ova, are the female reproductive cells. The second function is to create offspring. This process begins with **fertilization**. It's the joining of a male sperm cell with a female egg cell to form a fertilized egg. The female reproductive system must also nourish the developing baby until it is born. Giving birth is the system's third function.

The Female Reproductive System

The female reproductive system is made up of several organs, including two **ovaries**. The ovaries are a female endocrine gland that produce mature eggs and two different hormones. These hormones are called estrogen and progesterone and control the female reproductive system and sexual development.

Once a month, the ovaries release an egg cell into the fallopian tubes. If an egg is fertilized, it continues along the tube. It then attaches itself to the wall of the **uterus**. The uterus is a pear-shaped organ, located within the pelvis. It is where the developing baby is nourished and protected. The fertilized egg continues to develop for about nine months. At birth, the baby moves out of the uterus and passes out of the body through the vagina.

The Menstrual Cycle

The **menstrual cycle** begins when a female experiences puberty. Puberty occurs when the female reproductive hormones begin to be produced by the ovaries. The cycle lasts from the beginning of one menstruation to the next. For some females, the cycle may be as short as 23 days. For others, it may take as long as 35 days. A female's menstrual cycle may not become regular for one or two years after she first begins menstruating.

The cycle begins with **ovulation**. It's the process by which the ovaries release mature eggs, usually one each menstrual cycle. When the ovaries release an egg, it moves through the fallopian tubes toward the uterus. At the same time, the lining of the uterus begins to thicken. The lining thickens as the uterus prepares to receive the egg if it is fertilized.

Reading Check

Explain What happens to the lining of the uterus if fertilization does not take place after ovulation?

Ovaries
The ovaries hold the female's eggs. The ovaries also make the hormones estrogen and progesterone. These control female sexual development and other reproductive organs.

Cervix
This is the narrow part of the bottom of the uterus. The opening of the cervix enlarges to allow a baby to leave the uterus during birth.

This diagram shows the parts of the female reproductive system. **Identify the functions of the ovaries.**

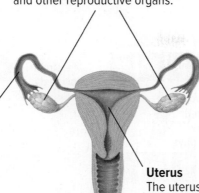

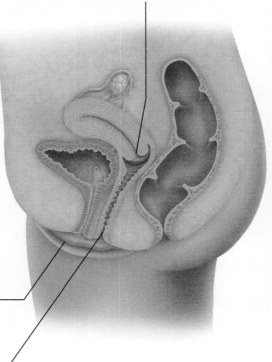

Uterus
The uterus is the organ in which a developing child is nourished.

Fallopian tubes
Eggs travel from the ovaries to the uterus through the fallopian tubes. Eggs are usually fertilized in these tubes.

Labia
Labia are folds of skin that cover the opening of the vagina.

Vagina
The vagina is the passageway that leads from the cervix to the outside of the body. Menstrual flow leaves the body through the vagina. Sperm enter the female reproductive system through the vagina. During birth, a baby leaves the mother's body through the vagina.

If the egg is not fertilized when it reaches the uterus, **menstruation** occurs. It's the flow from the body of blood, tissues, and fluids that result from the breakdown of the lining of the uterus. The lining of the uterus then leaves the body through the vagina. Menstruation typically lasts from three to five days. Then, about two weeks later, the ovaries release another egg, and the cycle starts over again.

Female Health Problems

MAIN IDEA The female reproductive system can be affected by a wide variety of problems, ranging from minor to life-threatening.

Before their menstrual cycles, many females experience a set of symptoms known as premenstrual syndrome, or PMS. Some of these symptoms are physical. For example, females may develop headaches or acne. They may also feel tired or crave certain foods. Emotional symptoms can also occur, such as depression or irritability. This is one of the most common problems affecting the female reproductive system. Other problems are less common, but may be more serious.

- Many females have abdominal cramps during the first few days of their menstrual cycles.
- A yeast infection in the vagina can cause itching, discharge, and sometimes pain. Medicine can clear up these symptoms.

Myth vs Fact

Myth Females can practice good hygiene by douching, or rinsing out the vagina with a solution of water and vinegar.

Fact Douching can be harmful. The body has its own natural system for flushing bacteria out of the vaginal area. Douching is more likely to cause irritation or infections than to prevent them.

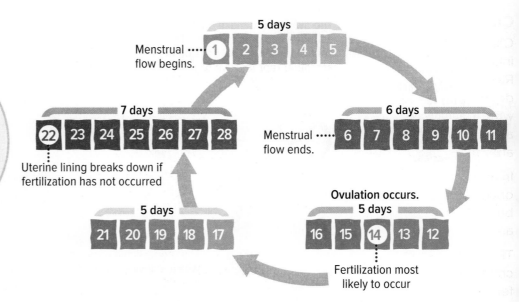

5 days
1 | 2 | 3 | 4 | 5

Menstrual flow begins.

6 days
6 | 7 | 8 | 9 | 10 | 11

Menstrual flow ends.

Ovulation occurs.
5 days
16 | 15 | 14 | 13 | 12

Fertilization most likely to occur

5 days
21 | 20 | 19 | 18 | 17

7 days
22 | 23 | 24 | 25 | 26 | 27 | 28

Uterine lining breaks down if fertilization has not occurred

The menstrual cycle is usually around 28 days. However, it may be shorter or longer for some females, especially during the first two years of menstruation. **What process begins the menstrual cycle?**

- Toxic shock syndrome is a rare bacterial infection that causes fever and vomiting. If untreated, it can cause death. It may occur if a tampon is left in the body too long.
- Ovarian cysts are fluid-filled sacs that form on the ovaries. These are usually harmless, but they may become painful if they grow too large.
- Sexually transmitted diseases, or STDs, can harm the female reproductive system.
- Women can get cancers of the reproductive system. Cancer may affect the breasts, ovaries, uterus, or **cervix**, which is the entrance to the uterus. Certain STDs can increase the risk of cervical cancer.
- Infertility is the inability to become pregnant. Some STDs can cause this problem if they go undetected or are not treated. Other causes include blocked fallopian tubes and fibroids, which are noncancerous growths in the uterus.

FEMALES CAN DETECT SIGNS OF CANCER BY DOING A BREAST SELF-EXAM EVERY MONTH.

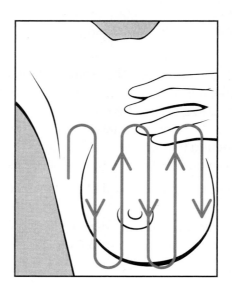

In a vertical pattern, check from the underarm to the sternum and from the collarbone to the ribs. **Explain why it is important for females to start performing regular breast self-exams during the teen years.**

Caring for the Female Reproductive System

Caring for the reproductive system is part of a female's overall health. It includes good hygiene. This means taking a daily shower or bath. Regular medical checkups are also important. In addition to their regular doctor, females should see a **gynecologist**. A gynecologist is a doctor who specializes in the female reproductive system. Females should see a gynecologist for the first time at around age 14 and then once a year after that.

In addition to medical exams, females should examine their own breasts once a month for signs of cancer. Breast cancer is rare among teens, but self-examinations can help girls become familiar with both the look and feel of their developing breasts.

These self-exams allow females to learn what is normal for them. Some changes to look for include lumps, swelling, or irritation of the skin. If a female does notice anything unusual or different during a breast self-examination, it is important to see a doctor.

Females should make sure that they change their tampons or sanitary napkins at least every four hours during their menstrual cycles. This practice helps maintain good hygiene. Changing tampons regularly can also help prevent potentially serious infections such as toxic shock syndrome. Finally, females can keep track of their menstrual cycles. This will help them notice problems such as missed cycles. A female should see a doctor if she misses several menstrual cycles in a row. She should also see a doctor if she experiences severe pain or bleeding in between cycles.

Reading Check

List What are four ways to care for the female reproductive system?

Lesson 3 Review

What I Learned

1. **VOCABULARY** Define *fertilization*.

2. **LIST** What are the three functions of the female reproductive system?

3. **RECALL** How often should females perform breast self-exams?

Thinking Critically

4. **INFER** How does fertilization affect the menstrual cycle?

5. **SYNTHESIZE** How are the male and female reproductive systems similar? How are they different?

Applying Health Skills

6. **ADVOCACY** By practicing abstinence as a teen, you can avoid STDs that could damage your reproductive system. Design a bumper sticker or button with a slogan that promotes abstinence. You may also use graphics or illustrations to support your ideas.

Infant and Child Development

Before You Read

Quick Write Describe a time when you or someone you know played with a baby or helped care for one. What could the baby do without your help? What did you have to help the baby do?

Vocabulary

cell
tissues
organs
body system
embryo
fetus
developmental task
infancy
toddler
preschooler

Reading Check

Explain Why some couples might choose to use family planning?

BIG IDEA Each human being passes through several stages of development before reaching adolescence.

Choosing to Start a Family

MAIN IDEA A couple will decide together when they are ready to take on the responsibility of starting a family.

Once you have formed a committed relationship with another person, you and that person may decide to start a family. Many older adults look back at the years they raised children as being meaningful. They may also admit that having children can be stressful and expensive. For these reasons, some couples choose to delay having children or to never to have children. They are engaging in family planning.

Family planning is the practice of controlling the number of children in a family, or to deliberately choose to space out the births of children in the family. Many couples choose family planning in order make sure their lives are established before bringing children into the world. One partner may want to become established in a good job. The couple may want to buy a house. They may want to make sure they have enough money saved to take care of children. In the U.S. a middle-income family should expect that each child will add about $15,000 to the family budget.

Couples who choose to plan when they will have children will see a doctor to talk about their decision. The doctor will tell them about methods they can use to prevent an unintended pregnancy until they are ready for a family. They may make the choice to use contraceptives in order to avoid pregnancy.

How Life Begins

MAIN IDEA Every human being starts out as a single cell that develops into a baby over the course of nine months.

As strange as it may seem, the human body began as a single **cell**. A cell is the basic unit of life. This cell formed when an egg cell from the mother joined with a sperm cell from the father. During the next few days, this single cell divided many times before reaching the mother's uterus.

Over the course of nine months, this tiny clump of cells continued to divide and form new structures. Eventually, these structures developed into all the parts of a human body. To understand how this process took place, first you need to know a little more about how the human body is constructed.

Building Blocks of Life

Have you ever thought about how a computer or a household appliance works? If you have, you know that it contains many different parts, each with its own job to do. However, all these different parts work together to make the object function properly. In the same way, your body contains many different parts that work together to enable you to breathe, eat, walk, speak, and do everything else you need to do in the course of a day.

Machines can have hundreds or thousands of parts. The body is made up of trillions of tiny parts known as cells. Cells that have similar functions in the body work together in larger units called **tissues**. Tissues are groups of similar cells that do a particular job. Some examples include muscle tissue, brain tissue, and connective tissue. Tissues, in turn, can form an organ. Your heart, brain, liver, and lungs are all **organs**. They are body parts made up of different tissues joined to perform a particular function. When many organs work together to carry out a certain job, they form a **body system**. A body system is a group of organs that work together to carry out related tasks.

YOUR WHOLE BODY BEGAN AS A SINGLE CELL.

The human body is made up of trillions of cells. **Name the larger units that cells form.**

Signs of Pregnancy

When a couple decides to have children, the female will prepare for pregnancy. She should see her doctor for a check-up to make sure she is healthy. It is also a good time to examine lifestyle choices. The female should bring a list of all the medications and vitamins that she is taking. She should stop using tobacco, alcohol, or other drugs.

During the visit, the doctor may suggest that the female gain or lose weight. Studies show that underweight women have a higher chance of giving birth to small babies. The doctor may also recommend eating a well-balanced diet that includes folic acid, calcium, and iron. This helps ensure proper nutrition before becoming pregnant. The doctor may also prescribe prenatal vitamins. The exam is a good time to talk about any worries that the future parents may have. It is a good time to talk about diseases that occur in families, such as diabetes. The doctor can explain the risks and provide advice. A doctor will also describe the first signs of pregnancy. These include:

- missed menstrual cycle,
- tiredness,
- nausea,
- headaches,
- constipation,
- breast tenderness,
- and mood swings.

Development Before Birth

The process of a single cell dividing to form all the body's tissues, organs, and systems is long and complex. Fertilization usually occurs inside the mother's body. The egg cell and the sperm cell join in one of her fallopian tubes. The cell begins to divide, preparing to form a new person.

The fertilized egg cell continues to divide as it moves to the uterus. The developing organism is now called an embryo. It is nourished by the uterus. The cells of the embryo continue to multiply and change, forming tissues. The tissues combine to form organs. After about eight weeks, the embryo is called a **fetus**. A fetus is a developing organism from the end of the eighth week until birth. All major organs have begun to take shape. These include the heart, brain, and lungs.

The fetus floats in a fluid-filled sac that cushions and protects it. It gets its nourishment from the mother's blood through the placenta, a thick, rich tissue that lines the walls of the uterus. A tube called the umbilical cord connects the mother's placenta to the fetus. Through it, the fetus can take in oxygen and nutrients. However, substances such as alcohol or tobacco can also reach the fetus through the placenta. Women need to be careful to avoid these harmful substances.

Reading Check

Name What is the smallest basic unit of life?

Prenatal Care

During pregnancy, prenatal care helps the baby develop in a healthful way. To give birth to a healthy fetus, the mother needs regular care from a doctor. The doctor will check that mother's weight is within a normal range. The doctor will also do tests to make sure that the baby is developing normal. An expecting mother will also take vitamins that help keep herself and her baby healthy.

All expectant mothers are advised to take folic acid. It prevents serious birth defects. Studies show that expecting mothers also need to get the proper amount of protein, fats, and other nutrients. Good nutrition is important to their health as well as the health of the fetus. The Food and Drug Administration (FDA) recommends that expecting mothers avoid certain foods. These include raw fish and raw eggs. Both can contain high levels of bacteria. The USDA's MyPlate also provides guidelines for pregnant women.

A new human life is formed during a nine month period. **Identify the month when the fetus begins to breathe.**

End of First Month

The embryo is less than ¼ inch long. The spinal cord is forming, and the embryo's heart is beating.

End of Second Month

The fetus weighs about 1/3 ounce and is about 2 inches long. It has arms, legs, eyes, and ears, and it is beginning to form red blood cells.

End of Sixth Month

The fetus is regularly sleeping and waking inside the uterus. It has hair on its head and nails on its fingers. It measures over 9 inches from head to backside and weighs as much as 2 pounds.

End of Seventh Month

The fetus's eyes are open and its bones are fully formed. It has toenails, and it is beginning to breathe. The soft hair that has covered its skin begins to fall off. It may measure 11 inches from head to backside and weigh 3 ¼ pounds.

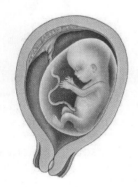

End of Third Month

The fetus has functioning organs, nerves, and muscles. It is beginning to develop bones and hair. It is possible to tell whether it will be a boy or a girl. The fetus is about 3 ½ inches long and weighs about 1 ½ ounces.

End of Fourth Month

The fetus's movements can be seen on an ultrasound image. It can hear sounds and make sucking motions with its mouth. It is about 6 inches long from its head to its backside, and it weighs over 7 ounces.

End of Fifth Month

The mother can feel the fetus moving. Its skin is wrinkled and covered with fine hair. It can swallow. Its fingerprints and footprints are forming. It may be 7 ½ to 8 inches long from head to backside and weigh more than a pound.

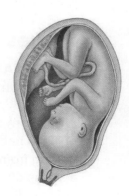

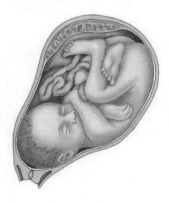

End of Eighth Month

The fetus will put on about half a pound in weight each week for the next month. Its eyes can detect light, and its body has become round.

End of Ninth Month

The fetus is fully developed. Its head now points toward the mother's cervix. A full-term fetus may measure 18 to 20 inches from head to backside and weighs 6 pounds or more.

1 Infancy
Description: The child depends on others to meet his or her needs.
Developmental task: learning trust

2 Early Childhood (Toddler Stage)
Description: The child is learning to walk, talk, and do other things on his or her own.
Developmental task: developing autonomy (the ability to do things for yourself)

3 Middle Childhood (Preschool Years)
Description: The child begins to come up with ideas on his or her own and to develop a sense of right and wrong.
Developmental task: developing initiative (the ability to plan and act on your own)

4 Late Childhood (Early School Years)
Description: The child starts school and learns new skills.
Developmental task: developing a sense of self-worth

5 Adolescence
Description: The teen explores many different roles.
Developmental task: developing identity (a sense of who you are)

6 Early (Young) Adulthood
Description: The young adult tries to establish close personal relationships.
Developmental task: developing intimacy (a strong connection with another person)

7 Middle Adulthood
Description: The adult seeks a sense of satisfaction through career, raising a family, and being active in society.
Developmental task: making a contribution to the world

8 Late Adulthood (Maturity)
Description: The adult looks back on the achievements of his or her life.
Developmental task: to be satisfied with the life one has lived.

Each stage of life has a different developmental task. **Name the developmental task during adolescence.**

Stages of Life

MAIN IDEA A human life can be divided into a series of stages, each with its own tasks to be mastered.

A newborn baby can do very little by itself. For the first several months of life, a baby depends on others for tasks as simple as eating and getting from place to place. To reach adulthood, a baby will go through a series of stages or steps with different needs. The psychologist Erik Erikson divided the human life cycle into eight separate stages. Erikson's stages of life include infancy, early childhood, middle childhood, late childhood, adolescence, early adulthood, middle adulthood, and late adulthood. At each stage, Erikson explained, you need to master a **developmental task**. These are events that need to happen in order for you to continue growing toward becoming a healthy, mature adult. The developmental stage figure shows the stages of life and some tasks for each stage.

Infancy

During **infancy**, or the first year of life after birth, babies learn about their own bodies. Tasks such as focusing the eyes or sitting up are big achievements for a baby. Infants also begin to observe the world around them. Babies learn to recognize familiar people and objects. They begin to learn language, making sounds and recognizing others' voices. During this first stage of life, infants also develop bonds of love and trust with their parents and others close to them.

Early Childhood

The next stage is a **toddler**. A toddler is a child between the ages of one and three. During these years, they will become more aware of themselves and their surroundings. Toddlers learn to do things on their own, such as feeding themselves and using the toilet. They can walk and are interested in exploring new objects and people. Young toddlers may also speak in short sentences and be able to follow simple directions.

TODDLERS WILL OFTEN IMITATE BEHAVIORS OF PARENTS OR SIBLINGS.

In early childhood, children learn to walk on their own. **What are some other skills children learn at this age?**

After age 2, toddlers start to learn more complicated skills. Physically, they may be able to run, jump, or kick a ball. Mentally, toddlers can usually recognize different shapes and follow directions that have two or three steps. Socially, toddlers will often imitate the behavior of their parents or siblings. However, they are also becoming more interested in being independent. Many parents of toddlers will say that their child's favorite word is "no."

Middle Childhood

If a toddler's favorite word is "no," a **preschooler** may like the word "why?" A preschooler is a child between the ages of three and five. At this stage of life, children tend to ask a lot of questions and use their imaginations. Preschoolers may enjoy telling stories or singing songs. Children between the ages of three and five become more interested in people outside their immediate families. They learn to cooperate with other children. Preschoolers' physical skills also improve. They can do exercises like hopping on one foot or riding a tricycle. They may be able to dress and undress themselves or choose foods at mealtimes.

During your school years, friendships become more important. **Describe some other differences between middle and late childhood.**

Late Childhood

From ages six to eleven, school plays an important part in children's lives. It brings them into regular contact with other children, as well as adults. Children become more independent from their families, and friends become more important to them. Through schoolwork, friends, and activities, children at this stage develop increased confidence in themselves.

By the time you reach your teen years, you have already learned and accomplished quite a lot in your life. You can do physical tasks such as riding a bike or throwing a ball. You can read and write, and you have learned to recognize and express your emotions. You can make decisions and solve problems on your own, but you have also formed new friendships and learned to work with others. All these skills will help you as you move through adolescence and into adulthood.

Adolescence

During the time between ages 12 and 18, you are no longer a child, but you are not yet an adult. During adolescence, puberty occurs. A child in this stage will experience physical, mental/emotional, and social changes that prepare the child for adulthood.

Myth vs. Fact

Myth All babies should be starting to walk and talk by the time they reach their first birthday.

Fact All children develop at their own pace. However, most one-year-olds can say one or two words, like "mama" or "dada." They should also be able to stand up with help from a parent. If parents are worried that their child is taking too long to learn these skills, they can check with a doctor to make sure nothing is wrong.

Reading Check

Recall What is the first stage of life?

Lesson 4 Review

What I Learned

1. **VOCABULARY** What is a *developmental task*?

2. **IDENTIFY** What are the parts that make up a body system, from smallest to largest?

3. **LIST** What are the eight stages of life?

Thinking Critically

4. **ANALYZE** What is the difference between an *embryo* and a *fetus*?

5. **INFER** What might happen to an infant who did not receive proper care or attention from adults?

Applying Health Skills

6. **GOAL SETTING** Vanessa's older sister, Kayla, has tried several times to quit smoking without success. Now she and her husband want to have a baby. She has asked Vanessa for her advice. Use the steps of the goal-setting process to show how Kayla could reach her goal of becoming tobacco free before she gets pregnant.

Staying Healthy as You Age

Before You Read

Quick Write Pretend that you are 50 years older than you are now. Write a letter to your teen self. What kinds of stories or advice would the older you want to share with the younger you?

Vocabulary

chronological age
biological age
social age

Developing Good Character

Respect Their bodies may not move as fast as yours, but older adults have a lot of wisdom and experience to share. You can show respect for older adults by listening and speaking in a polite manner. Make a list of topics you think would be interesting to discuss with older adults. Describe some other ways to show respect to older adults.

BIG IDEA Practicing good health skills will allow you to remain healthy throughout early, middle, and late adulthood.

Stages of Adulthood

MAIN IDEA People continue to learn and develop throughout adulthood.

Right now, as a teen, you go through a lot of changes as you move toward adulthood. However, this does not mean that you will stop changing once you become an adult. Your body will change gradually as you age. Your mental abilities will continue to change as you learn and develop new skills throughout your life. Your social life will also change as you form new relationships, which might include marriage or parenthood.

Just like childhood, adulthood can be divided into several stages. Each stage may involve milestones such as starting a career, getting married, or raising a family. However, these milestones will not be the same for everyone. Some people, for instance, marry late in life or they may never marry at all. Some adults choose to have many children, while others have none. Some people work in one career throughout adulthood, while others may change careers many times. Each person's path through adulthood is unique because no two people are exactly alike.

Early Adulthood

Many people, as they leave their teen years, are still working on their education. Others have finished school and are starting their first jobs. Young adults may try several different jobs before settling on a career path. People this age also work on forming close relationships with others. This may include marrying or starting a family. However, some adults do not marry until they are older, while others never marry. It is also important for people in early adulthood to continue habits that will contribute to good health later in life. These positive steps include a healthful diet and staying active.

> **EACH PERSON'S PATH THROUGH ADULTHOOD IS UNIQUE BECAUSE NO TWO PEOPLE ARE EXACTLY ALIKE.**

Buying a first home is one millstone many people reach in middle adulthood. **What are some other changes people may go through during this time?**

Middle Adulthood

In middle adulthood, many people have settled down in their work and personal lives. At this point, they may start to think about what they want to contribute to the world. People in middle adulthood often feel a need to achieve and be recognized. They may focus on advancing in their careers, raising their families, or helping their communities.

At some point in middle adulthood, people often think about what they have done with the first half of their lives and what they want to do next. Some people decide to make major changes in their lives at this time. Others focus on getting the most out of the paths they have already chosen. Many adults at this stage of life also renew their commitment to a healthful diet and remaining physically active.

Late Adulthood

Late adulthood begins around age 65. People at this age often decide to retire from their jobs and pursue new interests. Many become more active than when they were younger. Retirement may allow older adults to travel. Someone who has always lived and worked in a city may start visiting parks. In late adulthood, it is also important to maintain good physical health.

Some older adults pursue new careers or volunteer work. Late adulthood can also be a time of reflection. Older adults may ask whether they have fulfilled their potential. If they are satisfied, they can be content at this stage.

Reading Check

Compare What are some of the differences between early and middle adulthood?

Ways to Measure Age

MAIN IDEA Your age in years may be different from the age suggested by your physical health or your social behavior.

According to one saying, "You're only as old as you feel." Many people in late adulthood feel young for their age because they have maintained good health. Because of this, there are several different ways of measuring age.

- **Chronological age** is your age measured in years. It does not reflect your health. People who are the same age may be very different physically, mentally, and socially.
- **Biological age** is determined by how well various body parts are working. It focuses the physical condition of a person's body along with overall health. For instance, a 50-year-old with arthritis would have an older biological age than a 50-year-old in good health. Heredity, environment, and behavior all play a role in this type of age. People who have made healthful choices may have a younger biological age.
- **Social age** is measured by your lifestyle and the connections you have with others. Consider two young adults, both 22 years old. One of them is still attending college and living at home. The other has taken a job, gotten married, and started a family. The second person might have an older social age.

Good Health for Older Adults

MAIN IDEA Staying active can help older adults maintain good health.

Wellness, or balanced health, is as important for older people as it is for teens. Adults who take good care of their physical health may have a biological age that is younger than their age in years. Being physically active and eating healthy foods are two ways to do this.

Good mental health is also a part of biological age. Older people may be at risk for mental problems such as depression. Getting regular medical care can be a key to finding and treating these problems. Being involved in social and physical activities can also help.

Fit and active people tend to be healthier throughout their lives. **Predict how your choices as a teen may affect your biological age later in life.**

STAYING ACTIVE IS IMPORTANT FOR SOCIAL HEALTH.

Developmental tasks for middle and older adulthood differ. **Name two differences between middle and older adulthood.**

Staying active is important for social health as well. Older people will have better total health if they stay involved with friends and continue to participate in activities they enjoy. Some sports that older adults might play are tennis, softball, hiking, biking, dancing, or walking. Staying active can help older adults maintain good health. Getting regular physical activity can help an older adult reduce their social age and biological age.

Reading Check

Report What are some ways that older adults can stay physically active throughout their lifetime?

Lesson 5 Review

What I Learned

1. **VOCABULARY** Explain what the term *social age* means.

2. **NAME** What are three things that people may focus on during early adulthood?

3. **EXPLAIN** Why is regular medical care important for mental health in late adulthood?

Thinking Critically

4. **ANALYZE** How are chronological age and biological age related? How are they different?

5. **ANALYZE** Ophelia retired last year. She was looking forward to spending lots of time on her favorite hobby. However, she soon began to feel dissatisfied. She missed the daily social contact of her job. What could Ophelia do to bring her health triangle back into balance?

Applying Health Skills

6. **ANALYZING INFLUENCES** Sylvester is 65 years old and recently retired. He has just decided to get an Internet connection in his home for the first time. How might this decision affect his total health?

Personal Health Care

LESSONS

1 Personal Hygiene and Consumer Choices

2 Taking Care of Your Skin and Hair

3 Caring for Your Mouth and Teeth

4 Protecting Your Eyes and Ears

11

Personal Hygiene and Consumer Choices

Before You Read

Quick Write Write two ways that washing your hands with soap helps keep you and others healthy.

Vocabulary

hygiene
body odor
cuticle
hangnail
ingrown toenail
consumer
consumer skills
advertisement
media literacy
endorsement
infomercial
comparison shopping
generic products
coupons
unit price
guarantee
warranty
fraud
health fraud

Reading Check

Identify What are two ways body odor can be eliminated?

BIG IDEA It is important for you to know how to maintain good hygiene as well as be a smart consumer.

What is Hygiene?

MAIN IDEA Your personal hygiene affects all sides of your health triangle.

Think about your appearance. Have you bathed or showered today? Are your clothes neat and clean? Is your hair combed? Did you brush your teeth this morning? Caring for your appearance includes paying attention to your personal **hygiene**, or cleanliness.

Keeping your body clean is an example of good hygiene. Your hygiene and your appearance affect all three sides of your health triangle. When you look your best, you feel good about yourself. This improves your mental/emotional health. You are more confident around others, strengthening your social health. Of course, good hygiene also keeps your body physically healthy. For example, washing your hands helps prevent the spread of germs that cause illness and disease.

Keeping It Clean

Cleanliness is the key to good hygiene. Bathing or showering every day helps keep your skin and body clean. Use soap to wash away dirt, sweat, oils, and bacteria that collect on your skin. Having clean, healthy skin is part of your overall appearance.

During the teen years, **body odor**, or smell from the body, may become more noticeable. Body odor is caused by bacteria on the surface of your skin. When the bacteria mix with sweat, it creates a bad odor. Bathe or shower regularly to help eliminate body odor. Deodorants or antiperspirants can also help.

Another important part of maintaining good hygiene is washing your hands thoroughly and often. Use plenty of warm water and soap on your hands. Remember to wash the palms and backs of your hands and between your fingers. Washing your hands helps prevent illness and the spread of germs. Always wash your hands before you prepare and eat food. Remember to wash your hands after you use the bathroom, play with pets, visit a sick person, or touch garbage or another source of germs.

Facial and Body Hair

As you get a little older, you will develop facial and body hair. The growth of hair in new areas of your body, such as under your arms or on your face, is caused by hormones. As facial and body hair becomes thicker, males may choose to shave their facial hair. Females may choose to shave the hair on their legs or under their arms. Shaving your body hair is a personal choice.

Caring for Your Nails

Nails protect sensitive fingertips and the tips of toes. Your fingernails and toenails are made of a tough substance called keratin. The **cuticle** is a fold of epidermis around the fingernails and toenails. Sometimes, the cuticle can become torn or cut. Nail care includes caring for the cuticle. Common problems with fingernails and toenails include:

- **Hangnails.** A **hangnail** is a split in the cuticle along the edge of a fingernail. You can treat hangnails by carefully trimming the skin. The cuticle should heal in a few days.
- **Ingrown toenails.** An **ingrown toenail** is a condition in which the nail pushes into the skin on the side of the toe. This can result from trimming the nail on a curve rather than straight across or wearing shoes that are too tight. See a doctor if your toe becomes inflamed and sore, as it may be infected.

To care for your nails, wash your hands regularly. Use hand lotion to keep nails and skin moist. Trim your nails using a nail clipper or small scissors. Cut your nails straight across, so the nail is at or just beyond the skin. Use an emery board or nail file to round out the ends of your fingernails slightly and smooth out rough edges. Soften your nails with warm water. Use a cuticle stick to push back the cuticle. Avoid biting, tearing, or picking at your nails. Putting your fingers in your mouth can also spread germs.

Proper nail care keeps your nails looking clean and healthy. **Identify some steps you can take to improve the appearance of your nails.**

The changes your body undergoes during adolescence make hygiene especially important. **Explain how having good hygiene can make you feel more confident.**

Developing Consumer Skills

MAIN IDEA Knowing how to evaluate products and services will help you become a smart consumer.

A **consumer** is a person who buys products and services. As a consumer, you have likely walked through store aisles many times. How do you decide which items to buy? How do you know what will be best for you? Learning **consumer skills** will help you become a smart shopper. Consumer skills are techniques that enable you to make wise, informed purchases. Smart shoppers know how to compare products in terms of quality, effectiveness, safety, and cost. They also know how to resolve problems with purchases.

Informed shoppers know that their choices are influenced by **advertisements**. These are messages designed to influence people to buy a product or service. Advertisements are created to make products look appealing. **Media literacy** is the ability to understand the goals of advertising and the media. Media literacy skills can help you decide if an ad is truthful.

Type of Advertising

Some advertisements suggest that a product will make you beautiful or successful, like the people in the ad. Many ads feature a celebrity **endorsement**, which is a statement of approval. Advertisers pay athletes or actors for endorsing a product.

An **infomercial** looks like a news story or television show. However, infomercials are really advertisements. They always feature a product for sale.

Consumer skills are techniques that help you become a smart shopper. **Name two qualities that smart shoppers use to compare products.**

Advertisements may also contain hidden messages. Hidden messages uses words and pictures that make the product look appealing. The goal of hidden messages is to convince you to buy a product. The chart titled *Hidden Messages in Advertising* lists several types of hidden messages and their meanings.

Reading Check

Describe List some factors that influence your choices as a consumer.

HIDDEN MESSAGES IN ADVERTISING

Technique	Example	Hidden Message
Bandwagon	A group of people using the product or service.	*Everyone is using it, and you should too.*
Rich and Famous	Product displayed in expensive home or being used by famous people.	*It will make you feel rich and famous too.*
Free Gifts	A redeemable coupon is oered with the purchase of merchandise.	*It's too good a deal to pass up.*
Great Outdoors	A product is displayed in scenes of nature.	*If it's associated with nature, it must be healthy.*
Good Times	People smiling and laughing while using a product.	*This product will make your life fun.*
Testimonials	People who say they have used the product talk about Its benefits.	*It worked for them, so it'll work for me too.*

Shopping Smart

The first step in becoming a smart shopper is to understand what you are buying. With many products, this means reading the product label. By law, labels for foods, medicines, and many health products must include certain kinds of information. A label tells you the product name and what the product is intended to do. The label also gives directions, lists the ingredients, and may feature warnings.

Directions tell you how much of the product to use and how often to use it. Use a product only as directed. A product may not be effective if it is not used as directed. If any problems occur when you use a product, stop using it immediately. The product may contain an ingredient that is causing the problem.

Product labels can also help you compare similar products. Different brands of the same type of product may contain slightly different ingredients. The same brand may feature a range of products designed for slightly different uses. Also, product labels may make similar claims, but they may not do exactly the same thing.

Consumers can sometimes be persuaded to try products that have fancy packaging with catchy phrases. Packages are designed to make you want to buy the product. They may feature bright colors, attractive logos, and certain words designed to attract your attention in an ad or on a store shelf. Companies spend lots of money on designing packaging that will help sell their products.

THE FIRST STEP IN BECOMING A SMART SHOPPER IS TO UNDERSTAND WHAT YOU ARE BUYING.

Reading Check

List What factors are evaluated in comparison shopping?

ONE IMPORTANT FACTOR TO CONSIDER WHEN COMPARING PRODUCTS IS UNIT PRICE.

Comparison Shopping

When you compare two or more similar products, you are **comparison shopping**. It is collecting information, comparing products, evaluating their benefits, and choosing products with the best value. As you compare products, consider the benefits of one brand over another. Which brand offers more of what you are looking for? Does one brand fit your needs better than another? You should also consider the brand's reputation. Do you know anyone who has used and liked it? How about the cost?

Some brands may offer **generic products**. These are products sold in plain packages at lower prices than brand name products. Clipping **coupons** is another way to save money. They are slips of paper that reduce the price of a product. They can be found in newspapers, store flyers, and online.

One important factor to consider when comparing products is unit price. A product's **unit price** is the cost per unit of weight or volume. It is usually placed on a tab or sticker underneath it. You can calculate it yourself, if necessary.

1. Find the weight or volume given on each product container.
2. Divide the price of the product by its weight or volume.
3. The result is the unit price.

For example, a 5-fluid-ounce bottle of facial cleanser costs $4.50. Dividing $4.50 by 5 equals 90¢. The unit price is 90¢ per fluid ounce.

Finally, check to see if the item carries a **guarantee** or a **warranty**. A guarantee is a promise to refund your money if the product doesn't work as claimed. A warranty is the store's written agreement to repair a product or refund your money if the product does not function properly.

Managing Consumer Problems

Products usually work as advertised. However, some businesses sell products or services that don't work. When this is done to deceive the buyer, it is called **fraud**. Fraud is a calculated effort to trick or fool others. It is illegal and punishable by law. One serious type of fraud is **health fraud**. It is the selling of products or services to prevent diseases or cure health problems which have not been scientifically proven safe or effective for such purposes. If you have been a victim of fraud, contact the Food and Drug Administration.

Problems with Products

Some products are defective. Keep the receipt until you have used the product. If it is defective, try to return it to the store where you purchased it. Ask about the store's return policy when buying the product.

Online Shopping

Buying products online is easy and convenient. When buying online products, learn about the seller. Anyone can set up an online store. Some online stores sell counterfeit products. These are products that are similar to a real product but may contain less expensive ingredients. This is a type of fraud.

To check an online store, read reviews from other purchasers. Ask friends who may have purchased similar products. Search for referrals for online businesses. It's also important to confirm the seller's address and phone number to make sure you can contact the seller if needed. Always use a credit card to pay for your purchases. Sometimes a credit card company will help you if fraud occurs. Use comparison shopping to determine if the price is fair. Finally, learn about the seller's return policy. Some sellers will pay for return shipping while others will not.

Reading Check

Explain What steps can you take if you buy a defective product?

Developing Good Character

Responsibility Skilled consumers understand what they are buying. They also understand why they make certain choices. Being aware of advertising is an important skill. Ads make products seem appealing to get people to buy them. It is also important to be aware of a product's impact on the world around you. You can show responsibility by learning more about the things you buy.

Online shopping is easy and convenient. **Name one caution to use when shopping online.**

Lesson 1 Review

What I Learned

1. **VOCABULARY** What does *hygiene* mean?

2. **EXPLAIN** How does good hygiene affect all parts of your health triangle?

3. **IDENTIFY** What are two kinds of information found on health product labels?

Thinking Critically

4. **ANALYZE** What are the characteristics of informed health choices?

5. **SUGGEST** Turtles and other reptiles carry salmonella. This bacterium often makes people sick if they ingest it. What would be a good way to make sure you do not get sick from your friend's pet reptile?

Applying Health Skills

6. **ANALYZING INFLUENCES** Imagine that you are selecting a deodorant. Compare your wants and needs to the product's claims. What other factors would influence your decision?

Taking Care of Your Skin and Hair

Before You Read

Quick Write Write a short description of how you care for your skin and hair each day.

Vocabulary

epidermis
dermis
subcutaneous layer
sunscreen
acne
pores
dermatologist
ultraviolet (UV) rays
melanin
hair follicles
dandruff

Reading Check

List What are the three layers of the skin?

BIG IDEA Caring for your skin and hair is important to your overall physical health.

Healthy Skin

MAIN IDEA The skin is an organ with several important functions.

When you think of an organ, you might think of a body part such as your heart or your stomach. You might overlook your body's largest organ, which is not found beneath your skin—it is your skin! The average person's skin has a surface area of 2 square meters and weighs about 3 kilograms.

Your skin has many jobs. It acts as a waterproof shield that defends against germs and keeps them from getting into your body. Your skin helps maintain your body temperature. When you sweat, your skin gets rid of water and salts and cools your body. Your skin allows you to feel and sense pressure, temperature, and pain. Your skin also uses energy from sunlight to make vitamin D, which helps keep bones and teeth healthy.

The skin is composed of three main layers. The **epidermis** is the outermost layer of the skin. The palms of your hands and the soles of your feet are where the epidermis is thickest on your body. This outer layer of skin is at its thinnest where it covers the outside of your eyelids.

The **dermis** is the skin's inner layer. It is underneath the epidermis. The dermis contains nerve endings, blood vessels, and oil and sweat glands. The dermis is the part of your skin that enables your sense of touch. This layer is also the part of your skin that helps keep you cool by getting rid of waste when you sweat.

The innermost layer of skin, the **subcutaneous layer**, (subh-kyoo-TAY-nee-uhs) is the layer of fat under your skin. It connects your skin to muscle and bone. The subcutaneous layer is also the part of the skin that holds the roots of your hair follicles.

> **YOUR SKIN IS THE LARGEST ORGAN IN YOUR BODY.**

Caring for Your Skin

MAIN IDEA You can take several steps to help keep your skin healthy.

Keeping your skin healthy can help you feel, look, and smell good. Having healthy skin is also an important part of your overall physical health. Try these tips for taking care of your skin:

- Bathe or shower every day. Use soap to wash away dirt, sweat, oils, and bacteria that collect on your skin.
- If you have dry or delicate skin, you may find it helpful to moisturize it with lotion. Dry or cracked skin can itch or become irritated.
- Limit the amount of time you spend in the sun, especially between the hours of 10:00 a.m. and 4:00 p.m. This is when the sun's UV rays are strongest and most direct. Whenever you do spend time in the sun, wear protective clothing. You should also apply **sunscreen**, which is a cream or lotion that filters out some UV rays. Find a sunscreen with a sun protection factor (SPF) of 15 or higher. Reapply sunscreen about every two hours and after swimming.
- Avoid tanning beds, which can be harmful or damaging to the skin in ways similar to exposure to the sun.
- Avoid tattoos and piercings. Permanent body decoration can put you and your skin at risk for disease and scarring.

You will see advertisements for many types of skin care products. Your family and friends may make recommendations. Entire store aisles are filled with various cleansers, lotions, creams, and other items that promise to make your skin healthier and look better.

You may be tempted to try many of these products. You may find some that work as promised, while others appear to have no effect. However, remember to use your skills as a wise consumer. You can make informed choices about the types of products to buy and use on your skin.

Good skin care protects your overall health. **Name two ways to take care of your skin.**

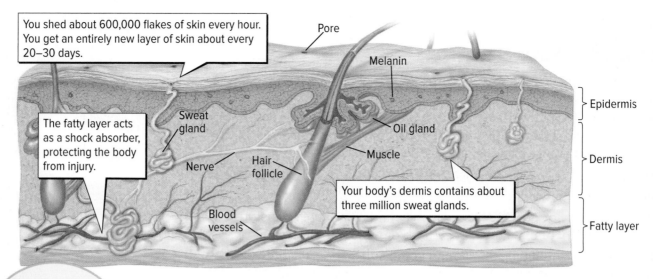

You shed about 600,000 flakes of skin every hour. You get an entirely new layer of skin about every 20–30 days.

The fatty layer acts as a shock absorber, protecting the body from injury.

Your body's dermis contains about three million sweat glands.

Pore
Melanin
Oil gland
Muscle
Sweat gland
Nerve
Hair follicle
Blood vessels

Epidermis
Dermis
Fatty layer

Your skin is a very complex organ. It has many parts. **Describe the function of the sweat glands.**

Reading Check

Describe How can bacteria and viruses affect the skin?

Whiteheads, blackheads, and pimples are all forms of acne that commonly affect teens. **Identify the structure in the skin that becomes clogged to from acne.**

Skin Problems

Did you ever wake up, look in the mirror and see a huge pimple? One skin problem experienced by many teens and some adults is **acne**. It's a skin condition caused by active oil glands that clog hair follicles. The openings of hair follicles onto the skin are called **pores**, which are tiny openings in the skin that allow perspiration to escape. Bacteria gather in the clogged pores, making them swell up. A graphic in the lesson shows how acne forms.

If you get acne, avoid picking it or trying to pop the pimple. In fact, try not to touch your skin at all. Instead, wash the area with a mild soap and warm water. Be gentle, and don't scrub too hard. Washing helps remove the dirt, oil, sweat, or makeup that can cause pores to clog. Over-the-counter acne medications can often help clear up breakouts. If you have a bad case of acne, you may need to see a **dermatologist**, who is a physician who treats skin disorders.

Viruses are the smallest and simplest pathogens. A virus can also affect the skin. Cold sores and warts are both caused by viruses. Both can be treated but are contagious. Contagious means the virus can be spread to others. A person with a wart or cold sore should avoid skin contact with others. Washing the hands thoroughly and often is also important.

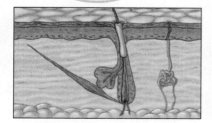

Whiteheads
A whitehead forms when a pore gets plugged up with the skin's natural oils and dead skin cells.

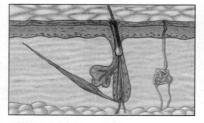

Blackheads
When a whitehead reacts to the air and darkens, it becomes a blackhead.

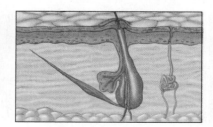

Pimples
A clogged follicle can burst, which lets in bacteria. The follicle becomes infected and a pimple forms.

Body Piercing and Tattooing

For some adults, piercings or tattoos are ways to express themselves. However, unlike different clothes or hair styles, body piercings and tattoos are permanent. Both carry potential health risks because they break the physical barrier of the skin. This can allow germs to enter the body. Germs that are passed along by tattoo or piercing needles can cause infection.

If the needles used for tattooing are not sterile, they can spread bacteria and viruses. Piercings can also spread disease and cause scarring. An oral piercing can damage your mouth and teeth. A tattoo or piercing may also impact your social health. They can limit future opportunities and relationships.

Additionally, many states in the U.S. have limits on the age at which a person can get a tattoo or body piercing. In many states, teens under the age of 18 need parental approval to get a tattoo or body piercing. This typically also applies ear piercings.

Sun Damage

Factors in the environment can affect your personal health. For example, the sun gives off **ultraviolet (UV) rays**, which are an invisible form of radiation that can enter skin cells and change their structure. Sunburn happens when UV rays damage skin cells. You might think soaking up lots of UV rays will give you a healthy-looking tan. In reality, a tan indicates that skin damage has occurred. One way that the result of tanning will show up on your skin is through wrinkles, or premature aging of the skin. Even worse, too much time in the sun increases the risk of skin cancer.

You can protect yourself from UV rays by limiting sun exposure and using sunscreen. Check the weather forecast for the UV Index. This is a scale that categorizes levels of UV radiation reaching Earth's surface. The figure in the lesson shows the UV Index.

Special cells in the epidermis make **melanin**, which is a substance that gives skin its color. Darker skin has more melanin than paler skin. Melanin can block some, but not all, UV rays from reaching the lower layers of skin.

What about indoor tanning beds? Many people use these devices to stay tan year-round. However, tanning beds also emit UV rays. They can damage the skin, lead to skin cancer, and injure the eyes and immune system.

Some people choose spray tans to make the skin look darker. Spray tans are safer than being exposed to UV rays through direct sunlight or tanning bed. However, a spray tan does not offer any additional protection from the sun's UV rays.

Ultraviolet rays from the sun can harm your skin. **How can you protect yourself from ultraviolet rays and reduce your risk for developing skin cancer?**

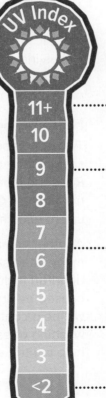

Levels 11 and higher: Extremely high risk. Avoid the sun between 10 a.m. and 4 p.m. Use sunscreen with an SPF of 15 or higher, and wear protective clothing and sunglasses.

Levels 8 through 10: Very high risk. Avoid the sun between 10 a.m. and 4 p.m. Use sunscreen with an SPF of 15 or higher, and wear protective clothing and sunglasses.

Levels 6 and 7: High risk. Cover up, wear a hat with a wide brim, and use sunscreen with an SPF of 15 or higher.

Levels 3 through 5: Moderate risk. Stay in the shade around midday.

Level 2 or below: Low risk. Wear sunglasses on bright days, cover up, and use sunscreen.

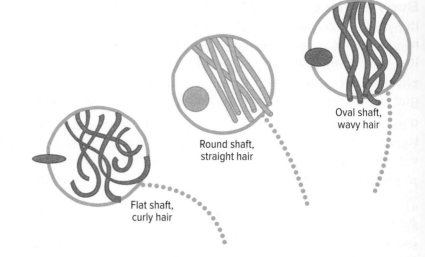

Hair can be curly, straight, or wavy depending on the shape of the hair shaft. **Identify the type of hair a person will have if his or her hair shafts have a round shape.**

Flat shaft, curly hair

Round shaft, straight hair

Oval shaft, wavy hair

Healthy Hair

MAIN IDEA Hair grows from follicles in the dermis and is made of keratin.

How much do you know about your hair? Hair grows from **hair follicles** in the dermis. Hair follicles are small sacs on the dermis from which hair grows. As new hair cells are formed, old ones are forced out. Your hair contains a protein called keratin, the same substance that forms fingernails and toenails. Keratin is the substance that gives hair strength and allows it to bend and blow in the wind without breaking.

The part of the hair that you can see is the shaft. The shape of the hair shafts determines whether your hair is wavy, curly, or straight. The graphic in the lesson shows these three types of hair. As is the case with living skin, hair gets its color from the pigment melanin. The color of your hair is determined by heredity. You inherited genes from your parents that determined your hair color.

Hair and Scalp Problems

Two conditions that can affect the health of your hair are dry or oily scalp. Either problem can be corrected by choosing the right shampoo. Read the label carefully. Different products are made for dry, oily, and normal hair. Chlorine in swimming pools can dry out your hair and even cause it to change color. Special shampoos can remove chlorine.

Another common scalp problem is **dandruff**. Dandruff occurs when too many dead skin cells flake off the outer layer of the scalp. Symptoms of dandruff include itchy scalp and flaking of the scalp. Dandruff can be caused by dry skin or certain skin diseases. Washing your hair regularly will help control itching of the scalp and flakes of dandruff. If regular hair washing does not work, using a special medicated shampoo can also help control and prevent dandruff.

Brushing or combing your hair regularly can keep it looking healthy. **List some ways you can take care of your hair.**

Sometimes, an itchy scalp is caused by head lice. Lice are tiny insects that live in hair and on the scalp and feed on human blood. Having lice is not an indication that a person is unclean. However, lice spread easily from one person to another.

To help prevent lice from spreading, avoid sharing hats, combs, and brushes. If you get head lice, use a medicated shampoo that is developed for treating lice. You will also need to wash all your bedding, towels, combs, brushes, and clothing. Since lice spread easily, everyone else in your home will need to take these same steps.

Caring for Your Hair

Keep your hair healthy by washing it regularly with a gentle shampoo and using conditioner if needed. Keep in mind that when shopping for shampoos and conditioners, you will find many different types and brands on store shelves. Use comparison shopping to choose the hair care products that are best for you. If possible, let your hair dry naturally. If you use a blow dryer, use low heat. Styling irons and high heat from hair dryers can make hair dry, brittle, and faded.

Brushing or combing your hair helps you improve your appearance. It also helps keep your hair healthy and whole. Brushing and combing removes dirt from your hair and scalp in between shampooing. It also helps spread natural scalp oils down along the hair shaft. The natural oils produced in the scalp help keep your hair strong and in good condition.

Reading Check

Describe What are two problems that can affect the scalp and hair?

Myth vs Fact

Myth Head lice can hop or fly from one person to another.

Fact The head louse (*Pediculus humanus capitis*) cannot fly or hop. It moves by crawling. The most common way that head lice are transmitted from one person to another is through direct contact, such as head-to-head contact. Less commonly, transmission occurs through clothing, such as hats, scarves, and coats, or personal items, such as combs, brushes, and towels.

Lesson 2 Review

What I Learned

1. **VOCABULARY** Define the terms *epidermis* and *dermis*.

2. **NAME** What are three functions of the skin?

3. **EXPLAIN** What protein does hair contain? How does hair grow?

Thinking Critically

4. **APPLY** A friend of yours is planning to get a pierced lip. What potential risks does this plan involve?

5. **EVALUATE** Your friend has told you that for the last several days her scalp has been itchy. This is not normal for her. What are some possible causes for this symptom? For each cause, suggest a course of action your friend could take.

Applying Health Skills

6. **PRACTICING HEALTHFUL BEHAVIORS** Tisha's friends have invited her to join them at the beach. They plan on lying out in the sun to tan. How can Tisha protect her skin from the sun's UV rays?

Caring for Your Mouth and Teeth

Before You Read

Quick Write Describe the steps you take to keep your teeth and gums healthy.

Vocabulary

plaque
tartar
gingivitis
fluoride
orthodontist

BIG IDEA Your teeth perform important functions and keeping them healthy is part of responsible healthful behaviors.

Your Teeth and Gums

MAIN IDEA Your teeth have several important functions.

Your teeth are vital to your health and appearance. They allow you to chew and grind food. They aid in forming certain speech sounds. Your teeth help give structure to your mouth.

The art in the lesson shows the parts of a tooth. Each tooth has a root that goes into your jaw. Roots are surrounded by pink flesh called gums. About three-fourths of each tooth is located below the gum line. The top of the tooth is the crown, which is covered with a layer of hard, white enamel. The neck of a tooth connects the crown to the root.

> **TWO IMPORTANT KEYS TO HAVING HEALTHY TEETH AND GUMS ARE PROPER BRUSHING AND FLOSSING.**

Tooth and Gum Problems

Proper care of your teeth and gums can prevent tooth decay. Tooth decay begins with **plaque**. Plaque is a thin, sticky film that builds up on teeth and leads to tooth decay. Bacteria in plaque feed on the carbohydrates—sugars and starches—in the foods you eat. These bacteria produce acids that can break down tooth enamel and lead to cavities, or holes in your teeth. The figure in the lesson shows how tooth decay occurs. The germs in plaque can also cause bad breath.

If plaque is not removed, then it becomes **tartar**, which is a hardened plaque that hurts gum health. You cannot brush away tartar. A dentist must remove it with special tools that clean and polish surfaces of the teeth.

The gums serve as anchors for your teeth. The teeth you can see in a mirror are only the top parts of your teeth. Most of each tooth is rooted in your gums. A common gum problem is **gingivitis**, a common disorder in which the gums are red and sore and bleed easily. If it is left untreated, gingivitis can lead to tooth loss or infection.

Reading Check

Explain How does tartar form?

Keeping Teeth and Gums Healthy

MAIN IDEA Ways to keep your teeth and gums healthy include brushing and flossing every day and having dental checkups twice a year.

Two important keys to having healthy teeth and gums are proper brushing and flossing. Brush after every meal (or at least twice a day) with toothpaste and a soft-bristled toothbrush. Brushing cleans the teeth and stimulates the gums. Brushing your teeth after eating also removes plaque from the surface of the teeth, before bacteria can produce acid that harms the teeth. It is important to brush and remove plaque before it becomes tartar, which only a dentist can remove. Whenever you do brush your teeth, dental experts recommend using a toothpaste or mouthwash containing **fluoride**. It's a chemical that helps prevent too decay.

In many areas, fluoride is added to tap water. The fluoride in the water also protects your teeth. However, fluoride from water is not a replacement for brushing or flossing. Remember to use the consumer skills you have learned to choose the type and brand of toothbrush, toothpaste, and mouthwash that best fit your needs.

Brushing your teeth after eating can also help prevent halitosis, or bad breath. Using mouthwash can also help kill bacteria and control bad breath. Halitosis can be caused by eating certain foods, poor oral hygiene, tobacco or alcohol use, bacteria on the tongue, decayed teeth, and gum disease.

The three main sections of a tooth are the crown, the neck, and the root. **Identify the part of the tooth you see when you smile and look in the mirror.**

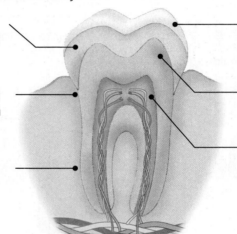

Crown—the part of the tooth visible to the eye

Neck—the part that connects the crown to the root

Root—the part that holds the tooth in the gum

Enamel—the hard material that covers the crown of the tooth

Dentin—bonelike material surrounding the pulp

Pulp—soft sensitive tissue containing nerves and blood vessels

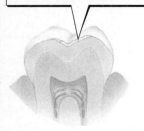

Stage 1
The bacteria in plaque combine with sugars to form a harmful acid. This acid eats into the enamel, the hard outer surface of the tooth.

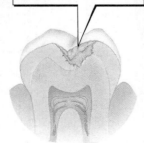

Stage 2
Repeated acid attacks on the enamel cause a cavity, or hole, to form.

Stage 3
If the cavity grows and reaches the sensitive inner parts of the tooth, it can cause a toothache.

Tooth decay occurs when bacteria in the mouth convert food into acids that dissolve enamel. **Explain what can happen if tooth decay goes untreated.**

Reading Check

Analyze Why is it important to visit your dentist at least twice per year?

In addition to brushing your teeth, flossing is also important. Dental floss is a thin plastic, nylon, or coated silk thread. When you slide it between each tooth after brushing, floss removes food particles from the sides of your teeth. It also removes plaque from between the teeth and under the gum line that a toothbrush cannot reach. Flossing keeps gums healthy and prevents gum disease. Flossing also helps to clean underneath braces. The graphic in the lesson shows how to brush and floss to prevent tooth decay.

A third key to maintaining your dental health is your diet. Choose foods that are high in the mineral calcium, such as yogurt, cheese, and milk. Your body uses calcium to build teeth and bones. You should also limit foods that are high in sugar, which can cause tooth decay. If you do eat sugary foods, brush your teeth as soon as you can.

Another important way to protect your teeth and gums is to have dental checkups twice a year. The dentist or dental hygienist will clean your teeth to help prevent tooth decay and gum disease. The dentist will also examine your teeth for cavities or other problems.

Brushing and flossing regularly helps prevent tooth decay and other health problems. **Explain why you should brush your teeth at least three times a day.**

Brush the outer surfaces of your upper and lower teeth. Use a combination of up-and-down strokes and circular strokes.

Thoroughly brush all chewing surfaces with a soft-bristle brush to protect your gums.

Brush the inside surfaces of your upper and lower teeth.

Brush your tongue and rinse your mouth.

Take about 18 inches of floss and wrap each end around the middle finger of each hand.

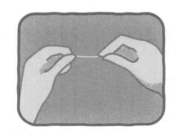

Grip the floss firmly between your thumb and forefinger.

Slide the floss back and forth between teeth toward the gum line until it touches the gum line.

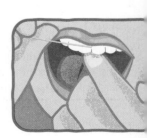

Curve the floss around the sides of each tooth. Keep sliding the floss back and forth gently as you move it up and down.

Other Dental Issues

Other problems of the teeth and mouth include misalignment and impacted wisdom teeth. Improperly aligned upper and lower teeth are also known as having a "bad bite." This can be caused by crowded teeth, extra teeth, thumb sucking, injury, or heredity. If left untreated, a bad bite can result in lost teeth.

Your dentist will also check for wisdom teeth, or extra molars in the back of your mouth. These typically appear in your late teens. An impacted wisdom tooth can cause swelling and be very painful. Wisdom teeth may crowd or push on other teeth or become infected. If you have problems when your wisdom teeth come in, your dentist may have to do surgery to remove them.

If your teeth need straightening, your dentist may refer you to an **orthodontist**. An orthodontist is a specialist who prevents or corrects problems with the alignment or spacing of teeth. An orthodontist may apply braces in order to straighten your teeth.

Braces can help improve your appearance and make your teeth easier to clean. Some teens may not like having braces on their teeth, because they do change your appearance for a while. Braces can be made of either metal or a clear material. Clear braces have become popular because they are less noticeable. Many adults choose this option if they did not have their teeth straightened earlier in life.

Your smile is an important part of your appearance. **List some steps you can take to help keep your teeth healthy.**

Reading Check

Explain How is your diet related to the health of your teeth and gums?

Lesson 3 Review

What I Learned

1. **VOCABULARY** What is plaque? Why should plaque be removed?

2. **LIST** What are the functions of the teeth?

3. **NAME** What are two healthful behaviors that keep your teeth and gums healthy?

4. **EXPLAIN** How do effective dental hygiene practices help prevent illness?

Thinking Critically

5. **HYPOTHESIZE** What can happen to your teeth and gums if you do not floss regularly?

6. **EVALUATE** Why is good gum care important for the health of your teeth?

Applying Health Skills

7. **ADVOCACY** Create a booklet that explains the importance of proper tooth and gum care. Include original art, if you like, with step-by-step instructions. Distribute copies to students in other classes.

Protecting Your Eyes and Ears

Before You Read

Quick Write Make a list of five ways you used your sense of hearing today. Describe how these activities would be different if you could not hear very well.

Vocabulary

cornea
pupil
farsightedness
nearsightedness
astigmatism
cataracts
glaucoma
optometrist
ophthalmologist
deafness
decibel
tinnitus

Reading Check

Identify What are the two most common vision problems?

BIG IDEA Caring for your eyes and ears will keep them healthy.

Healthy Eyes

MAIN IDEA The function of your eyes is to focus light on the retina, which sends information to your brain about what you see.

Your eyes are your windows to the world. Like a camera, your eyes focus light in order to give your brain a picture of the world around you. Eyes allow you to see shapes, colors, and motion. Each part of the eye plays an important role in how you see. Light enters the eye through the pupil. The **cornea** is a clear, protective structure of the eye that lets in light. It directs the light rays as they enter the **pupil**, which is the dark opening in the center of the iris. The lens focuses the light on the retina. The retina contains nerve cells that send signals to the brain though the optic nerve. The figure in the lesson shows the parts of the eye and how they work together.

Vision Problems

Two common vision problems are **farsightedness** and **nearsightedness**. Farsightedness is a condition in which faraway objects appear clear while near objects look blurry. In a person with farsightedness, the words on this screen may look unclear. That person might need reading glasses to see this page clearly. Nearsightedness occurs when objects that are close appear clear while those far away look blurry. A nearsighted person can clearly see objects only if they are up close.

A third common eye condition is **astigmatism**. It's an eye condition in which images appear wavy or blurry. Eye problems are usually corrected with eyeglasses or contact lenses. Both help the lens of the eye focus light on the retina. An eye doctor can determine if you need corrective lenses. He or she also identifies and treats more serious eye problems, such as **cataracts** or **glaucoma**. Having cataracts is an eye condition in which the lens becomes cloudy as a person ages. Glaucoma is an eye condition in which fluid pressure build up inside the eye.

> **YOUR EYES ARE YOUR WINDOWS TO THE WORLD.**

Caring for Your Eyes

MAIN IDEA You can keep your eyes healthy by taking care of them.

Your eyes need light to see. However, too much light can hurt them. The sun can be very hard on your eyes. For example, you should never look directly at the sun. In addition, UV rays can damage eye cells. You can solve this problem by wearing sunglasses that block UV rays.

Sitting too close to the television can cause eyestrain and headaches. Sit at least 6 feet away. If you get eyestrain while sitting at a computer, change the monitor position to cut down on the glare. Reading in dim light can also cause eyestrain. Be sure you have adequate lighting while you read, work, or watch TV. Light should come from above your reading material. Taking frequent breaks while using your computer, watching TV, or reading can help prevent eyestrain.

Protect your eyes when you are doing yard work, handling power tools, or playing sports that involve flying objects. Wear glasses or goggles designed for these activities. Wear goggles when you work with strong chemicals, such as when you are working in a science laboratory.

Reading Check

Explain Why is it important to protect your eyes while in the sun?

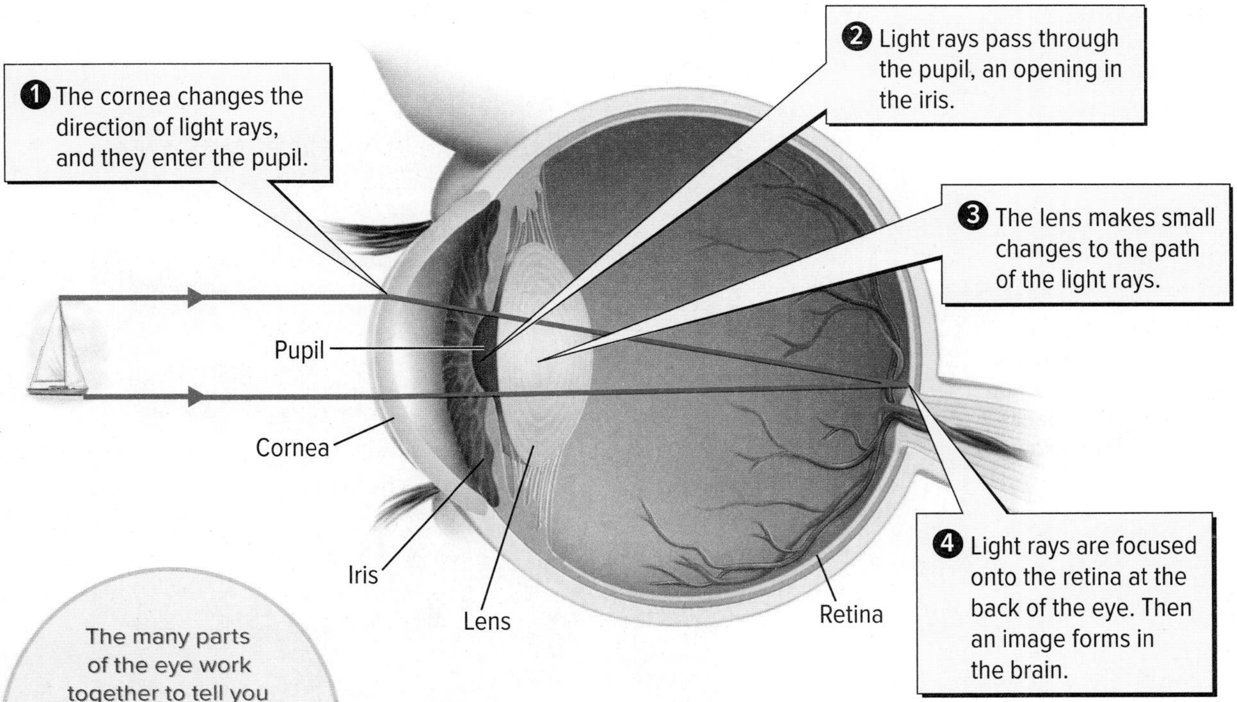

❶ The cornea changes the direction of light rays, and they enter the pupil.

❷ Light rays pass through the pupil, an opening in the iris.

❸ The lens makes small changes to the path of the light rays.

❹ Light rays are focused onto the retina at the back of the eye. Then an image forms in the brain.

Pupil

Cornea

Iris

Lens

Retina

The many parts of the eye work together to tell you about the world around you. **Describe how the various parts of the eye interact to make vision possible.**

If something gets in your eye, do not rub it. Instead, flush particles out with clean water or eye drops. Sharing eye makeup or eye care products can spread germs. Diseases such as conjunctivitis, or pinkeye, can make eyes red and painful. If your eyes hurt for more than a short time, see a doctor right away.

Keeping your eyes healthy includes getting regular exams or vision screenings. If you already wear glasses or contacts, you should regularly visit an **optometrist**. An optometrist is a health care professional who is trained to examine the eyes for vision problems and to prescribe corrective lenses. Schedule a visit at least once a year. An optometrist may do initial tests for eye diseases. He or she may then send you to an ophthalmologist. An **ophthalmologist** is a physician who specializes in the structure, functions, and diseases of the eye. An ophthalmologist also fits people with corrective lenses.

> KEEPING YOUR EYES HEALTHY INCLUDES GETTING REAGULAR EXAMS OR VISION SCREENINGS.

Caring for your eyes includes having a yearly eye exam. **Identify what an optometrist checks during a yearly exam.**

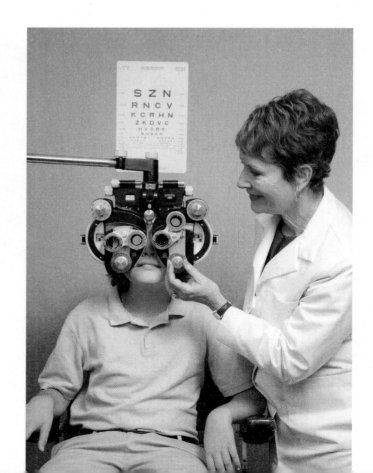

Healthy Ears

MAIN IDEA Ears gather sound and help you stay balanced.

Like your eyes, your ears allow you to receive information. Your ears gather sound. Sound waves travel through the ear canal and strike the eardrum, causing it to vibrate. The vibrations cause tiny hairs to move in the inner ear. When these hairs move, the auditory nerve sends messages to the brain. Your brain interprets the messages as sounds.

Your ears also enable you to control your balance. Balance is controlled by tube-like structures in the inner ear. Fluid and tiny hair cells inside these tube-like structures send messages to your brain when you move. The brain interprets the messages and tells your body what adjustments it needs to make.

The ears carry sound to the brain and help you stay balanced. **Name the part of the ear responsible for these two main functions.**

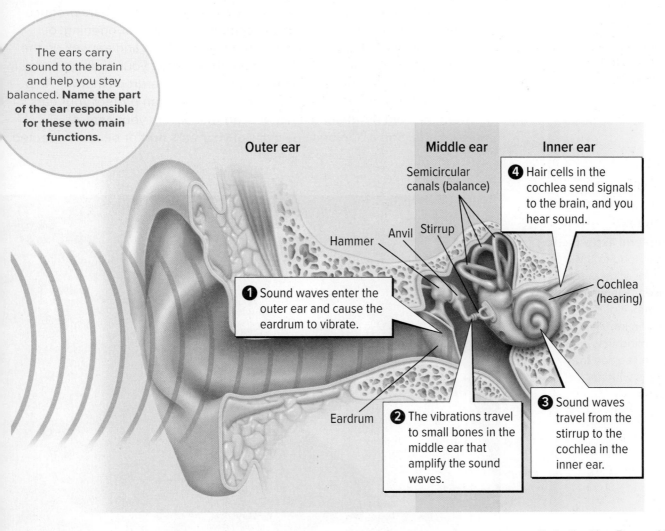

Outer ear

Middle ear

Inner ear

Semicircular canals (balance)

❹ Hair cells in the cochlea send signals to the brain, and you hear sound.

Anvil Stirrup

Hammer

Cochlea (hearing)

❶ Sound waves enter the outer ear and cause the eardrum to vibrate.

Eardrum

❷ The vibrations travel to small bones in the middle ear that amplify the sound waves.

❸ Sound waves travel from the stirrup to the cochlea in the inner ear.

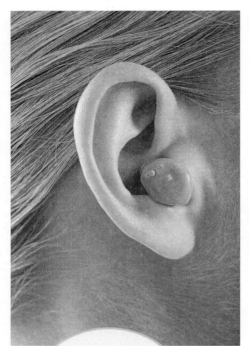

People with hearing loss may wear hearing aids. These increase the loudness of sounds. **Identify a common way that people with loss can communicate.**

Ear Problems

Ear infections are the most common ear problems. Germs can spread into the ear from the nose or throat. Ear infections can be treated by a doctor.

The most serious ear problem is hearing loss. **Deafness** is a condition in which someone has difficulty hearing sounds or has complete hearing loss. It can result from injury, disease, or birth defect. The unit for measuring the loudness of sound is called a **decibel**. Normal conversation is about 60 decibels. Noises above 85 decibels can damage the ear and cause hearing loss.

Tinnitus is a constant ringing in the ears. It is often a result of exposure to noise for a long period of time. For some people with tinnitus, the ringing sensation is constant. Ongoing tinnitus is an early warning sign of inner-ear nerve damage.

Caring for Your Ears

MAIN IDEA Protecting your ears from loud sounds is the best way to care for them.

To care for your ears, wash and dry them regularly. Your ears produce earwax, which helps trap dirt and carry it out of the ear opening, or auditory canal. Use a wet washcloth to wipe off dirt and earwax on the outside of your ears. Do not insert anything inside your auditory canal. If you get water in your ears, you can use special eardrops that will help dry out the water and prevent an infection. You can protect your ears from the cold by wearing a hat or scarf that covers your ears, or earmuffs. See a doctor if any parts of your ears hurt or become infected.

Myth vs Fact

Myth Sun exposure will not damage my eyes.

Fact Long-term exposure to UV rays can lead to the development of cataracts, an eye disease in which the lens becomes cloudy. Cataracts can impair vision and cause blindness. Surgery can replace the affected lens with an artificial lens.

Protect your hearing from loud sounds by keeping the volume low or wearing ear muffs. **Identify the decibel level at which hearing damage may occur.**

The best way to care for your ears is to protect them from loud sounds. Repeated exposure to sounds above 85 decibels can cause hearing loss. When listening to music, especially with earbuds or headphones, keep the volume down. How can you tell how loud is too loud? If other people can hear sound coming from your headphones or earbuds, you may have your music turned up too loud.

Reading Check

Explain Why is it important to limit your exposure to loud sounds?

Lesson 4 Review

What I Learned

1. **VOCABULARY** What is the difference between an optometrist and an ophthalmologist?

2. **LIST** What are three habits that you would recommend to promote eye health and protect vision?

3. **EXPLAIN** What happens when sound waves reach your outer ear?

Thinking Critically

4. **ANALYZE** Tracy is nearsighted. What type of vision correction do you think might be best for Tracy? Explain your reasoning.

5. **EVALUATE** A woodshop produces noise that reaches 100 decibels. What can you do to reduce the risk of injury to your ears? Explain.

Applying Health Skills

6. **PRACTICING HEALTHFUL BEHAVIORS** Imagine that you are planning a backpacking trip in the mountains. The weather will be sunny but very cold. What kind of gear or clothing would you pack to protect your eyes and ears?

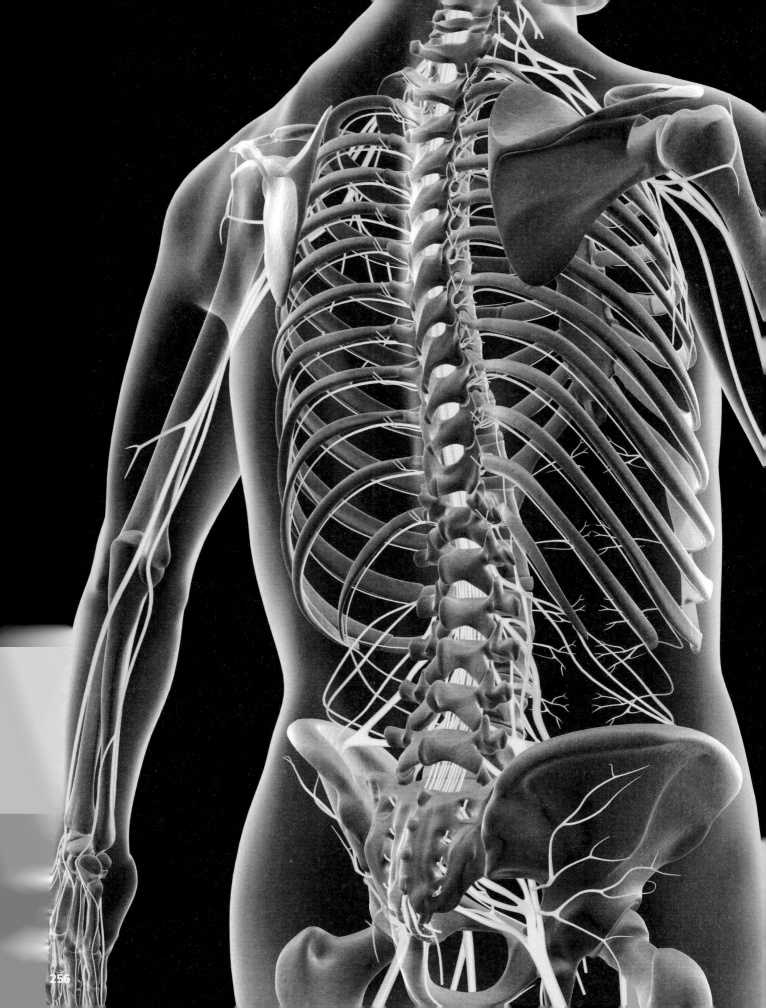

Your Body Systems

LESSONS

12

Your Skeletal and Muscular Systems

Before You Read

Quick Write List some bones and muscles that you are familiar with.

Vocabulary

skeletal system

marrow

ligament

cartilage

tendon

muscular system

skeletal muscle

cardiac muscle

smooth muscle

BIG IDEA Your skeletal and muscular system work together to make your body move.

Your Skeletal System

MAIN IDEA Your skeletal system provides your body with a framework.

Your **skeletal system** is the framework of bones and other tissues that supports the body. It includes bones, joints, and various connective tissue. You can feel bones in your hands, arms, legs, and feet. All of your bones make up your skeleton. Your body has more than 200 bones. Bones are attached to muscles. Each movement you make is caused by your bones and muscles working together.

The skeletal system has important functions. Bones provide support. Are you sitting right now? If so, your bones and muscles are working to hold you in your sitting position. If you raise your hand to speak in class, your bones and muscles cause the movement. Touch your head. The hard part on the top of your head protects your brain. Other bones protect your spinal cord, lungs, and other internal organs.

Another function of bones is to produce and store materials needed by your body. Red blood **marrow**, a tissue in the center of some bones, makes millions of blood cells each day. Bones store fat and calcium. Calcium is needed for strong bones and teeth and for many cellular processes.

Several kinds of connective tissues help move and protect your bones. Bones work together at joints. Joints provide flexibility and enable the skeleton to move. A **ligament** is a type of connecting tissue that holds bones to other bones at the joint. When the bones in joints move, ligaments stretch and work to keep the bones together.

Ligaments connect bones but do not protect them. Bones are protected by **cartilage**, which is a strong flexible tissue that allows joints to move easily, cushions bones, and supports soft tissues. Cartilage protects bones, such as those connected by your knee joint. Your elbows and shoulders are also protected by cartilage.

A **tendon** is a type of connecting tissue that joins muscles to bones and muscles to muscles. It can also protect your bones. Tendons help to stabilize joints and keep them from moving out of place.

Reading Check

Compare Identify two types of joints and give an example of each.

Your skeletal system contains two types of joints—immovable joints and movable joints. Immovable joints do not move. For example, your skull contains several immovable joints. Movable joints allow you to move your hands and feet and bend parts of your body such as your knees and elbows.

Your Muscular System

MAIN IDEA Your muscular system allows your body to move and helps keep it stable.

You use muscles when you walk and stretch. The tissues that move parts of the body and control the organs are the **muscular system**. Muscles also provide your body with stability and protection.

Muscles attached to bones support your body to provide stability and balance. If you stumble and lose your balance, your muscles pull you back to a stable position. Muscles cover most of your skeleton like a layer of padding. Muscles cover your abdomen, chest, and back to protect your internal organs. Muscles also work to keep your body at its normal temperature of around 37°C.

When you are cold, your muscles contract quickly and cause you to shiver. When you are too warm or have exercised, your body may sweat. In either case, when your body is too cold or too warm, your muscles work to turn chemical energy into thermal energy to keep your body at a safe temperature.

THE SKELETAL SYSTEM IS MADE UP OF BONES, JOINTS, AND CONNECTIVE TISSUE.

Reading Check

Explain How do muscles work?

Joints provide flexibility and enable the skeleton to move. **Describe how ligaments work to help the body move.**

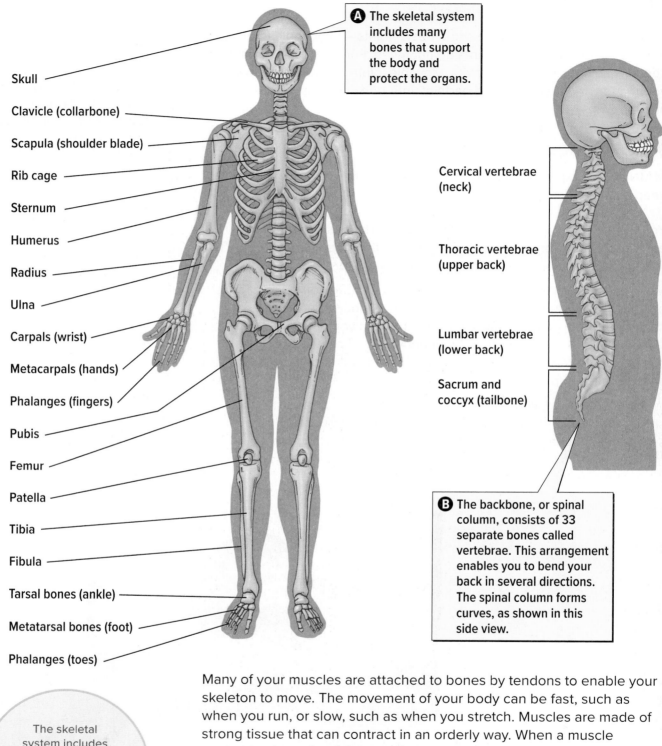

Skull

Clavicle (collarbone)

Scapula (shoulder blade)

Rib cage

Sternum

Humerus

Radius

Ulna

Carpals (wrist)

Metacarpals (hands)

Phalanges (fingers)

Pubis

Femur

Patella

Tibia

Fibula

Tarsal bones (ankle)

Metatarsal bones (foot)

Phalanges (toes)

A The skeletal system includes many bones that support the body and protect the organs.

Cervical vertebrae (neck)

Thoracic vertebrae (upper back)

Lumbar vertebrae (lower back)

Sacrum and coccyx (tailbone)

B The backbone, or spinal column, consists of 33 separate bones called vertebrae. This arrangement enables you to bend your back in several directions. The spinal column forms curves, as shown in this side view.

The skeletal system includes many bones that support the body and protect the organs. **Which bones protect your heart?**

Many of your muscles are attached to bones by tendons to enable your skeleton to move. The movement of your body can be fast, such as when you run, or slow, such as when you stretch. Muscles are made of strong tissue that can contract in an orderly way. When a muscle contracts, the cells of the muscle become shorter. When the muscle relaxes, those cells return to their original length.

Many of your muscles are not attached to bones. As these muscles contract, they cause blood and food to move through your body. These are the muscles that cause your heart to beat. These types of muscles also make the hair on your arms stand on end when you get goose bumps.

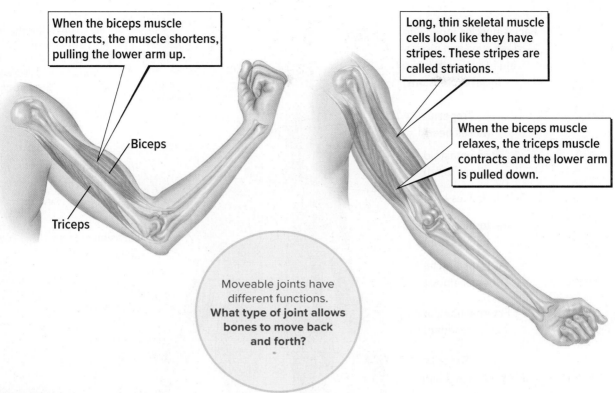

When the biceps muscle contracts, the muscle shortens, pulling the lower arm up.

Biceps

Triceps

Long, thin skeletal muscle cells look like they have stripes. These stripes are called striations.

When the biceps muscle relaxes, the triceps muscle contracts and the lower arm is pulled down.

Moveable joints have different functions. **What type of joint allows bones to move back and forth?**

Types of Muscles

Your body has three different types of muscles: skeletal, cardiac, and smooth. Each of these muscle tissues has a specific function. **Skeletal muscle** is attached to bones that enable you to move your body. It's a voluntary muscle. This means that you control the skeletal muscles to make your body move.

Your heart is made of cardiac muscle. **Cardiac muscle** is found in the walls of your heart. It's an involuntary muscle. Cardiac muscles work on their own, without your control. When cardiac muscles contract and relax, they pump blood through your heart and blood vessels throughout your body.

Smooth muscle is a type of muscle found in organs and in blood vessels and glands. It's an involuntary muscle named for its smooth appearance. Blood vessels in your body are lined with smooth muscles. Your stomach, bladder, and intestines also contain smooth muscles. Contraction of the smooth muscles controls the movement of blood through the vessels. They also move other materials through the body, such as food in the stomach.

Reading Check

Differentiate Explain the difference between voluntary and involuntary muscles.

MUSCLES PROVIDE STABILITY, PROTECTION, AND MAINTAIN BODY TEMPERATURE.

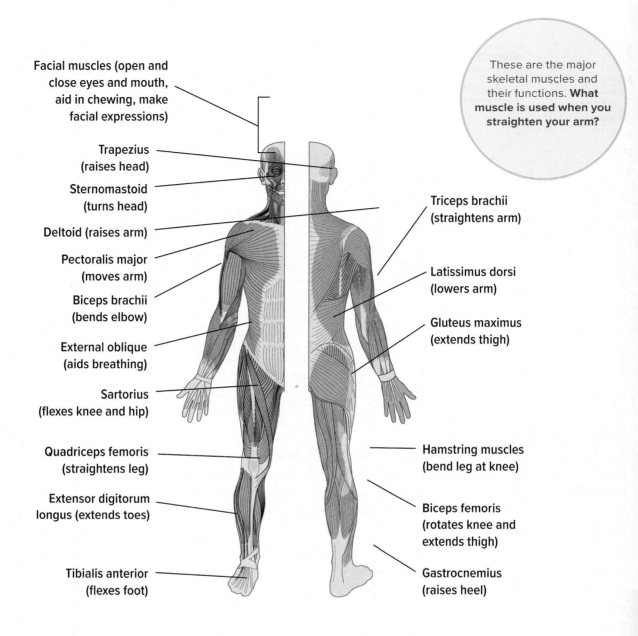

Facial muscles (open and close eyes and mouth, aid in chewing, make facial expressions)

Trapezius (raises head)

Sternomastoid (turns head)

Deltoid (raises arm)

Pectoralis major (moves arm)

Biceps brachii (bends elbow)

External oblique (aids breathing)

Sartorius (flexes knee and hip)

Quadriceps femoris (straightens leg)

Extensor digitorum longus (extends toes)

Tibialis anterior (flexes foot)

Triceps brachii (straightens arm)

Latissimus dorsi (lowers arm)

Gluteus maximus (extends thigh)

Hamstring muscles (bend leg at knee)

Biceps femoris (rotates knee and extends thigh)

Gastrocnemius (raises heel)

These are the major skeletal muscles and their functions. **What muscle is used when you straighten your arm?**

YOUR BODY HAS THREE DIFFERENT TYPES OF MUSCLES: SKELETAL, CARDIAC, AND SMOOTH.

How Muscles Work Together

Skeletal muscles work by pulling bones. They do not push on your bones. Each movement that you make involves your muscles pulling on your bones. Muscles often work together to help your body move. Muscles contract and expand to pull on bones and create movement at a joint. The process of two muscles working together is called paired movement. For example, when you pull up your lower arm at the elbow, your biceps muscle contracts. When you then lower your arm, your biceps muscle relaxes and the triceps muscle contracts in order to pull the arm down.

Problems with Bones and Muscles

MAIN IDEA Your bones and muscles can develop problems.

Your bones, muscles, and connective tissues are strong, but they need your care. Problems can develop because of injury, infection, poor posture, and lack of nutritious foods. Some problems of the skeletal system can include:

- **Fracture.** A fracture is a break in a bone caused by an injury.
- **Dislocation.** This occurs when a bone is pushed out of its joint. Dislocation can stretch or tear a ligament.
- **Sprain.** A sprain is an injury to the ligament connecting bones at a joint. This occurs when a ligament is stretched or twisted and causes swelling.
- **Strain.** This is a small tear in a muscle or tendon. Strains can occur when a muscle has been overstretched. A strain may be referred to as a pulled muscle.
- **Overuse injuries.** Injuries as a result of overuse occur over a period of time. An example of overuse is a shin splint, which can develop in runners.
- **Osteoporosis.** This condition results in brittle or porous bones. Osteoporosis can be caused by long-term lack of nutrition or exercise.
- **Scoliosis.** This is a curving of the backbone. The spine curves to one side of the body in an S-shape or C-shape.
- **Muscular dystrophy.** This disorder weakens muscles over time. It is usually inherited and causes skeletal muscle tissue to gradually waste away.

Reading Check

Summarize What are the most frequent causes of injuries to bones and muscles?

> YOUR BONES, MUSCLES, AND CONNECTIVE TISSUES ARE STRONG, BUT THEY NEED YOUR CARE.

Bones and muscles are vulnerable to injury which can affect your mobility and health. **List some of the injuries that affect the skeletal system.**

Caring for Your Bones and Muscles

MAIN IDEA You can help keep your bones and muscles healthy.

Your good health habits can keep your bones and muscles strong and healthy. Bone and muscle health requires energy from the foods you eat. A diet that is rich in nutrients such as protein, fiber, potassium, and vitamin C can help keep your muscles strong.

Physical activity also helps keep muscles healthy and strong. Decreased muscle strength can increase the risk of heart disease and injury, and make joints less stable. Do flexibility exercises so you can move more easily and work out more safely. Choose activities that strengthen your muscles and bones. Improve your cardiovascular endurance to give your heart and lungs more power. Warm up before and cool down after physical activity. If you feel pain, stop and give your body time to recover.

Also important is your posture, or the way you hold your body. Good posture means the bones and joints in your back stay in place and your muscles are used properly. To prevent too much strain on your back, avoid carrying a heavy backpack. Bend and use your legs, not your back, when you lift something heavy. Running, walking, cycling, and swimming can help keep your back strong and healthy.

Reading Check

Explain Tell how good posture contributes to bone and muscle strength.

These runners exercise to keep fit. **How are these teens helping their bones and muscles?**

763

1020

Ergonomics is the study of designing and arranging objects people use. The goal is to make the objects efficient and safe to use. To keep bones and muscles healthy, use proper ergonomics when sitting at a desk or working on a computer.

Desk ergonomics include having your chair set at the correct height. Your feet should be flat on the floor. Your knees should be level with your hips. Your hands should be at or below your elbows as you type. Your wrists should be straight. The computer monitors should be an arm's length away from your eyes. Practicing proper desk ergonomics is part of maintaining good posture.

What Teens Want to Know

What causes paralysis? The nervous system can be affected by injury or disease. A spinal cord injury causes the nervous system to lose the ability to send messages to the muscles. Muscular disorders such as muscular dystrophy cause the muscular system to lose the ability to respond to messages from the nervous system.

Lesson 1 Review

What I Learned

1. **VOCABULARY** Define the skeletal system.

2. **EXPLAIN** Describe two functions of the skeletal system.

3. **COMPARE AND CONTRAST** What is the difference between voluntary and involuntary muscles?

Thinking Critically

4. **SYNTHESIZE** Imagine you are going to start running to prepare for a marathon in six months. What steps can you take to protect your bones and muscles?

5. **ANALYZE** Why do you think poor posture can cause backaches?

Applying Health Skills

6. **PRACTICING HEALTHFUL BEHAVIORS** List the physical activities in which you participate. Evaluate your activities to determine how well you are strengthening your bones and muscles. Do you think you should add any activities? Tell how you might change your physical activities to improve bone and muscle strength.

Your Nervous System

Before You Read

Quick Write Describe some ways that your brain and your nerves tell your body what to do.

Vocabulary

nervous system
brain
neurons
central nervous system (CNS)
peripheral nervous system (PNS)
spinal cord
somatic system
autonomic system

Reading Check

Define What is another phrase for neurons?

BIG IDEA Your nervous system controls and sends messages throughout your body.

Parts of the Nervous System

MAIN IDEA Your movements and body processes are controlled by the nervous system.

Your **nervous system** is your body's message and control center. It gathers, processes, and responds to information received by the **brain**. The brain is the command center, or coordinator, of the nervous system. This happens very quickly. Your nervous system can receive information, process it, and respond to it in less than a second.

The nervous system controls all body processes, such as digestion, breathing, and blood flow through the body. Your nervous system processes your physical and emotional feelings and reactions to stimuli. A stimulus is a change in environment that causes a response. Examples of stimuli include, heat, cold, and actions such as seeing and catching a thrown ball.

The nervous system receives messages from your five senses—vision, hearing, smell, taste, and touch. Your senses send a message to your brain, and your body reacts. For example, think about how you react when you touch something hot—you quickly sense the heat and pull away.

The cells that make up the nervous system are called **neurons**. Neurons are also called nerve cells. Neurons are the message carries that help the different parts of your body communicate with one another.

> **YOUR NERVOUS SYSTEM CARRIES MESSAGES THROUGHOUT YOUR BODY.**

The Central Nervous System

Your nervous system has two parts—the **central nervous system (CNS)** which includes the brain and the spinal cord, and the **peripheral nervous system (PNS)**. The PNS includes the nerves that connect the central nervous system to all parts of the body. The brain controls your thoughts, speech, memory, and muscle movement. The brain controls voluntary muscle movement, or things you have to think about doing, such as standing, running, waving, and speaking. The brain also controls involuntary muscle movement, or things you do without thinking, such as your heartbeat, swallowing, blinking, coughing, and sneezing.

The largest and most complex part of the brain is the cerebrum. It controls memory, language, and thoughts. The part of the brain that controls voluntary muscle movement is called the cerebellum. It also stores muscle movements you have learned, such as tying your shoes or riding a bike. The brain stem is the area of the brain that controls involuntary muscle movement.

Your **spinal cord** is a long bundle of neurons that sends messages to and from the brain and all parts of the body. It is part of your central nervous system. The neurons in the spinal cord reach out to other parts of the body. The brain responds to information through the neurons in the spinal cord.

The nervous system receives messages from your five senses. **Name the five senses.**

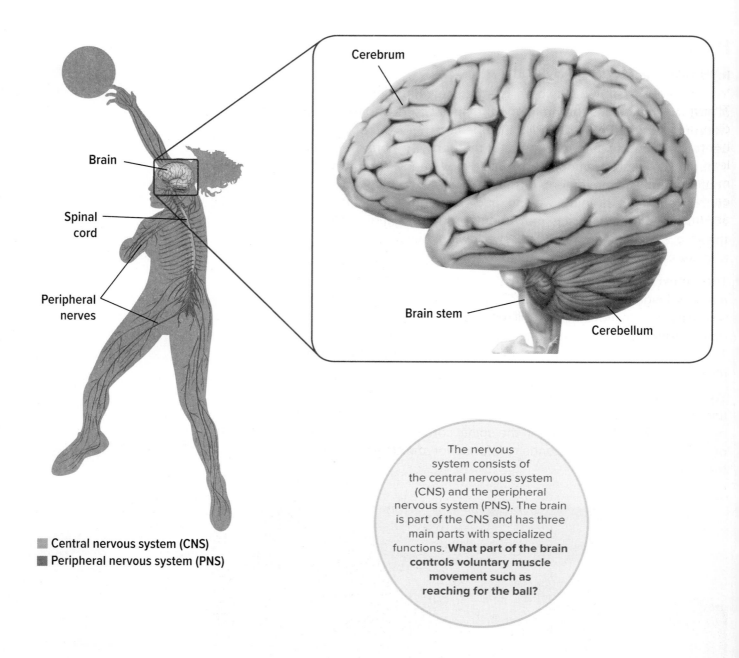

Brain

Spinal cord

Peripheral nerves

Cerebrum

Brain stem

Cerebellum

Central nervous system (CNS)
Peripheral nervous system (PNS)

The nervous system consists of the central nervous system (CNS) and the peripheral nervous system (PNS). The brain is part of the CNS and has three main parts with specialized functions. **What part of the brain controls voluntary muscle movement such as reaching for the ball?**

The Peripheral Nervous System

If a bee stings you, it hurts because your nerves sense pain. They can send a message through the peripheral nervous system to the central nervous system and on to the brain—in under a second!

The peripheral nervous system has two parts. The **somatic system** is part of the nervous system that deals with actions that you control. The **autonomic system** is the part of the nervous system that deals with actions you do not usually control. Raising your hand is a voluntary movement, or one you control. The beating of your heart is an involuntary movement, or one you do not control.

Problems Affecting the Nervous System

MAIN IDEA Injury or disease can harm the nervous system.

The nervous system can become injured or be affected by diseases or disorders. One of the most common causes of damage is injury to the head, neck, or back. For example, an injury to the spinal cord could lead to paralysis. This means the loss of feeling in or being unable to move some body parts. Since the brain is your control center, an injury can cause memory loss, brain damage, or the loss of some physical abilities. Other issues that affect the nervous system include:

- **Multiple sclerosis,** or MS, which damages the outer part of some nerves. MS can cause problems with thinking and memory. Some people lose muscle control or become unable to walk because of MS.
- **Cerebral palsy,** which is a disease of the nervous system that either inherited or caused by brain damage.

Garrett has cerebral palsy, a nervous system disorder. He gets good grades and is active in his community. **Name two other nervous system disorders.**

Reading Check

Explain How can alcohol affect the nervous system?

Reading Check

Identify Name two other ways to protect your nervous system.

Developing Good Character

Citizenship You can demonstrate good citizenship by sharing what you learn about protecting your health. For example, encourage family members to protect their brains by always wearing a helmet when riding a bike. *What are some other ways you could promote healthy choices in your family or neighborhood?*

- **Alzheimer's disease,** which most often affects older adults, harms the brain and causes loss of memory.
- **Parkinson's disease,** which is a brain disorder that causes shaking and stiffness of the arms and legs.
- **Epilepsy,** which occurs when signals in the brain do not send messages in the normal way. Epilepsy can cause a person to briefly lose muscle control or have seizures.
- **Viruses,** such as polio, rabies, meningitis, encephalitis, and West Niles virus. Alcohol can affect the brain. Alcohol can destroy millions of brain cells that cannot be replaced.
- **Alcohol,** which can destroy brain cells. Alcohol also affects your thinking, your balance, and the way your body moves.
- **Other drugs,** which can harm the part of the brain that helps control your heart rate, breathing, and sleeping. Drugs also either speed up or slow down the nervous system.

Caring for Your Nervous System

MAIN IDEA Healthy behaviors can protect the nervous system.

A healthful lifestyle will help protect your nervous system. Make healthful food choices, drink plenty of water, and get plenty of sleep. Stay physically fit and maintain a healthy weight. You can also guard against illnesses that affect the nervous system. Protect yourself against insects and avoid animals that may carry disease. Wash your hands thoroughly and often.

Use safety equipment when you participate in physical activities. Protect your brain by wearing a safety helmet when riding a bike or skating. If you participate in gymnastics, have a person nearby who can spot you. If you lift weights, it is also helpful to have a spotter on hand. You can protect your back and spinal cord by lifting properly.

> MAKE HEALTHFUL FOOD CHOICES, DRINK PLENTY OF WATER, AND GET PLENTY OF SLEEP. STAY PHYSICALLY FIT AND MAINTAIN A HEALTHY WEIGHT.

This teen wears a protective helmet when she rides her bike. **Why is a protective helmet important when riding a bike?**

Be sure to follow basic safety rules. Always wear a seat belt when riding in a car. Pay attention to signs telling where you may or may not ride a skateboard. When you skate or ride a bike, watch for traffic, people and animals in your path.

You practice positive health behaviors when you decide not to use alcohol or other drugs. These destroy brain cells and affect your thoughts, emotions, and judgment. You want to keep your brain cells healthy!

Lesson 2 Review

What I Learned

1. **VOCABULARY** What are neurons?

2. **DEFINE** Define *CNS* and *PNS*.

3. **DESCRIBE** What are stimuli?

Thinking Critically

4. **SYNTHESIZE** List five voluntary muscle movements that you make each day.

5. **ANALYZE** Choose a sports figure or athlete that you know about. It could be a member of a team, a dancer, or a skateboarder. Describe how that person practices safe habits to protect the body's health.

Applying Health Skills

6. **ACCESSING INFORMATION** Epilepsy is a nervous system disorder in which a person has seizures. During a seizure, the person may lose consciousness, twitch, and shake. Use online and library resources to investigate what happens in the brain of a person who has epilepsy. Write a paragraph describing what you find.

Your Circulatory and Respiratory Systems

Before You Read

Quick Write In a few sentences, explain what you already know about how the heart works and what it does.

Vocabulary

circulatory system
cardiovascular system
heart
arteries
veins
capillaries
cell respiration
pulmonary circulation
systemic circulation
plasma
respiratory system
lungs
respiration
larynx
trachea
epiglottis
bronchi
diaphragm
Air Quality Index (AQI)

Reading Check

Recall What are the main parts of the circulatory system?

BIG IDEA Your heart is the center of your circulatory system, and your lungs are the center of your respiratory system.

Your Circulatory System

MAIN IDEA Your circulatory system is like a transportation system inside your body.

The body works all the time, even when you are asleep. Your **circulatory system** keeps your body working. The circulatory system is the group of organs and tissues that carry needed materials to cells and remove their waste products. The circulatory system includes the heart, different types of blood vessels, and the blood. It is also called the **cardiovascular system**. This includes organs and tissues that transport essential materials to body cells and remove their waste products.

Cardio refers to the heart, and *vascular* refers to the blood vessels. The circulatory system moves blood to and from tissues in the body. The blood delivers oxygen, food, and other materials to cells. It also carries waste products away from cells.

Think about a network of busy roads, all traveling in different directions. The circulatory system is like a highway system, and your blood cells are like the semi trucks that travel the roads, carrying materials around, through, and out of your body.

Parts of the Circulatory System

Your **heart** is a muscle that acts as the pump for the circulatory system. It pushes blood through tubes called blood vessels. There are three different types of blood vessels, **arteries**, **veins**, and **capillaries**. Arteries are blood vessels that carry blood away from the heart to various parts of the body. Veins are blood vessels that carry blood from all parts of the body back to the heart. The capillaries are tiny blood vessels that carry blood to and from almost all body cells and connect arteries and veins. They deliver oxygen and nutrients in the blood directly to the body's cells. The process in which the body's cells are nourished is called **cell respiration**. It's a process in which the body's cells are nourished and energized.

You cannot live more than a few minutes without oxygen. Your heart and lungs work together to deliver oxygen to your body's cells. **Pulmonary circulation** takes place when blood travels from the heart, through the lungs, and back to the heart. When blood travels this path, it gets rid of carbon dioxide. It also fills up with oxygen. Then **systemic circulation** begins when oxygen-rich blood travels to all body tissues except the lungs. At the same time, blood also delivers other nutrients to the cells and picks up waste products.

The heart has four chambers. The top chambers are called atria, or "rooms." Blood enters the heart through the two atria. The lower chambers are called ventricles. Blood leaves the heart through the two ventricles. Blood pressure is highest when the heart contracts, or pushes out blood. It is lowest between heartbeats, when the heart relaxes.

> **THE BODY WORKS ALL THE TIME, EVEN WHEN YOU ARE ASLEEP.**

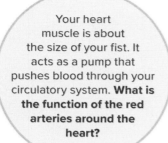

Your heart muscle is about the size of your fist. It acts as a pump that pushes blood through your circulatory system. **What is the function of the red arteries around the heart?**

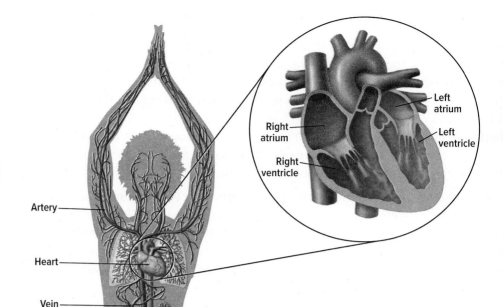

Right atrium

Right ventricle

Left atrium

Left ventricle

Artery

Heart

Vein

Reading Check

Recall Which blood cells carry oxygen from the lungs to all parts of the body?

Your Blood

Blood supplies your body with nutrients. It also helps fight off illness. Blood is made up of several parts—both liquids and solids. The yellowish, watery part of blood is called **plasma**. Plasma makes up about half the volume of blood in the body and is largely water. Its job is to transport blood cells and dissolve food. The solid parts of blood include red and white blood cells and platelets.

- **Red blood cells** carry oxygen to all other cells in the body. They also carry away some waste products.

- **White blood cells** help the various body systems destroy disease-causing germs.

- **Platelets** are small, disk-shaped structures that help your blood clot. Clotting helps keep you from losing too much blood when you have a cut or other injury.

Red blood cells are one of four specific types: A, B, AB, or O. Your blood type is inherited from your parents and remains the same throughout your life.

It is good to know your blood type. Some blood types are compatible. This means they can be safely mixed if a person needs blood. Mixing blood types that are not compatible can be harmful or even fatal.

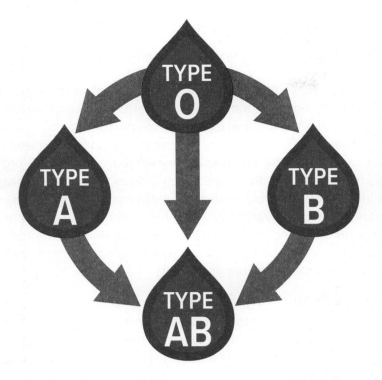

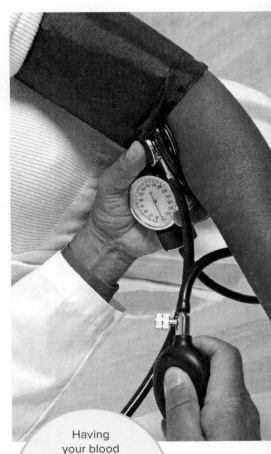

Donated blood saves many lives each year. **Which blood type is compatible with the other three blood types?**

People with any blood type can receive type O. As a result, people with type O blood are called "universal donors." People with type AB blood can receive any blood type but can only give to others with type AB. They are known as "universal recipients."

Blood is given during surgery or when a person needs blood due to a serious injury or illness. Blood may also carry an Rh factor, or a protein found on the surface of red blood cells. The Rh factor is another inherited trait. Blood is either Rh-positive or Rh-negative. People with Rh-positive blood can receive blood from donors who are either Rh-positive or Rh-negative. People with Rh-negative blood can only receive blood from donors who are also Rh-negative. Both the blood type and the Rh factor must be compatible in order for blood to be received safely.

Problems Affecting the Circulatory System

Some circulatory problems affect the heart or blood vessels. Others mainly affect the blood, while some affect other body systems. Circulatory problems include:

- **Hypertension,** which is also called high blood pressure. It can lead to kidney failure, heart attack, or stroke.
- **Heart attack,** or the blockage of blood flow to the heart.
- **Stroke,** which usually results from blood clots in the brain, or from a torn blood vessel.
- **Arteriosclerosis,** or a condition in which arteries harden and reduce blood flow.
- **Anemia,** which is an abnormally low level of hemoglobin, a protein that binds to oxygen in the red blood cells.
- **Leukemia,** or a type of cancer in which abnormal white blood interfere with production of other blood cells.

Having your blood pressure checked is part of a physical exam. **Why is it important to maintain normal blood pressure?**

Your Respiratory System

MAIN IDEA Your respiratory system controls your breathing.

Oxygen is essential to the body for survival. You get oxygen by breathing. Breathing is the movement of air into and out of the **lungs**. They are two large organs that exchange oxygen and carbon dioxide. Breathing enables your respiratory system to take in oxygen and eliminate carbon dioxide. The **respiratory system** includes the organs that supply your blood with oxygen. Breathing in, or inhaling, brings oxygen into your lungs. Your blood circulates through your lungs, exchanging carbon dioxide for oxygen. The exchange of gases between your body and the air is called **respiration**. Exhaling, or breathing out, is the action of your lungs getting rid of carbon dioxide and other waste materials from your body.

Parts of the Respiratory System

When you breathe in, air enters through the nose and mouth. In the nose, air is warmed and moistened. Hairs and sticky mucus in the nose help track dust and dirt from the air. Air passes through the nose and mouth into the throat.

The pharynx is a tube-like passageway at the top of the throat that receives air, food, and liquids from the mouth or nose. The epiglottis is a flap of tissue at the lower end of the pharynx. It keeps food and liquids from entering the respiratory system.

Reading Check

Recall Name the main organs of the respiratory system.

OXYGEN IS ESSENTIAL TO THE BODY FOR SURVIVAL.

Air moves into and out of the lungs through the respiratory system. **Which part of the respiratory system contains the bronchi?**

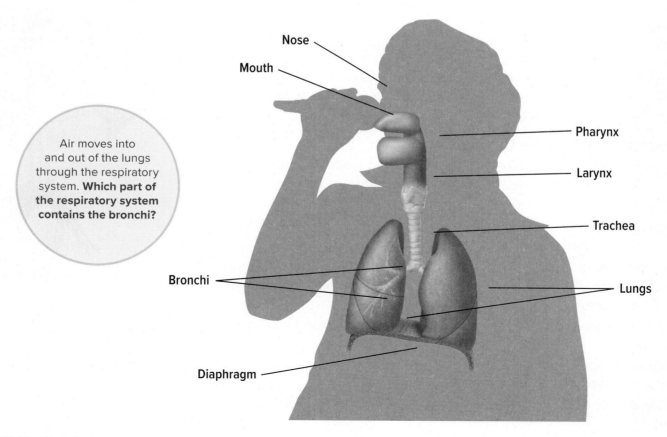

Nose

Mouth

Pharynx

Larynx

Trachea

Bronchi

Lungs

Diaphragm

Air passes from the pharynx into a triangle-shaped area called a voice box, or **larynx**. It's the upper part of the respiratory system, which contains the vocal cords. Two thick folds of tissue in the larynx—the vocal cords—vibrate and make sounds as air passes over them. The tissues of the larynx allow a person to speak. Air then enters the **trachea**, which is a passageway in your throat that takes air into and out of your lungs. The **epiglottis** is a flap of tissue in the back of your mouth that keeps food out of your trachea. It covers the trachea when you eat and uncovers it when you breathe. Two passageways that branch from the trachea, one to each lung are called **bronchi**. Inside the lungs, the bronchi continue to branch off into even smaller tubes.

How You Breathe

When carbon dioxide is in your blood, your nervous system signals your body to let it out, or exhale. The change in air pressure inside your chest causes breathing to occur. Breathing begins with the **diaphragm**. It's a large, dome-shaped muscle below the lungs that expands and compresses the lungs, enabling breathing.

When you breathe in, the diaphragm contracts. This allows the lungs to expand and fill with air. When you breathe out, the diaphragm expands. As it gets larger, it pushes on the lungs, forcing out the air.

Breathing involves both voluntary and involuntary muscle movements. You do not have to think about breathing. However, you can hold your breath or control the rate of your breathing.

Reading Check

Classify Tell whether the diaphragm is an organ or a muscle.

Coughing and difficulty breathing are symptoms of respiratory problems. **Name other symptoms of a cold.**

TOBACCO SMOKE, CHEMICALS, GERMS, AND AIR POLLUTION ARE HARMFUL TO YOUR HEALTH.

Problems Affecting the Respiratory System

Tobacco smoke, chemicals, germs, and air pollution are harmful to your health because they can damage the many parts of your respiratory system. Tobacco use may not cause all the problems that can affect your respiratory system, but it can make many problems worse. Some respiratory illnesses can make breathing difficult. Others can become life-threatening. See the table to learn causes and symptoms of some respiratory illnesses.

Respiratory problems can be prevented. **Which of these respiratory illnesses are caused by risk behaviors?**

RESPIRATORY ILLNESSES

Illness	Causes	Symptoms
Colds, flu	viruses	congestion, runny nose, watery eyes, coughing, sneezing
Bronchitis (brahn KI tus)	viruses, bacteria	coughing and fatigue due to mucus blocking the bronchi and bronchioles slows air movement
Pneumonia (noo MOH nyuh)	viruses, bacteria	difficulty breathing due to fluid in the alveoli that slows gas exchange
Asthma (AZ muh)	dust, smoke, pollen, pollution	difficulty breathing due to swollen airways and increased mucus
Emphysema (em fuh SEE muh)	smoking	coughing, fatigue, loss of appetite, and weight loss due to destruction of alveoli
Lung cancer	smoking	coughing, difficulty breathing, and chest pain

Keeping Your Circulatory and Respiratory Systems Healthy

MAIN IDEA You can help keep your heart, blood vessels, and lungs healthy and strong.

The health of your circulatory system has a major effect on your current and future health. Your blood carries oxygen to your cells and carbon dioxide away from them. Your blood also carries vital nutrients to your organs, muscular system, and skeletal system.

Caring for Your Circulatory System

The best way to help keep your heart healthy is to be physically active. Teens should set a goal of getting 60 minutes of physical activity each day. Regular activity strengthens your heart muscle and allows it to pump more blood with each heartbeat.

What you eat can have either a positive or negative effect on your circulatory system. Try to limit the amount of fat you eat. Fats, especially saturated and trans fats, can cause deposits to form in your arteries. These deposits increase blood pressure. As you lower your fat intake, increase your intake of dietary fiber. Whole grains and fresh vegetables are a great source of fiber. Whole-grain cereals, raw vegetables, and breads made with whole grains also make filling and satisfying snacks.

Another way to keep your heart and blood vessels healthy is to avoid tobacco. Tobacco use can cause lung cancer, emphysema, and other lung diseases. The nicotine in tobacco can constrict your blood vessels. When your blood vessels are constricted, or narrower, your heart has to work harder. This can result in high blood pressure and lead to heart disease.

Finally, learn to manage the stress in your life. Stress can cause high blood pressure, which puts a strain on the entire cardiovascular system. You can learn strategies to deal with stress in healthful ways. Regular physical activity is a very effective way to help relieve stress. Staying active can also help you maintain a heart-healthy weight.

> **Reading Check**
>
> **Recall** What is one good way to help keep your circulatory system in good health?

> **SET A GOAL TO GET 60 MINUTES OF PHYSICAL ACTIVITY EACH DAY.**

• • • • • • • • • •

Reading Check

Recall What is one good way to protect your respiratory system?

• • • • • • • • • •

Caring for Your Respiratory System

Your whole body depends on having a healthy respiratory system. However, you can take positive action to help keep your lungs breathing strong. Here are some things you can do to benefit your respiratory system:

- **Avoid tobacco use.** Smoking can cause cancer. All tobacco products contain substances that can cause cancer.
- **Stay away from people who smoke.** Avoid places where the air is smoky. Breathing secondhand smoke, or air that has been contaminated by others' tobacco use, can be just as harmful as smoking.
- **Take care of your body.** Give your body a chance to heal and recover when you have a cold or the flu. See a health professional if an illness does not go away.
- **Drink plenty of fluids.** Whether you feel ill or healthy, you always need plenty of fluids. Drink more water when you participate in physical activity or exercise.
- **Be physically active on a regular basis.** Keep your body systems active and strong.

Marcus knows how to enjoy the outdoors. **How is Marcus caring for his body on his walk?**

- **Eat a healthful diet.** As with all body systems, your respiratory system needs a proper balance of nutrients.
- **Pay attention to weather alerts for your area.** Allergy, ozone, and pollution alerts can prepare you for when outside air may be less healthful. Knowing your area's **Air Quality Index (AQI)** can help you maintain your respiratory health. AQI is the measure of ozone, sulfur dioxide, carbon monoxide, and fine particles close to the ground.
- **Manage stress.** Use strategies to maintain your stress levels. As you have learned, stress can have an impact on all sides of your health triangle.
- **Protect yourself from infections.** Wash your hands thoroughly and frequently with soap and water. Keep your body covered and protected when walking in the woods. Eat a healthful diet and get plenty of sleep.

YOUR WHOLE BODY DEPENDS ON HAVING A HEALTHY RESPIRATORY SYSTEM.

Myth vs Fact

Myth The hiccups are a mystery.

Fact We know what causes hiccups. The hiccups are caused by the diaphragm. Occasionally, the diaphragm has spasms that cause air to be taken in or pushed out rapidly. The sound of a hiccup is caused by the sudden rush of air being stopped by the vocal cords. Some hiccups can be caused by eating a big meal, swallowing air, stress, or excitement. The mystery is how to get rid of the hiccups!

Lesson 3 Review

What I Learned

1. **VOCABULARY** What are the main organs of the respiratory system?

2. **EXPLAIN** How are sounds made when a person speaks?

3. **DESCRIBE** Tell how the diaphragm helps you breathe.

Thinking Critically

4. **SYNTHESIZE** Think about the movement of your chest as your lungs take in air. Is this voluntary or involuntary movement? What changes when you do deep-breathing exercises?

5. **ANALYZE** When Nora's father went to donate blood, he was asked his blood type. He wasn't sure. How can Nick learn his blood type? Why is it important?

Applying Health Skills

6. **ANALYZING INFLUENCES** A number of factors in the environment might influence respiratory health. Make a list of these factors and discuss their role in the health of the community.

Your Digestive and Excretory Systems

BIG IDEA Your digestive and excretory systems process the food you eat for use by your body.

Your Digestive System

MAIN IDEA Digestion is the first step in the way your body processes the food you eat.

Think about the foods you ate during your last meal. Do you know what happens to food after you eat it? As soon as food enters your mouth, it begins its journey through your **digestive system**. The digestive system is a group of organs that work together to break down foods into substances that your cells can use. No matter what you eat, your food goes through four steps—ingestion, digestion, absorption, and **excretion**. Excretion is the process the body uses to get rid of waste. The first step, ingestion, is the act of eating, or putting food in your mouth, or eating.

The Process of Digestion

The digestive system begins in your mouth. With your first bite of food, your teeth begin to smash and grind the food into small bits. The food mixes with your **saliva**. It's a digestive juice produced by the salivary glands in your mouth. Substances that aid in the body's chemical rections to food are called **enzymes**. They are in your saliva and help break down the food. Then **digestion** begins. It's the process by which the body breaks down food into smaller pieces that can be absorbed by the blood and sent to each cell in your body.

Once you chew and swallow something you eat, the food first enters your throat. Throat muscles contract and expand to push the food down the esophagus into the stomach. The esophagus is a muscular tube that connects the mouth to the stomach. Waves of muscle contractions allow food to move through the esophagus and the rest of the digestive tract. Once the partially digested food leaves the esophagus, it enters the stomach. The stomach is a large, hollow organ that stores food temporarily.

> **DO YOU KNOW WHAT HAPPENS TO FOOD AFTER YOU EAT IT?**

The stomach also aids in chemical digestion. Chemical digestion is when chemical reactions in the body break down pieces of food into small molecules. In the stomach, food mixes with gastric juices until it forms a watery liquid. During this digestive process, the food may stay in the stomach up to four hours. The food then passes from the stomach to the **small intestine**.

The small intestine is a coiled tube from 20 to 23 feet long, in which about 90 percent of digestion takes place. Most chemical digestion occurs in the small intestine. Nutrients in the small intestine enter the blood through blood vessels. This is known as absorption: the body begins to absorb the nutrients from the food.

Organs that Aid in Digestion

The liver and pancreas are both organs that produce substances that enter the small intestine and help with chemical digestion. The **liver** is a digestive gland that secretes a substance called *bile*, which helps to digest fats. The **gallbladder** is small, saclike organ that stores bile until it is needed in the small intestine. The **pancreas** is a gland that helps the small intestine by producing pancreatic juice, a blend of enzymes that breaks down proteins, carbohydrates, and fats.

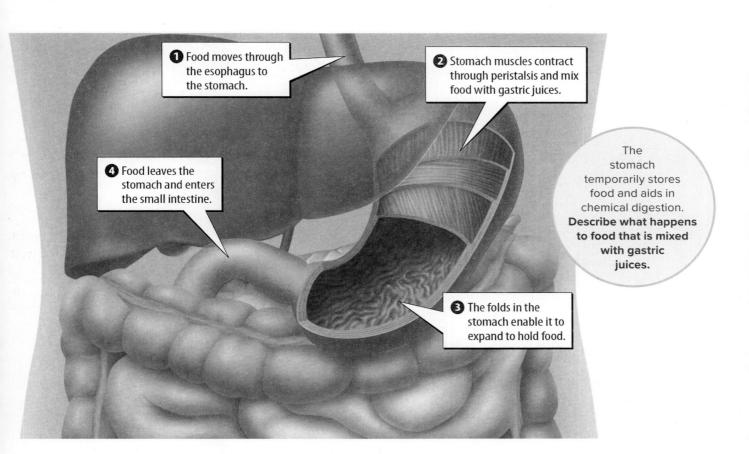

❶ Food moves through the esophagus to the stomach.

❷ Stomach muscles contract through peristalsis and mix food with gastric juices.

❹ Food leaves the stomach and enters the small intestine.

❸ The folds in the stomach enable it to expand to hold food.

The stomach temporarily stores food and aids in chemical digestion. **Describe what happens to food that is mixed with gastric juices.**

Your Excretory System

MAIN IDEA Waste from the food you eat is processed by your excretory system.

The group of organs that work together to remove wastes are the **excretory system**. They include the **kidneys**, the bladder, and the **colon**. The kidneys are organs that remove waste material, including salts from the blood. They produce urine. The bladder stores urine until it is ready to be passed out, or removed, from your body.

Foods that are not absorbed in the small intestine move into the colon. It's a tube five to six feet in length that plays a part in both digestion and excretion. The materials that pass through the large intestine are the waste products of digestion. The waste products become more solid as water is absorbed. The waste products are pushed into the final section of the large intestine, or the rectum. Muscles in the rectum and anus control the release of solid waste, or feces.

The excretory system controls your body's water levels. Your skin and lungs also help to remove waste from your body. Your skin gets rid of some wastes in the form of sweat. Your lungs get rid of carbon dioxide when you exhale or breathe out.

YOUR EXCRETORY SYSTEM WORKS TO REMOVE WASTE FROM YOUR BODY.

Many wastes are excreted through the kidneys and bladder. **What body system sends a signal that the bladder is full?**

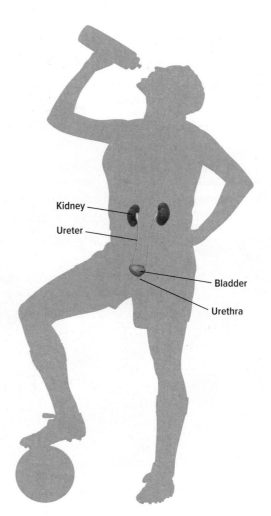

Kidney

Ureter

Bladder

Urethra

Digestive and Excretory Problems

MAIN IDEA Some problems can occur in the digestive and excretory systems.

Have you ever had a stomach ache or felt uncomfortable after eating? Several factors can cause problems of the digestive system. As you might imagine, many issues may be related to what you eat, but other problems can be symptoms of serious illness.

A common warning from your digestive system is indigestion. The mildest symptom of indigestion is often a bloated or unusually full feeling after eating. It can also include belching, painful gas, nausea, or a burning sensation in the stomach area. Indigestion is often your digestive system's way of telling you to eat more slowly and healthfully.

Another common problem is heartburn, or a burning sensation in the chest or throat. It is caused by stomach acids flowing back into the esophagus. Heartburn can be caused by diet. However, if it lasts too long, medical attention is required.

Contaminated food or water is often the cause of diarrhea, or watery feces. Diarrhea might also be a symptom of a disease of the colon. A person with severe diarrhea for a long time should also see a medical professional.

Pain in the stomach area might be caused by an **ulcer**. It's an open sore in the stomach lining. Ulcers may be caused by bacteria, and alcohol use can also be a factor.

Sometimes a person can feel severe pain caused by mineral crystals, or stones, that develop in the digestive system. Gallstones, which form in the gall bladder, and kidney stones, which form in the kidneys, often require medical care.

Appendicitis is the inflammation of the appendix. The appendix is a tube about four inches long, located near where the small intestine and large intestine meet. If your appendix becomes inflamed, you feel pain in the lower right side of your body. Appendicitis is very serious and requires emergency surgery.

Hemorrhoids are swollen veins at the opening of the anus. They may be itchy or painful and sometimes cause bleeding.

Reading Check

Paraphrase What is a common sign that you might need to eat more healthfully?

HAVE YOU EVER HAD A STOMACHACHE AFTER EATING OR FELT UNCOMFORTABLE?

Caring for Your Digestive and Excretory Systems

MAIN IDEA A healthful diet and lifestyle are important to digestive and excretory health.

Your digestive and excretory systems are important to your overall physical well-being. As with all body systems, a healthy lifestyle can keep your digestive and excretory systems healthy. Here are some steps you can take:

- **Eat a healthful diet with plenty of fiber.** Choose low-fat and high-fiber foods from all food groups. Include plenty of fresh fruits and vegetables.
- **Take time to eat and chew food thoroughly.** Avoid rushing your meals, which can overload your digestive system. Eating and chewing slowly will also help prevent you from eating too much.

Drinking water is essential to your health. **How else can you provide your body with water?**

- **Drink plenty of water.** Your digestive system needs water to work properly. Drink six to eight 8-ounce glasses of water each day. Unsweetened fruit juices, low-fat milk, soup, and many fruits and vegetables are also sources of water.
- **Take care of your teeth and gums.** Your teeth begin the digestive process by helping to chew, mash, and grind your food into small pieces to swallow. Brush your teeth at least twice a day with fluoride toothpaste and floss daily. Get regular dental checkups.
- **Wash your hands.** Make a habit of washing your hands thoroughly with soap and water. This is especially important before preparing or eating foods. Regular hand washing will help prevent the spread of bacteria that can upset the digestive system.
- **Avoid risk behaviors.** Alcohol use can interfere with the way your digestive system absorbs nutrients. It can also contribute to ulcers and indigestion. Tobacco use has been linked to ulcers and other digestive problems such as heartburn, gallstones, and kidney stones.
- **Be physically active.** As with all body systems, keeping your body fit and maintaining a healthy weight will have positive effects on your digestive and excretory health.

> **A HEALTHY LIFESTYLE CAN KEEP YOUR DIGESTIVE AND EXCRETORY SYSTEMS HEALTHY.**

Reading Check

Recall Why are your teeth important to the digestion process?

Myth vs Fact

Myth All forms of bacteria are bad.

Fact Your digestive system contains between 10 and 100 trillion bacteria. That's ten times the number of cells in your body! Certain bacteria are necessary for the digestion of food. Without "friendly" bacteria, you could eat all you wanted, but the food could pass through your intestines mostly undigested. The trick is to stay away from "unfriendly," or harmful, bacteria.

Lesson 4 Review

What I Learned

1. **VOCABULARY** Define *digestive system*.

2. **EXPLAIN** In what body part does most of your digestion take place?

3. **DESCRIBE** What role does the stomach play in the digestive process?

Thinking Critically

4. **SYNTHESIZE** Describe the path of food from the mouth to the colon.

5. **ANALYZE** Imagine you just ate a huge meal from a fast-food restaurant. Now you have a stomachache. What could be the problem and the cause?

Applying Health Skills

6. **GOAL SETTING** Identify a behavior that can promote digestive health but which you are not currently practicing. Use the skill of goal setting to help you make this behavior a habit. Share the steps in your action plan with your classmates.

Your Endocrine System

Before You Read

Quick Write Think about when you may have experienced a growth spurt or seen one in someone else. Write one or two sentences to describe the growth spurt.

Vocabulary
endocrine system
gland
metabolism
pituitary gland

Reading Check

Recall What is metabolism?

BIG IDEA Your body has glands and organs that allow it to function and grow.

Your Endocrine System

MAIN IDEA Your endocrine system produces chemicals that regulate body functions.

When you start back at school each year, do you notice that many of your classmates have grown? The body system responsible for growth and other body changes is the **endocrine system**. It's the system of glands throughout the body that regulate body functions.

The endocrine system sends messages to the body through the blood in the form of hormones. A **gland** is a group of cells, or an organ, that secretes a chemical substance. Those chemical substances are hormones. For example, the thyroid gland controls **metabolism**. The process by which the body gets energy from food.

Glands and Hormones

One of the major glands of the endocrine system include the **pituitary gland**. It is a gland that signals other endocrine glands to produce hormones when needed. Other glands of the endocrine system are the thyroid, parathyroid, adrenals, hypothalamus, thymus, and the pancreas. The glands of the reproductive system are also part of the endocrine system. Glands produce specific hormones, which travel through the bloodstream to cells that need them. Some hormones are produced continuously, while others are produced at certain times.

When you feel nervous or stressed, your heart rate and blood flow to the brain may increase. Your blood sugar and blood pressure may rise. Sweat production increases and air passages expand. Digestion and other bodily processes may slow down to conserve energy. Your adrenal glands release the hormone adrenaline, which allows your body to respond to stress. The figure shows the glands and how the hormones they produce control body functions.

Problems of the Endocrine System

The most common problem of the endocrine system is diabetes. Diabetes can cause heart disease, kidney failure, blindness, and circulatory problems. Diabetes is the seventh leading cause of death in the U.S.

People with diabetes have too much sugar in their blood. Type 1 diabetes occurs when the pancreas cannot produce enough insulin. People with type 1 diabetes need to take extra insulin to help control their blood sugar. Type 2 diabetes occurs when the body cannot properly use the insulin it produces. However, type 2 diabetes is preventable. Maintaining a healthy weight and staying physically active can help prevent type 2 diabetes.

Staying hydrated will help keep all your body systems healthy. **Why is it important to drink plenty of water?**

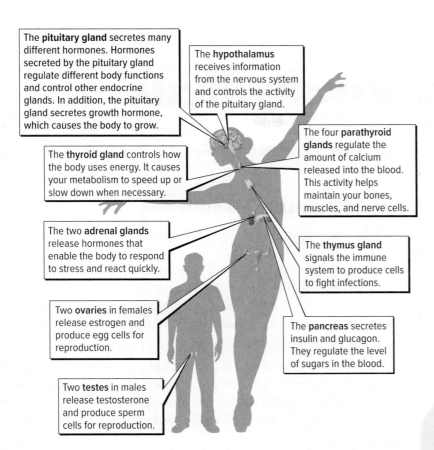

The **pituitary gland** secretes many different hormones. Hormones secreted by the pituitary gland regulate different body functions and control other endocrine glands. In addition, the pituitary gland secretes growth hormone, which causes the body to grow.

The **hypothalamus** receives information from the nervous system and controls the activity of the pituitary gland.

The four **parathyroid glands** regulate the amount of calcium released into the blood. This activity helps maintain your bones, muscles, and nerve cells.

The **thyroid gland** controls how the body uses energy. It causes your metabolism to speed up or slow down when necessary.

The two **adrenal glands** release hormones that enable the body to respond to stress and react quickly.

The **thymus gland** signals the immune system to produce cells to fight infections.

Two **ovaries** in females release estrogen and produce egg cells for reproduction.

The **pancreas** secretes insulin and glucagon. They regulate the level of sugars in the blood.

Two **testes** in males release testosterone and produce sperm cells for reproduction.

DO YOU NOTICE THAT MANY OF YOUR CLASSMATES HAVE GROWN?

Another endocrine system problem is an overactive or underactive thyroid. Having an overactive thyroid may result in the gland making too much hormone. Symptoms of hyperthyroidism include swelling in the front of the neck, nervousness, increased sweating, and weight loss.

An underactive thyroid gland is also known as hypothyroidism. In the case of an underactive thyroid, the gland is not making enough hormones to regulate metabolism. Symptoms of an underactive thyroid gland include tiredness, depression, weight gain, hair loss, and pain in the muscles and joints.

Lesson 5 Review

What I Learned

1. **VOCABULARY** Define *hormones*.

2. **DESCRIBE** What does the endocrine system do?

3. **EXPLAIN** What is diabetes?

Thinking Critically

4. **ANALYZE** If a person's thymus gland is not functioning properly, what could happen the next time they are exposed to the virus that causes a cold?

5. **APPLY** Anthony is feeling stressed and nervous about taking a test. How is Anthony's endocrine system responding to this stress?

Applying Health Skills

6. **DECISION MAKING** Several people in Lanie's family have type 2 diabetes. How can Lanie use her decision-making skills to reduce her risk of developing type 2 diabetes?

Your Immune System

Before You Read

Quick Write Write one or two sentences describing the last time you had a cold or fever.

Vocabulary

immune system
immunity
inflammation
lymphatic system
lymphocytes
antigens
antibodies
vaccine

Reading Check

Recall What parts of the digestive system provide barriers to pathogens entering the body?

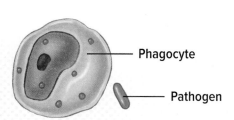

Phagocyte

Pathogen

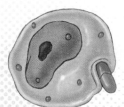

A phagocyte destroys a pathogen by surrounding it and breaking it down. **How does a phagocyte know to go to an inflammation?**

BIG IDEA Your immune system helps your body defend itself against infections.

Your Immune Responses

MAIN IDEA Your body has several defenses against pathogens.

Pathogens, germs that cause diseases, are everywhere. They are in the air you breathe, the water you drink, and on the objects you touch. Most bacteria, viruses, and other pathogens do not make you sick. Your body has natural barriers between you and pathogens.

Your first line of defense against infection is your body's natural barriers in other body systems. Natural barriers are your skin, the saliva and stomach acid in the digestive system, and mucous membranes in the respiratory system. The circulatory system and nervous systems also work together to raise the body's temperature, killing the pathogens with fever.

When pathogens get past the body's natural barriers, your **immune system** responds. It's a combination of body defenses made up of the cells, tissues, and organs that fight pathogens in the body. It has two main responses— nonspecific and specific. Together these responses provide **immunity**, or the ability to resist the pathogens that cause a particular disease.

If a pathogen gets through, the immune system takes action with a specific response. It recognizes specific pathogens that have attacked before. Once your immune system has created a specific response, those response cells remain in your body. When the same pathogen attacks again, your immune system is prepared to fight it and reacts right away.

Inflammation

White blood cells flow through the circulatory system. Their job is to fight germ-causing diseases and pathogens. They do most of their work attacking pathogens in the fluids outside your blood vessels. They fight infection several different ways.

Some white blood cells can surround and destroy bacteria. Other white blood cells release chemicals that make it easier to destroy pathogens. Some white blood cells can produce proteins that destroy viruses and other foreign substances in the body.

When pathogens get past the body's first-line defenses, your immune system reacts with what is called a nonspecific immune response. A nonspecific response typically begins with **inflammation**. It's the body's response to injury or disease, resulting in a condition of swelling, pain, heat, and redness. The brain sends signals that tell white blood cells to rush to the affected area and destroy the pathogens. Circulation to the area slows down. The symptoms of inflammation are caused by white blood cells surrounding pathogens and destroying them.

When you have an infection, the body starts producing a protein to stimulate the body's immune system. If pathogens spread, your body temperature may rise and cause a fever. A higher body temperature makes it harder for pathogens to reproduce. A fever also signals the body to produce more white blood cells to destroy pathogens. When you have a fever, you know that your white blood cells are fighting pathogens.

The Lymphatic System

If the body's first-line and second-line defenses do not destroy all the invading pathogens, another type of immune response occurs. This is a specific immune response that calls on the **lymphatic system**. It's a secondary circulatory system that helps the body fight pathogens and maintains its fluid balance. The fluid that circulates through the body's lymphatic system is known as lymph.

Two main **lymphocytes** are B cells and T cells. These are special white blood cells in the lymphatic system. Lymph includes macrophages. Macrophages attach themselves to invading pathogens and destroy them. Macrophages help lymphocytes recognize the invading pathogens and prepare for future attacks.

Different cells work together to fight invading pathogens. **What is the purpose of memory B cells and memory T cells?**

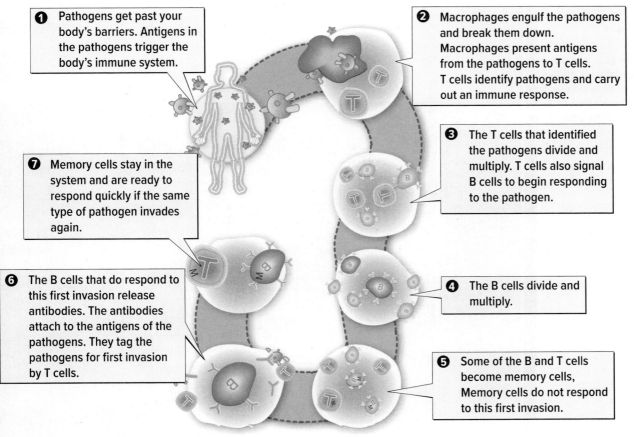

❶ Pathogens get past your body's barriers. Antigens in the pathogens trigger the body's immune system.

❷ Macrophages engulf the pathogens and break them down. Macrophages present antigens from the pathogens to T cells. T cells identify pathogens and carry out an immune response.

❸ The T cells that identified the pathogens divide and multiply. T cells also signal B cells to begin responding to the pathogen.

❼ Memory cells stay in the system and are ready to respond quickly if the same type of pathogen invades again.

❻ The B cells that do respond to this first invasion release antibodies. The antibodies attach to the antigens of the pathogens. They tag the pathogens for first invasion by T cells.

❹ The B cells divide and multiply.

❺ Some of the B and T cells become memory cells. Memory cells do not respond to this first invasion.

Antigens and Antibodies

Lymphocytes react to **antigens**, or substances that send the immune system into action. The immune system responds to these antigens by producing **antibodies**. These are proteins that attack to antigens, keeping them from harming the body. Lymphocytes known as B cells produce a specific antibody for each specific antigen. If the same type of pathogen invades the body again, these specific antibodies are ready to attack.

T cells in the lymphatic system can do two things. T cells known as helper cells stimulate the production of B cells to produce antibodies. T cells called killer cells attach to invading pathogens and destroy them.

Some of the new B cells and T cells do not react to pathogens immediately. These B cells and T cells wait and are ready to react if the same kind of pathogen invades the body again. These cells are called memory B cells and memory T cells.

These cells help your immune system stop diseases that have attacked before. For example, if you have had measles or if you have been vaccinated against measles, your immune system remembers. It will attack the antigens for the measles virus.

MANY KINDS OF CELLS WORK TOGETHER IN YOUR IMMUNE SYSTEM TO FIGHT INVADING PATHOGENS.

Your skin is a first line of defense against germs. **Why do you think it is important to wash your hands frequently?**

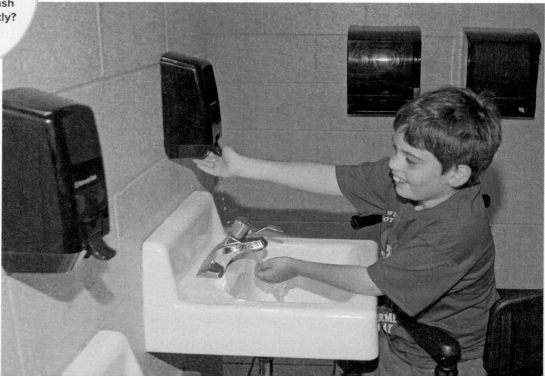

Immunity

MAIN IDEA The body develops immunity in different ways.

Everyone is born with natural immunity. Even before a mother gives birth, antibodies pass from her body to her developing fetus. However, these immunities last only a few months. The baby's immune system becomes active and produces antibodies on its own to fight pathogens.

A **vaccine** causes the immune system to produce antibodies for certain diseases. A vaccine is a preparation of dead or weakened pathogens that is introduced into the body to cause an immune response. This process is called immunization.

Vaccines have been developed for many diseases, such as polio, measles, mumps, chickenpox, hepatitis, and strains of the flu. Some vaccinations, or shots, are given in a series over several months. Others must be given repeatedly over a lifetime. If you cut yourself on a piece of rusty metal, a doctor may ask when you got your last tetanus shot. Rusty metal can introduce harmful pathogens into your body. To fight them, your immune system will need antibodies. To stay healthy, it is important to keep your vaccinations current. If you have not had a certain vaccination for a while, you may need what is called a booster shot.

Reading Check

Paraphrase How is a baby born with natural immunities?

> **EVERYONE IS BORN WITH NATURAL IMMUNITY.**

Lesson 6 Review

What I Learned

1. **VOCABULARY** Define immune system.

2. **RECALL** What does the lymphatic system do in the body?

3. **IDENTIFY** Name three ways the body achieves immunity against diseases.

Thinking Critically

4. **EXPLAIN** What is the difference between a nonspecific immune response and a specific immune response? Which kind of response does the immune system "remember"?

5. **APPLY** Why should you avoid drinking from the same container as a friend who has a cold?

Applying Health Skills

6. **ACCESSING INFORMATION** Each of the 50 states determines which vaccines they require for students entering school. Research the vaccine requirements in your state. Why do you think states require certain vaccines before students enter school?

Tobacco

LESSONS

13

Facts About Tobacco

Before You Read

Quick Write Make a list of as many tobacco products as you can think of prior to reading this lesson. Briefly describe each product.

Vocabulary

nicotine
addictive
smokeless tobacco
snuff
tar
carbon monoxide

BIG IDEA The substances contained in tobacco products are very harmful to your health.

What Is Tobacco?

MAIN IDEA Tobacco is a harmful and addictive substance in all its forms.

Tobacco is a woody, shrub-like plant with large leaves. It is grown throughout the world. Tobacco is considered to be a stimulant, meaning that it increases the heartbeat and raises blood pressure. The stimulant is **nicotine**, an addictive, or habit-forming, drug found in tobacco. An **addictive** drug is one that is capable of causing a user to develop intense cravings. It contains harmful substances that are released when a person smokes or chews it.

Tobacco companies add more harmful ingredients when they prepare tobacco to be sold. It is estimated that there are more than 600 ingredients in tobacco. When tobacco is burned, it creates more than 7,000 chemicals. About 70 of these chemicals are known to cause cancer. Some of the same ingredients found in cleaning products or pest poisons are added to tobacco products.

The chemicals in tobacco affect the person who uses it as well as others who are nearby. For this reason, the U.S. government has limited the sale of tobacco products to adults. It is illegal to sell tobacco products to anyone under the age of 21. It is also illegal for people under age 21 to use tobacco products.

Forms of Tobacco

Tobacco companies harvest leaves from tobacco plants. The leaves are then prepared for smoking or chewing. Tobacco products come in many forms. The most common include cigarettes, vaping or e-cigarettes, cigars, pipes, specialty cigarettes, and smokeless tobacco.

CIGARETTES These are the most commonly used forms of tobacco products. Cigarettes contain shredded tobacco leaves. They may also have filters intended to block some harmful chemicals. However, filters do not remove enough chemicals to make cigarettes less dangerous. Also, some tobacco users may try not to inhale the smoke but smoking in any form is not safe for your body.

VAPING or e-CIGARETTES Vaping is a newer tobacco-delivery method. The device creates a vapor that the user inhales. A cartridge is attached to an e-cigarette device. The e-cigarette device heats the liquid contents of a cartridge. The user inhales the vapor.

e-Cigarettes are thought to be safer to use than cigarettes because they do not burn tobacco. However, the devices do deliver tobacco, which is addictive. Other chemicals in vaping cartridges are:

- ultrafine particles that are inhaled in the lungs,
- flavorings that contain the chemical diacetyl have been linked to lung disease,
- cancer-causing chemicals, such as formaldehyde
- volatile organic compounds that can cause eye, nose, and throat irritation. The compounds can also damage the liver, kidneys, and nervous system, and
- heavy metals including nickel, tin, and lead.

CIGARS AND PIPES As is the case with cigarettes, the tobacco used in cigars and pipes is made up of shredded tobacco leaves. However, one large cigar can contain as much tobacco and nicotine as an entire pack of 20 cigarettes. Pipes and cigars also cause some of the same serious health problems that cigarettes do. Cigar smoke contains up to 90 times more cancer-causing chemicals than those found in cigarette smoke. People who smoke cigars or pipes are more likely to develop mouth, tongue, or lip cancer than people who do not use tobacco. Cigar and pipe smokers also face an increased risk of dying from heart disease compared to nonsmokers.

Reading Check

Identify What are the most common health risks associated with cigars and pipes?

Vaping was thought to be safer than smoking cigarettes. **Name some chemicals in vaping cartridges.**

SPECIALTY CIGARETTES This category includes flavored, unfiltered, and clove cigarettes. Clove cigarettes are also called Kreteks and bidis. The U.S. has tried to ban sales of flavored cigarettes in an effort to discourage young people from trying them. Flavored tobacco is also smoked in special water pipes called hookahs. These forms of tobacco contain higher concentrations of harmful chemicals than do regular cigarettes.

SMOKELESS TOBACCO Smokeless tobacco is ground tobacco that is chewed or inhaled. Chewing tobacco is often called "dip" or "spit tobacco." **Snuff** is the other form of smokeless tobacco. It is finely ground tobacco that is inhaled or held in the mouth or cheeks. While it does not affect the lungs the way smoking does, smokeless tobacco is not a safe alternative to cigarettes. It contains addictive nicotine and other chemicals and causes gum disease and oral cancer.

Chemicals in Tobacco

MAIN IDEA Tobacco contains many chemicals that can harm your body.

Harmful chemical compounds exist in all forms of tobacco. These are released when a person smokes or chews tobacco. Chemicals in tobacco can also affect nonsmokers who inhale others' smoke. Most of these chemicals hurt your body's ability to work properly.

- Nicotine is one of the harmful substances found in tobacco leaves and in all tobacco products. A person begins to depend on it after it has been in the body regularly. A person can become addicted to nicotine very quickly. Nicotine has other effects, too. It makes your heart beat faster and raises your blood pressure. It causes dizziness and an upset stomach and reduces the amount of oxygen your blood carries to the brain.

Nicotine affects the heart and lungs, affecting sports performance. **Explain how nicotine affects the heart.**

The tar in cigarettes can cause serious breathing problems. **Describe how tar affects breathing.**

- **Tar** is a thick, dark liquid that forms when tobacco burns. The smoker inhales this dangerous substance. Tar covers the airways and the bronchi. Lungs covered with tar can become diseased and cause serious breathing trouble.
- **Carbon Monoxide** Carbon monoxide is a colorless, odorless poisonous gas produced when tobacco burns. It harms the brain and the heart by reducing the amount of oxygen available to these organs. If too much carbon monoxide enters your body, it can kill you.

HARMFUL CHEMICAL COMPOUNDS EXIST IN ALL FORMS OF TOBACCO.

Reading Check

Summarize What are three harmful substances found in tobacco smoke?

Lesson 1 Review

What I Learned

1. **VOCABULARY** What does the term *addictive* mean? Use it in a complete sentence.

2. **SUMMARIZE** Why is tobacco harmful?

3. **IDENTIFY** Name three substances in tobacco smoke that are harmful to the body.

Thinking Critically

4. **ANALYZE** If many cigarettes have filters, why are they still not safe?

5. **EXPLAIN** Why are e-cigarettes harmful to a user's health?

Applying Health Skills

6. **COMMUNICATION SKILLS** Ryan has learned that his friend Spencer smokes cigars and uses smokeless tobacco. Spencer tells Ryan that this is not as bad as smoking cigarettes. What facts might you suggest that Ryan share with Spencer to convince him that his tobacco use is still harmful?

Health Risks of Tobacco Use

Before You Read

Quick Write List two future goals. Then write a short paragraph telling how a tobacco-related illness could affect each of these goals.

Vocabulary

alveoli

emphysema

chronic obstructive pulmonary disease (COPD)

. . . .

Reading Check

Recall Why is smoking especially hazardous to teen health?

. . . .

BIG IDEA Tobacco is a dangerous drug with serious health consequences.

Tobacco Use Is Harmful to Your Health

MAIN IDEA Health experts have been warning about the dangers of tobacco for many years.

The message that tobacco use is bad for your health is not new. In 1964, the Surgeon General issued a report saying that smoking may be hazardous. A year later, tobacco companies were ordered to add health warnings to cigarette packages. Since then, the warning labels have become more prominent. Other tobacco products now also carry similar warnings.

Tobacco use has serious consequences. The chemicals in tobacco and tobacco smoke can cause damage to most of the body's systems. In the United States, more than 480,000 people die every year from smoking-related illnesses.

E-Cigarettes were thought to be a healthier alternative to tobacco use. But only because e-cigarettes are not combustible. In mid-2019, a new concern with the use of vaping pens arose. Several teens and young adults who used vape pens developed a lung disease. As of September 2019, the disease caused seven deaths. More than 2,000 people became seriously ill. Health researchers believe that an ingredient in some vape cartridges cause the illness. However, the cause has not been confirmed and research is continuing.

Tobacco use is especially damaging to teens because their bodies are still growing. The chemicals in tobacco interfere with this process of growth and development. Tobacco use is also illegal for anyone under the age of 21. Since the 1960s, the CDC has done research on the harmful effects of tobacco use. It can cause disease in all body systems. These illnesses can be prevented by choosing to stay tobacco-free. To learn more about the CDC's research go to www.CDC.gov and search for tobacco use.

> THE CHEMICALS IN TOBACCO CAN CAUSE DAMAGE TO MOST BODY SYSTEMS.

Nicotine raises the heart rate and blood pressure. Tobacco users often cannot run as long or as fast as they did before they started smoking. They get sick more often and tend to stay sick longer. Tobacco use can cause disease of the mouth and lungs.

Tobacco use also damages the rest of the body. In science class you have learned that the human body is a system of interacting subsystems, such as the respiratory system, the circulatory system, the nervous system, and the excretory system. Tobacco use causes diseases in all of these systems. The figure in the lesson lists some of the effects of tobacco on your body systems. Many of these illnesses can be prevented if you choose the positive health behavior of staying tobacco free.

Respiratory System	Tobacco smoke damages the air sacs in the lungs. This damage can lead to a life-threatening disease that destroys these air sacs. Smokers are also between 12 and 22 times more likely than nonsmokers to develop lung cancer.
Digestive System	All forms of tobacco increase the risk of cavities and gum disease. Tobacco dulls the taste buds and can cause stomach ulcers. Tobacco use is linked to cancers of the mouth, throat, stomach, esophagus, and pancreas.
Nervous System	Tobacco use reduces the flow of oxygen to the brain, which can lead to a stroke.
Excretory System 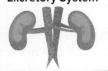	Smokers have at least twice the risk of developing bladder cancer as nonsmokers. Smokeless tobacco can also put users at risk of developing bladder cancer.
Circulatory System	Tobacco use is linked to heart disease. It increases the chances of a heart attack. Smoking also raises blood pressure and heart rate.

Using tobacco harms many body systems, causing many health problems or diseases. **Explain the effects of tobacco use on the digestive system.**

Respiratory System

Tobacco smoke contains tar, which coats the inside of the lungs. Smoke damages the **alveoli**, which are tiny air sacs in the lungs. When this happens, your lungs are less able to supply oxygen to your body. This damage can cause **emphysema**. It's a disease that results in the destruction of the alveoli in the lungs. When this disease affects a large part of the lungs, it can cause death.

The chemicals in tobacco also put smokers at a greatly increased risk of developing lung cancer. Tobacco use is the leading cause of lung cancer. A person who quits smoking completely can greatly reduce this risk of lung cancer.

Circulatory System

Tobacco use affects the circulatory or cardiovascular system, which includes the heart and blood vessels. As nicotine enters the circulatory system, blood vessels constrict, or squeeze together. Over time, the blood vessels can harden. When this happens, the blood vessels cannot carry enough oxygen and nutrients to all the parts of the body that need them. When you exercise, your muscles need more oxygen. If blood vessels cannot carry enough oxygen to muscles because they have hardened, it can make exercising harder.

Tobacco use also raises blood pressure and heart rate. Blood vessels narrow and harden due to nicotine and other factors. As a result, the heart has to work harder to move blood, oxygen, and nutrients through the body. When the heart has to work harder and blood vessels are narrower, blood pressure goes up. High blood pressure puts more stress on the heart and blood vessels. This increases the chance of a heart attack, stroke, or heart disease. When you exercise, your heart rate increases. If your resting heart rate is already higher due to tobacco use, your heart rate during exercise could rise to dangerous levels.

Nervous System

Your brain needs oxygen. The carbon monoxide in tobacco smoke can cut down the amount of oxygen that the blood can carry to the brain. Nicotine reaches the brain in only a few seconds and attaches to special receptors in brain cells. The brain then adapts by increasing the number of nicotine receptors. Tobacco users then have a strong need for more tobacco.

Tobacco use has serious consequences. **Identify the number of people in the U.S. who die from tobacco-related illnesses each year.**

A PERSON WHO COMPLETELY QUITS SMOKING CAN GREATLY REDUCE RISK OF LUNG CANCER.

Digestive System

Smoking can damage your digestive system. It can lead to mouth and stomach ulcers, which are painful, open sores. Smoking also harms teeth and gums, causing teeth to yellow. Smokers are more likely to get cavities and gum disease.

Excretory System

Tobacco can also harm your excretory system. Smokers and tobacco users are much more likely to develop bladder cancer than are nonsmokers. Chemicals in tobacco smoke are absorbed from the lungs and get into the blood. From the blood, the chemicals get into the kidneys and bladder. These chemicals damage the kidneys and the cells that line the inside of the bladder and increase the risk of cancer. Smoking tobacco is also a factor in the development of colorectal cancer, a cancer that affects the colon and the rectum.

Reproductive System

Using tobacco during pregnancy can harm a developing fetus. Smoking cigarettes during pregnancy increases the risk of a baby being born too early. It also can cause the baby to have a low birth weight, and/or birth defects. Tobacco use can also harm a fetus's brain and lungs. The CDC has also said some of the flavorings in e-Cigarettes may be harmful to a fetus.

Reading Check

Summarize List two harmful effects of smoking tobacco.

The lungs are part of your respiratory system. The photo on the left shows a healthy lung. The photo on the right shows a cancerous lung. **Name other ways that tobacco use affects the respiratory system.**

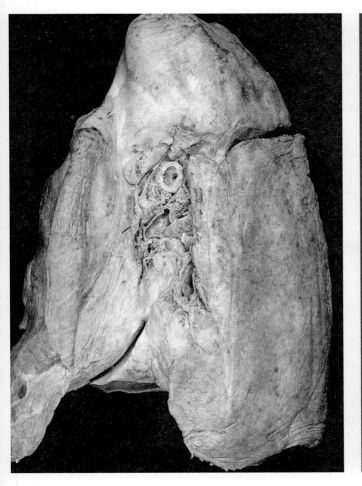

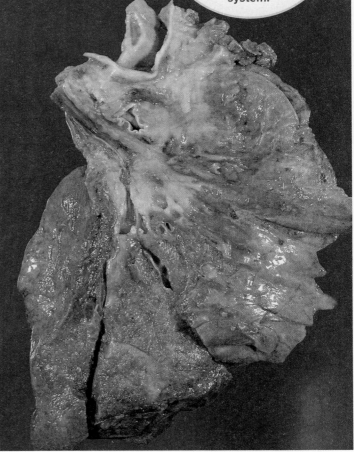

Effects of Tobacco Use

MAIN IDEA Tobacco use causes both short-term and long-term damage to the body.

Tobacco use causes changes in the body. Some of the effects of tobacco use are immediate. These short-term effects can often be felt right away.

1. **Cravings.** Nicotine is a very addictive drug, which means it causes the body to want more of it. A person who uses tobacco may feel a need for more very soon after using it.

2. **Breathing and heart rate.** For a smoker, it becomes harder to breathe during normal physical activity. It is more difficult for a tobacco user to work out for a long period of time. Nicotine also causes the heart to beat faster than normal.

3. **Taste and appetite.** Tobacco use dulls taste buds and reduces appetite. Tobacco users may lose much of their ability to enjoy food. However, when a person quits using tobacco, taste buds will heal.

4. **Unpleasant feelings.** Tobacco users may experience dizziness. Their hands and feet may also feel colder than normal.

5. **Unattractive effects.** Tobacco use causes bad breath, yellowed teeth, and smelly hair, skin, and clothes. It also ages the skin more quickly.

As you have learned, the chemicals in tobacco cause damage to many body systems. The impact of tobacco use is not limited only to smokers and other tobacco users. Simply being around other people who smoke can also cause health problems. Some of the long-term damage caused by tobacco use can even be life-threatening.

- **Bronchitis.** Tobacco smoke can damage the bronchi, or the passages through which air travels to the lungs. Also, a buildup of tar in the lungs can cause a smoker to have fits of uncontrollable coughing.

- **Emphysema.** This disease can make a person use most of his or her energy just to breathe. Emphysema is a common cause of death for smokers.

- **Chronic obstructive pulmonary disease (COPD).** This condition in which passages in the lungs become swollen and irritated, eventually losing their elasticity.

- **Lung cancer.** Nearly 90 percent of lung cancer deaths are caused by smoking.

- **Heart disease.** The nicotine in tobacco greatly increases the risk of heart attack or stroke.

- **Weakened immune system.** Long-term tobacco use harms the body's defenses against various diseases. Tobacco users are also more likely to get common illnesses such as coughs, colds, and allergies.

SIMPLY BEING AROUND SMOKERS CAN CAUSE HEALTH PROBLEMS.

The chemicals in tobacco affect many body systems. **Identify the percentage of lung cancer deaths caused by smoking.**

Reading Check

Compare and Contrast
How are the short-term effects of tobacco use similar and different from the long-term effects?

The Wellness Wheel

The Wellness Wheel shows different areas of wellness that can change as you go through adolescence. You can use the wellness wheel to list changes you may be experiencing. Some of those changes may be related to tobacco use on the left side of the wheel. On the right side of the wheel you can list strategies to use in response to each of these changes. For example, you may experience a change in your social life if your friends start to use tobacco and you do not. On the left side of the wheel you might write "I may not hang out with my friends as much because of their tobacco use." What strategy could you list on the right side of the wellness wheel in response to this change?

The Wellness Wheel shows different areas of wellness that affect your life. **List one way that tobacco use can affect your physical health and financial health.**

What Teens Want to Know

Will smoking make my skin look worse? Smoking cigarettes causes changes in the skin that speed up the aging process. Only sun exposure does more damage to the skin than smoking.

A research study found microscopic wrinkles developing on the skin of smokers as young as 20 years old. In addition, a smoker's skin tends to develop a yellowish or grayish look.

Lesson 2 Review

What I Learned

1. **VOCABULARY** What does *emphysema* mean? Use it in an original sentence.

2. **SUMMARIZE** Describe the ways in which smoking harms the systems in the body.

3. **IDENTIFY** What is the leading cause of death among people who smoke?

Thinking Critically

4. **ANALYZE** Which of the health risks associated with tobacco use do you consider the most serious? Explain your answer.

5. **APPLY** Your friend has been smoking cigarettes and now wants to try out for the track team. How would you explain the effects that smoking may have on his performance during try-outs?

Applying Health Skills

6. **DECISION MAKING** Use the decision-making process to analyze the consequences of tobacco use. Write a one paragraph analysis of the effects of tobacco use on health. Then write out each step of the decision-making process and explain how each step guided your analysis.

Costs of Tobacco Use

. . . .

Before You Read

Quick Write Make a list of habits you know are hard to break. Choose one that interests you and write a paragraph describing what you think it would be like to change that habit.

Vocabulary

psychological dependence
physical dependence
withdrawal
relapse
secondhand smoke
mainstream smoke
side stream smoke
passive smoker

. . . .

Reading Check

Describe Why is it especially risky for teens to try tobacco?

BIG IDEA Tobacco contains nicotine, which is an extremely powerful and addictive drug.

A Powerful Drug

MAIN IDEA Tobacco contains strong substances that make it difficult to stop using once a person has started.

Many people know the dangers of tobacco use, but they continue to use it over many years. As you have learned, tobacco contains nicotine. Nicotine is a powerful drug that causes addiction. Scientific studies have shown that nicotine is as addictive as powerful drugs such as cocaine or heroin. This addiction is both psychological and physical.

Teens who use tobacco are at a greater risk for forming an addiction to tobacco. An addiction to tobacco can be recognized by observing a tobacco users' behavior. Some of the behaviors that indicate a person may be addicted to tobacco use include the following:

- Needing tobacco at certain times of the day, such as after a meal.
- Becoming irritable or restless if tobacco use is not possible.
- Craving tobacco use under certain circumstances, such as when attending a party.

The Path to Addiction

When nicotine enters the body, it interacts with receptors in the tobacco user's brain. The brain sends a message to the body to speed up heart and breathing rates. As heart and breathing rates return to normal, the user wants more. Tobacco use soon becomes a habit, and the user can quickly become addicted.

Studies have shown that as many as 90 percent of adult smokers began using tobacco before the age of 18. The teen brain is not fully developed. This makes the teen brain much more likely to become addicted to substances such as nicotine. Teens are also more likely to develop a severe level of addiction than people who begin smoking at a later age. Teens who use tobacco are also much more likely to use other drugs. For example, a recent national survey shows that more than 90 percent of cocaine users smoked cigarettes before they started using cocaine. Another study has shown that nicotine addiction may lead to other addictions.

Tobacco use leads to nicotine addiction. Once a person is addicted to the nicotine in tobacco, it becomes very difficult to quit. Here are some facts about how people become addicted to tobacco use:

- The government has found that tobacco companies' market to young people. Some people start using tobacco as early as age 11 or 12.
- Research has shown that every day in the United States, more than 6,000 teens and preteens try their first cigarette or other form of tobacco. However, only 7 percent of teens say they use tobacco daily. Most teens do not use tobacco.
- Teens can feel symptoms of nicotine addiction only days or weeks after they first start using tobacco. The symptoms of addiction are felt even before teens start to use tobacco regularly. This can be especially harmful since the teen years are a time of rapid growth and development.
- The earlier in life someone tries tobacco, the higher the chances that person will become a regular tobacco user. Early tobacco use also lowers the chances a person will ever be able to quit.

Reading Check

Justify Why should a person collaborate with others to end an addiction to tobacco?

EARLY TOBACCO USE LOWERS THE CHANCES A PERSON WILL EVER BE ABLE TO QUIT.

The percentage of U.S. teens who use tobacco daily is about 7 percent. **Identify how long it can take for teens to feel the symptoms of nicotine addiction.**

Physical and Psychological Dependence

Psychological dependence occurs when the desire for tobacco becomes greater than the fear of its dangers. It is a person's belief that they need a drug to feel good or function normally. **Physical dependence** is an addiction in which the body develops a chemical need for a drug. It can happen quickly for a person who uses tobacco. Teens can develop a physical dependence for nicotine even more easily than adults.

As you have learned, nicotine is a drug. As with any drug, the body will develop a tolerance. This grows over time and causes a tobacco user to crave nicotine. Anyone who quits tobacco goes through **withdrawal**. It's a series of painful physical and mental symptoms that a person experiences when they stop using an addictive substance. Cravings for nicotine typically increase during withdrawal. Someone who is trying to quit may also have mood changes, feel nervous or irritable, or be extra hungry.

The body undergoes physical changes when a person no longer uses tobacco. For example, the nicotine and carbon monoxide from tobacco are replaced by more oxygen in the blood. The extra oxygen is healthier for the body, but it may affect the brain and result in headaches or dizziness for a while. The extra oxygen in the blood may also cause a person who has quit to feel tingling in the fingers or toes. Since nicotine acts as a stimulant, someone who has stopped using tobacco may suddenly feel extra tired or sleepy.

Sometimes, the symptoms of withdrawal are so bad that a person starts using tobacco again. Many people who stop smoking have a **relapse**, or a return to the use of a drug after attempting to stop. Relapses are common during the first few months after a tobacco user quits.

The physical and psychological effects of withdrawal may lead to a person to return to nicotine. However, if someone starts smoking again after working hard to quit tobacco, it can leave that person frustrated and angry. As a result, many people try to quit several times before they finally break the habit.

Very few people can quit using tobacco on their own. Remember that nicotine is addictive. A person who wishes to quit using tobacco may need help from a doctor or an addiction specialist.

The doctor or addiction specialist might give the person medicines that help end the addition. These include gums, patches, and other medicines. The first step is to talk to a doctor or someone who specializes in addiction recovery. The person wishing to quit using tobacco can collaborate with a doctor or addiction specialist. Together they can select the method that will work best for that person.

The best way to prevent tobacco addiction is to never start using tobacco. **Explain why it is difficult to stop smoking once you start.**

Tobacco's Other Costs

MAIN IDEA Tobacco use is costly to society.

You have learned about how harmful tobacco use can be to your health. Using tobacco has more than health costs. Every person in society pays a price. People who do not use tobacco may also experience health problems.

Tobacco companies spend more than $23 million each day on marketing. This marketing is designed to encourage people to use tobacco products. That advertising affects all U.S. taxpayers, even if they don't use tobacco products. Tax money is used to pay for the healthcare costs of smokers who don't have health insurance. Many other costs are associated with tobacco use.

Costs to Smokers

Tobacco use is expensive. Researches have determined that the average cigarette smoker uses one-and-a-half packs each day. The average price per pack in most states has risen to about $5.50. This means the typical smoker spends more than $8 per day on cigarettes. That adds up to $250 over the course of one month, or $3,000 per year. In 10 years', time, the average smoker in the U.S. will have spent more than $30,000 on cigarettes.

Secondhand smoke is very dangerous to everyone's health. **Evaluate why smoke-free restaurants are considered to be healthier.**

Tobacco users also spend more money on healthcare. They pay higher health insurance rates than nonsmokers. Tobacco users can also expect to live shorter lives and have more health problems than people who do not use tobacco. This is especially true for females. Tobacco use shortens a woman's life by an average of five years. Fires are another health risk posed by smoking. Tobacco left burning is often the cause of fires at home. Carelessly discarded cigarettes or matches can also spark wildfires.

Costs to Nonsmokers

Even if you do not smoke, being around those who do can be harmful. Whenever people smoke near you, you breathe their secondhand smoke, or environmental tobacco smoke (ETS). **Secondhand smoke** is filled with nicotine, carbon monoxide, and other harmful substances.

Secondhand smoke comes in two forms: **mainstream smoke** and **side stream smoke**. Mainstream smoke is the smoke that is inhaled and then exhaled by a smoker. Side stream smoke is the smoke that comes from the burning end of a cigarette, pipe, or cigar. The U.S. Environmental Protection Agency (EPA) has labeled secondhand smoke as a human carcinogen. This means it causes cancer.

A nonsmoker who breathes in secondhand smoke is called a **passive smoker**. Passive smokers are also at risk for disease. In fact, they have the same health problems as smokers. About 46,000 nonsmokers die each year from heart disease. About 3,000 more die of lung cancer every year.

Secondhand smoke also affects children and people with asthma. Children who breathe secondhand smoke are more likely to have respiratory and other health problems. These include allergies, asthma, ear infections, bronchitis, pneumonia, and heart disease.

Costs to Unborn Children

Unborn children can also be affected by secondhand smoke. They may be born too early, have a low birth weight, and can die from exposure to tobacco smoke. One disorder called Sudden Infant Death Syndrome (SIDS) has been linked to tobacco use by mothers while pregnant.

Costs to Society

Tobacco not only harms the body, it costs the user and society a lot of money. Smoking and other forms of tobacco use have hidden costs. Two additional costs involved in tobacco use are:

- **Lost productivity.** Productivity is how much a person is able to finish in the time he or she works. People who use tobacco have lower productivity levels on the job. They are sick more often than nonsmokers and get less done. Lost productivity costs businesses a lot of money. The nation as a whole pays a large price too. The government estimates that smoking costs the U.S. economy $96 billion per year in lost productivity.

- **Health care costs.** People who use tobacco tend to need more medical treatment than those who do not. If tobacco users have health insurance, it may help pay for some of their treatment. However, because health insurance companies face more costs to cover tobacco users, they charge higher rates for their insurance. If a tobacco user has no health insurance, the government helps cover the costs. This means that every U.S. family pays for tobacco use through their taxes.

Secondhand smoke contains many of the harmful substances that smokers inhale. **Name two substances contained in secondhand smoke.**

Reading Check

Identify Name three groups of people affected by tobacco use.

Reducing the Costs of Tobacco Use

MAIN IDEA Laws and education protect nonsmokers and lower the cost of tobacco use to society.

The CDC suggests that each pack of cigarettes sold ads more than $10 in health care and lost productivity costs. Some public action groups and Congress have investigated ways to lower these costs. New laws would restrict or even ban the manufacture and sale of tobacco products. Many states have bans that prevent people under the age of 21 from buying or using tobacco products. To learn the laws in your state, go to www.tobacco21.org and search for the *state-by-state* page. Click on your state to learn about tobacco laws.

Tobacco Taxes

Adding taxes to the sale of tobacco products has also reduced the cost of tobacco use to society. Taxes are now added to each pack of cigarettes sold in the United States. Some states add taxes too. These taxes make it more costly to buy tobacco products. They also give the government more money to educate people about the dangers of tobacco use.

Smoke-Free Environments

At the state and local levels, public action groups have worked to lower the cost to nonsmokers. One method is to enact laws that ban smoking in public spaces. These laws protect nonsmokers from secondhand smoke.

Most businesses and restaurants ban indoor smoking. Some also restrict outdoor smoking. Some may set up a smoking area to keep smokers away from the front of the building. Nonsmokers can then enter a building without having to walk through smoke or smell smoke while dining outside. Some towns and cities have made it illegal to smoke in certain outdoor locations where children and young people gather. Laws have been passed that restrict tobacco use on beaches, in playgrounds, and public gardens. These laws protect nonsmokers and children from the effects of secondhand smoke. The Federal government has also passed laws to protect the rights of nonsmokers. Since 1989, it has been illegal to smoke on all airplane flights in the United States.

No Smoking areas protect nonsmokers from secondhand smoke. **Name two locations that are typically no smoking areas.**

EVEN IF YOU DON'T SMOKE, BEING AROUND THOSE WHO DO CAN BE HARMFUL.

Federal law requires that cigarette packages have one of four different warning labels. **Describe the goal of tobacco warning labels.**

Labeling Laws and Advertising Limits

Other laws control how tobacco companies sell tobacco products. Cigarette packages are required to have clear warning labels, or disclaimers. These labels say that using the product is harmful to good health. Cans and pouches of smokeless tobacco must also display these warnings. The same laws apply to advertisements for tobacco.

Limits on tobacco advertising are stronger than ever before. In the United States, these laws protect young people from being influenced by tobacco advertisements. One example is where outdoor tobacco ads can be placed. Tobacco companies are restricted from placing outdoor ads within 1,000 feet of schools and playgrounds. They are banned from advertising on television or radio. They cannot give away or sell promotional hats, T-shirts, and other items. This is why tobacco companies pay to place their products in the media. It's also why they pay celebrities to be shown using their products. These advertising tactics help them to work around the laws.

Antismoking Campaigns

Anti-tobacco ads help to remind people of the dangers of tobacco use. The ads support that tobacco use is a risk behavior. The ads also explain that many negative health effects of tobacco use.

Smokers who see anti-tobacco ads may are reminded of the health dangers of tobacco use. A smoker may quit or seek treatment. The ads can help nonsmokers recognize the benefits of remaining tobacco free.

SURGEON GENERAL'S WARNING: Smoking Causes Lung Cancer, Heart Disease, Emphysema, And May Complicate Pregnancy.

Smoke free areas in the community help promote healthy lifestyles. **Describe how no smoking signs placed throughout the community can help promote healthy lifestyles.**

Reading Check

Summarize What actions have been taken to lower the cost of tobacco use to society?

State and local governments are also working to stop tobacco use by teens. They enforce laws that prevent selling tobacco to anyone under the age of 21. They prohibit smoking in schools.

States have also sued tobacco companies to recover costs to the public related to tobacco use. Some of the money from these lawsuits has helped fund anti-tobacco campaigns. Many communities promote healthy lifestyles.

Lesson 3 Review

What I Learned

1. **VOCABULARY** What does the term a *withdrawal* mean? Use it in a complete sentence.

2. **EXPLAIN** What is the difference between physical dependence and psychological dependence.

3. **IDENTIFY** What health care costs are involved with tobacco use?

Thinking Critically

4. **EXPLAIN** What solutions have been offered for the problems of tobacco use? Explain how these solutions seem to help lower the costs of tobacco use to society.

5. **ANALYZE** In 2018, about 7 out of every 100 (7 percent) middle school students were currently using tobacco products. That means that 93 percent of middle school students were not using tobacco products. Write one paragraph explaining why the majority of middle school students are choosing not to use tobacco products.

Applying Health Skills

6. **REFUSAL SKILLS** Some teens try tobacco for the first time because of peer pressure. With a small group, brainstorm effective ways to say no when peers offer or suggest that you use tobacco. Make a list of the best ideas and share them with your class.

Saying No to Tobacco Use

Before You Read

Quick Write Write a short dialogue between yourself and a peer who wants you to try tobacco. Show how you can politely but firmly refuse.

Vocabulary

target audience
product placement
point of sale promotion
cold turkey
nicotine replacement therapies
(NRT)

Reading Check

Describe Explain how teens are influenced by their peers to use tobacco.

BIG IDEA It is important to have strategies to resist the strong influences around tobacco use.

Why Do Teens Use Tobacco?

MAIN IDEA Many sources can influence teens to try tobacco.

Teens are paying attention to the information about the effects of tobacco use. The CDC reports that about 25 percent of all teens in the U.S. have tried cigarettes. The percentage of teens who use tobacco regularly is falling. However, the CDC also notes that the percentage to teens who vape is about 48 percent. The appeal of vaping is linked to peer and media influences. In this lesson, you will explore how peers and the media can influence you to use tobacco and vape pens.

Influences of Peers and Family

Teens may be influenced by peers to try tobacco or vaping. A teen may feel pressured to try a vape pen or cigarette if a friend uses them. If a friend uses it, it can't be bad, right? As you read earlier, teens can swiftly become addicted to nicotine. Vape pens offer a concentrated form of nicotine that can addict a teen even more quickly than cigarettes.

Some teens become addicted because they want to fit in with a group. They try a vape pen or smoke a cigarette a few times to be accepted by a group and become addicted. Others think that vaping or smoking make them seem more mature. This might make them feel more confident around other teens.

Teens who live in homes where parents vape or smoke are also more likely to try it themselves. A parent who uses nicotine may warn a teen about the dangers of using it. The teen may wonder, if it's so bad why do you use it? Most likely the adult is addicted to nicotine and cannot break free of the addiction. Having a sibling who smokes may also encourage tobacco use. Some teens try vaping and tobacco because they want to rebel against authority and become addicted to nicotine. Others are curious or think the health risks won't affect them.

Teens whose parents smoke are at higher risk of becoming smokers too. **Explain how living with a parent who smokes can influence you to use tobacco products.**

Influence of the Media and Advertising

Another negative influence on teen vaping and smoking is the media. Even though the media is prohibited from selling cigarettes to anyone under 18, the same is not true about vaping. Many teens who buy fruit-flavored vape cartridges do not know that one ingredient in the cartridge is nicotine. Some teens have gone from taking one hit a day on a vape pen, to using an entire cartridge every day. They have become addicted to the nicotine in the vape cartridge.

Advertisers also make smoking and vaping look attractive. Television shows, movies, and video games often show characters having fun while smoking or vaping. An estimated one-third of popular movies made for children and teens still show images of people smoking. Tobacco companies, and now vape companies, spend millions of dollars to advertise their products. They use colorful ads show happy, attractive people smoking.

Nine out of ten smokers start smoking by age 18. Eighty percent of underage smokers use the three most-advertised brands. Tobacco companies see teens as a **target audience,** or a group of people for which a product is intended.

One key strategy that tobacco companies use is **product placement.** It's a paid arrangement a company has made to show its products in media such as television or film. If you see a favorite celebrity smoking,

> **9 OUT OF 10 SMOKERS START SMOKING BEFORE AGE 18.**

> The CDC tracks current attitudes on tobacco use. **Describe how the effects of vaping are different from smoking?**

ATTITUDES ABOUT SMOKING

	Agree	Disagree	No Opinion or Do Not Know
Seeing someone smoke turns me off.	67%	22%	10%
I would only date people who don't smoke.	86%	8%	6%
It is safe to smoke for only a year or two.	7%	92%	1%
Smoking can help you when you're bored.	7%	92%	1%
Smoking helps reduce stress.	21%	78%	3%
Smoking helps keep your weight down.	18%	80%	2%
Chewing tobacco and snuff cause cancer.	95%	2%	3%
I strongly dislike being around smokers.	65%	22%	13%

Source: Centers for Disease Control and Prevention.

it could have a strong influence on you. This would be especially true if you did not know how harmful tobacco products are to your health. Tobacco companies also use another strategy called a **point-of-sale promotion**. It's an advertising campaign in which a product is promoted at a store's checkout counter. It's designed to get shoppers' attention as they wait to pay for other items.

Resisting Tobacco Use

MAIN IDEA Knowing how to resist tobacco will help you stay tobacco free.

You can protect your health now and in the future if you make a commitment to stay tobacco free. If you make that choice, you will enjoy better physical, mental/emotional, and social health. Ninety percent of adult smokers start smoking before age 18. If you avoid tobacco use as a teen, you will greatly decrease the chance that you will smoke as an adult.

Reasons to Say No

Being tobacco free is a safe behavior that includes many benefits. A healthier body is just one of those benefits. Read the list below to learn more.

- **Overall health**. People who smoke get sick more easily and more often than nonsmokers.
- **Clear, healthy skin**. If you use tobacco, your skin cells are less able to take in oxygen and other nutrients.
- **Fresh breath**. Cigarettes and smokeless tobacco products cause bad breath.
- **Clean, fresh-smelling clothes and hair**. Smokers usually smell like smoke. Stinky cigarette odors stick to clothes and hair and are not easy cleaned.

CIGARETTE ORDORS STICK TO CLOTHES AND HAIR AND ARE NOT EASILY CLEANED.

Friends can influence the choices you make. **Identify a strategy that you can use to refuse tobacco.**

- **Better sports performance**. Nonsmokers are usually better athletes than smokers because they can breathe more deeply and are healthy overall.
- **Money savings**. Tobacco is expensive. Increased taxes on tobacco mean the costs will keep going up. Teens who do not buy tobacco have more to spend on other items such as clothes and music.
- **Environmental health**. By staying tobacco free, you help reduce secondhand smoke. You are also protecting the people around you.

Ways to Say No

You take responsibility for your health when you choose not to use tobacco. Choose to spend time with others who are tobacco free. Be prepared to be asked if you want to try to tobacco. Practice your refusal skills to help you make the best decision. Try using the **S.T.O.P.** strategy: **S**ay "no" firmly. **T**ell why not. **O**ffer another idea. **P**romptly leave. You can find some examples of reasons you can give in the figure below. You can practice saying no in an assertive style that shows you are serious but also shows you respect others. Speak in a firm voice with your head up. This will tell others you mean what you say.

Your Rights as a Nonsmoker

You have the right to breathe air that is free of tobacco smoke. Many laws are in place to protect nonsmokers. It has become easier to find smoke-free places. As a nonsmoker, you can ask people not to smoke around you.

Reading Check

Give Examples Identify three ways you can say no to tobacco.

Most teens know someone who vapes or smokes. Here are some ways to say no. **List other ways to say no.**

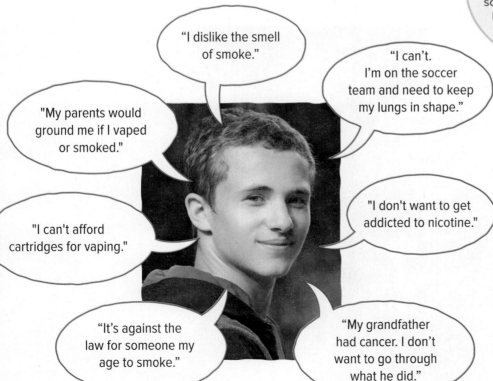

Breaking the Tobacco Habit

MAIN IDEA Many resources are available to people who want to be tobacco free.

Quitting tobacco use at any age will greatly benefit a person's health. The likelihood of developing cancers, heart disease, and respiratory disease will be greatly reduced. The benefits of quitting tobacco use are increased if a person quits while young. The process of quitting can be difficult, but there are many resources to help smokers stop using tobacco. The body goes through physical changes when a person no longer uses tobacco. Learning to live without tobacco takes time and a lot of willpower. Tobacco users often try a variety of methods to try to quit before they find one that works for them.

Some people may choose to stop by going **cold turkey**. The cold turkey method means stopping all use of tobacco products immediately. It's a method that can be difficult for people because they need help breaking the addiction to nicotine. They will experience withdrawal symptoms for up to six months. One source of help is **nicotine replacement therapies (NRT)**. These are products that assist a person in breaking a tobacco habit. They include nicotine gums, lozenges, and patches worn on the skin.

Many organizations also help users quit. For example, tobacco users can find tips and support groups through the American Lung Association, the American Heart Association, or the American Cancer

Reading Check

Define What are nicotine replacement therapies?

Quitting tobacco can protect health. **Name two methods used to stop using tobacco.**

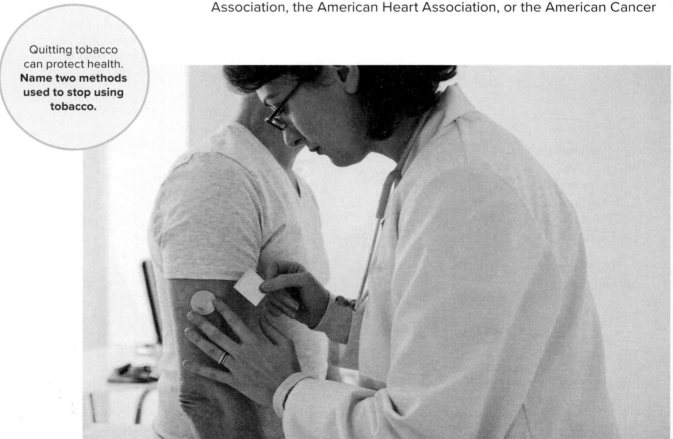

Society. Some schools now have programs to help teens who want to quit using tobacco. If you know someone who is trying to get rid of a tobacco habit, you can share the following information:

- List your reasons.
- Get support and encouragement from family or friends.
- Set small goals.
- Choose tobacco-free places to spend time.
- Change your tobacco-related habits.
- Be physically active.
- Keep trying.

In addition to the types of health organizations noted above, other good sources of information for people who want to quit include hospitals, web sites, and libraries. Doctors and nurses can also be helpful in helping people quit tobacco use. Medical professionals can offer advice about ways to deal with the symptoms of nicotine withdrawal. Doctors may also prescribe certain NRTs for those unable to quit on their own.

Fitness Zone

Fitness Calendar To stay healthy and tobacco free, it helps me to have an activity calendar. I can set goals for how much physical activity I want to get each day. When I choose new activities to try, it makes it more interesting. I also like to use an online diary to keep track of how well I am doing with my goal.

> **QUITTING TOBACCO USE AT ANY AGE WILL GREATLY BENEFIT A PERSON'S HEALTH.**

Lesson 4 Review

What I Learned

1. **VOCABULARY** Explain what the term *cold turkey* means. Use it in a complete sentence.

2. **LIST** What are two ways that advertisers target teens with vape products and tobacco?

3. **SUMMARIZE** How can people who wish to end a nicotine addiction get help.

Thinking Critically

4. **INFER** Why is it easier to say no to nicotine use than to quit once you are addicted?

5. **APPLY** How would you influence a peer to quit vaping or smoking?

Applying Health Skills

6. **GOAL SETTING** Make plan to help a friend end an addiction to nicotine. Do research to find available resources. Include a list of activities that the friend can do to replace the need for a vape hit or to smoke.

Alcohol

LESSONS

. . . .

1 Alcohol Use and Teens

2 Effects of Alcohol Use

3 Alcoholism and Alcohol Abuse

14

Alcohol Use and Teens

Before You Read

Quick Write Write a few sentences describing what you already know about alcohol use.

Vocabulary
alcohol
drug
depressant
inhibitions
binge drinking
minor
driving while intoxicated (DWI)

BIG IDEA Most teens do not use alcohol, but several factors influence teens to try it.

What is Alcohol?

MAIN IDEA Alcohol is a drug that affects the mind and body.

Have you ever seen spoiled food that has mold growing on it? This change is caused by a chemical reaction. A similar change leads to the creation of **alcohol**. It's a drug created by a chemical reaction in some foods, especially fruits and grains. The type of alcohol in beer, wine, and liquor is one of the most widely used and abused drugs in the United States.

Over time, using too much alcohol can damage body organs and cause disease. Alcohol also affects the brain and central nervous system, causing changes in behavior. Alcohol is a **drug** that acts as a **depressant**. A drug is a substance other than food that changes the structure or function of the body or mind. A depressant is a drug that slows down the body's functions and reactions, including heart and breathing rates. These physical changes may make it difficult to think and act responsibly.

Not all alcohol use is bad. For adults, a small amount of wine each day may help keep the cardiovascular system healthy. However, that amount is limited to one 5 oz. glass of wine per day for an adult female and two glasses for an adult male. For some people, though, even small amounts of alcohol can affect how they feel and behave.

Alcohol can cause people to lose their **inhibitions**. These are conscious or unconscious restraints on behaviors or actions. Some people may act in ways that are not typical for that person. Some people become relaxed and friendly. Others become depressed and angry. Under the influence of alcohol, people say and do things they will later regret.

> **OVER TIME, USING TOO MUCH ALCOHOL CAN DAMAGE BODY ORGANS AND CAUSE DISEASE.**

Why Do Some Teens Use Alcohol?

MAIN IDEA Teens face many influences that encourage them to try alcohol.

Studies show that most teens do not use alcohol. A 2018 study shows that only 8 percent of teens in grade 8 have tried alcohol. Why do some teens try it, even when they know alcohol is harmful to their health and also illegal? Curiosity is one reason a teen may try alcohol. Another reason is that they think it will make them more popular.

Some teens think alcohol use makes them feel relaxed or more grown up. Others use it to feel some relief from emotions that they have not yet learned how to handle. Some teens may believe that using alcohol will help them cope with problems. However, using alcohol is not a healthy way to cope with problems. It can stop people from developing healthy coping skills. Healthy coping skills include talking to a trusted adult, using stress management skills, and/or using refusal skills. Using alcohol to cope with problems can also lead to addiction to alcohol.

Alcohol in the Media

Television commercials, movies, Internet sites, and online ads often make using alcohol seem fun and exciting. You have probably seen an ad for some type of alcoholic drink. The people who appear in the ads look young and attractive. Companies that make and sell alcohol do this on purpose. They don't want people to see or think about the negative effects of their products. Media images may lead some teens to feel that drinking alcohol is okay. Some teens may also think that using alcohol will add more fun and excitement to their lives, like the people in the ads.

Reading Check

Characterize What are some typical reasons that a teen may try alcohol?

Media ads are designed to make products look fun and exciting. **Name the ways ads could encourage teens to try alcohol?**

Only about 8 percent of eighth graders have ever tried alcohol. **Name two reasons why some teens may try alcohol.**

Peer Pressure

"I want to fit in, too." many teens think when trying alcohol for the first time. Teens may be influenced to try alcohol because of negative peer pressure. Some may choose to use alcohol to try to fit in or to not be embarrassed in front of their friends. However, having even one drink can be harmful to your health. Simply trying alcohol can lead to risky situations. For example, sometimes teens may dare one another to consume a lot of alcohol very quickly. This type of **binge drinking**, or having several drinks in a short period of time, is very dangerous and can even cause death. It is not always easy to say no, but negative peer pressure is not a good reason to choose alcohol.

Reasons Not to Drink

MAIN IDEA The negative effects of alcohol use pose even greater risks for teens.

Teens have many reasons to avoid alcohol use. The use of alcohol can have harmful effects. This is especially true for teens. A teen's body and mind are still growing and developing. Alcohol use can impair brain development. The USDA recommends that adults who choose to use alcohol consume it only in moderation. This is defined as no more than one drink per day for an adult female and no more than two for an adult male. However, some people should never consume alcohol. According to the CDC, those who should not use alcohol at all include people who are:

- **Minors**, or people under the age of adult rights and responsibilities.
- Pregnant or women who are trying to become pregnant.
- Taking medications that can be harmful when they are mixed with alcohol.
- Recovering from alcoholism or unable to control the amount they drink.
- Facing any type of medical condition that can be made worse by alcohol use.
- Driving, planning to drive, or engaging in any other activity that requires skill, coordination, and alertness.

Alcohol use may also affect a teen's emotional development. Some people believe that alcohol use helps them make new friends. The alcohol may make a person behave in ways that are not normal for them.

Teens who consume alcohol also risk trouble with the law. Each state has laws related to teens and alcohol use. To find the laws in your state, go to the website for your state legislature. Use the search feature to find any laws related to teens and alcohol use.

> ALCOHOL USE CAN IMPAIR BRAIN DEVELOPMENT.

Reading Check

Define What is a minor? How does this term relate to alcohol?

Many teens are making the decision to stay alcohol free. **Explain how abstaining from alcohol can improve your health.**

Developing Good Character

Respect Choosing to be alcohol free shows that you respect yourself. Create, sign, and date a pledge listing your reasons for choosing to avoid alcohol. Remember that this pledge is a promise to yourself to make a healthful choice.

Identify a way you could encourage your peers to also make healthful choices regarding alcohol.

Buying or using alcohol can lead to an arrest, a fine, or time in a youth detention center. A person of any age who is convicted of **driving while intoxicated (DWI)** risks losing his or her license. DWI occurs when a person is driving a car with a blood alcohol concentration (BAC) of 0.08% or greater. It is also called driving under the influence (DUI). If a person under the influence causes an auto accident or injures another person, the driver may face more serious consequences.

Choosing not to use alcohol is a healthful decision. It shows you understand how risky alcohol use can be. As you have learned, some teens may believe that using alcohol will help them fit in with their peers. In reality, most teens do not use alcohol. By choosing not to drink, you will already be fitting in with most people your age. Many teens realize the negative effects that alcohol can have on all sides of their health triangle and are saying no to alcohol use.

Lesson 1 Review

What I Learned

1. **VOCABULARY** Define *alcohol*. Use it in a sentence.

2. **STATE** What are two reasons not to drink alcohol?

3. **EXPLAIN** Why should a person not use alcohol to cope with problems?

Thinking Critically

4. **APPLY** What do you think is the most important reason for a teen not to use alcohol?

5. **PREDICT** How can using alcohol affect a teen's development?

6. **APPLY** In 2018, only eight percent of 8th graders reported using alcohol in the past month. That means that 92 percent of 8th graders did not use alcohol in the past month. Write one paragraph explaining why the majority of 8th graders are choosing not to use alcohol?

Applying Health Skills

7. **GOAL SETTING** Think about personal goals you have, such as going to college or the kind of job you would like to have some day. Write one or two of these on a sheet of paper. Leave space under each one. Use that space to explain how alcohol use could prevent you from reaching your goals.

Effects of Alcohol Use

Before You Read

Quick Write Write a few sentences describing what you already know about the harmful effects of alcohol.

Vocabulary

intoxicated
blood alcohol concentration (BAC)
alcohol poisoning
fatty liver
cirrhosis
reaction time
fetal alcohol syndrome (FAS)

BIG IDEA Alcohol use has far-reaching effects to the body, other people, and personal relationships.

How Alcohol Affects the Body

MAIN IDEA Alcohol has many short- and long-term effects on your body.

Alcohol begins to affect body systems soon after it is consumed. It is quickly absorbed by the bloodstream. Alcohol affects the brain and central nervous system as soon as 30 seconds after it is consumed. Alcohol does not affect everyone in the same way, however. Some people can consume more than others before they become **intoxicated**. It's being physically and mentally impaired by the use of alcohol.

Being intoxicated is known as "being drunk." The person has consumed more alcohol than the body can tolerate. However, the amount of alcohol someone consumes is only one factor in understanding the effects of alcohol use.

How Alcohol's Effects Vary

Different people react to alcohol use in different ways. One of the biggest factors is the user's **blood alcohol concentration (BAC)**. BAC is the amount of alcohol in the blood. An alcohol user's BAC is expressed as a percentage.

A blood alcohol concentration of 0.02 percent will cause most people to feel light-headed. A BAC of 0.08 percent interferes with a person's ability to drive a car safely. Police officers use this percentage to determine whether a person is legally intoxicated. A BAC of 0.40 percent can lead to coma and death. A number of other factors can also influence how alcohol affects an individual. A person's size and gender make a difference. So, does how much alcohol someone consumes and how fast. Because of the effects of alcohol on safety, the U.S. has set the minimum legal drinking age at 21.

Short-Term Effects

MAIN IDEA The use of alcohol has an immediate effect on many parts of the body.

Alcohol has both short-term and long-term effects on the body. Recall that the body is a system of interacting subsystems, such as the circulatory system, the nervous system, and the excretory system. Alcohol affects all of these systems and more. You can see how systems interact with each other when you learn about how alcohol is processed by the body.

The use of alcohol not only causes immediate risks, but it can also cause serious health problems over time. Alcohol can affect the brain, stomach, liver, and kidneys right away. Using a lot of alcohol over time can cause serious damage to these organs. The Centers for Disease Control and Prevention (CDC) provides information on the harmful health effects of alcohol use by teens.

Alcohol and the Brain

Alcohol is absorbed into the bloodstream and reaches the brain very quickly. As a result, the brain and nervous system slow down immediately. Even one drink can make it difficult to think clearly. Alcohol blocks messages trying to get to the brain. After more drinks, it becomes harder to concentrate and remember. Also, it becomes difficult to speak clearly or walk in a straight line. Consuming alcohol before exercising can decrease endurance. It can also increase a person's risk for injury. Alcohol can slow reaction time and decrease coordination. Both of these changes could affect an athlete's performance. A person under the influence of alcohol is also more likely to engage in other risk behaviors such as driving while intoxicated, tobacco and other drug use, sexual activity, and acts of violence.

Brain
Immediate effects: impaired judgment, reasoning, memory, and concentration; slowed reaction time; decreased coordination; slurred speech; distorted vision and hearing; reduced inhibitions; alcohol poisoning, causing unconsciousness and even death

Long-term effects: Brain cell destruction, nervous system disorders, and memory loss

Heart
Immediate effects: Increased heart rate

Long-term effects: Irregular heartbeat, heart-muscle damage

Blood Vessels
Immediate effects: Widened blood vessels, creating a false sense of warmth

Long-term effects: High blood pressure, stroke

Liver
Immediate effects: Processes of the liver, which filters out over 90% of the alcohol in the body may become unbalanced.

Long-term effects: Scarring and destruction of liver tissue and liver cancer, which can both cause death.

Kidneys
Immediate effects: Increased urination, which can result in dehydration, headache, and dizziness.

Long-term effects: Kidney failure resulting from high blood pressure.

Stomach
Immediate effects: Vomiting, which can lead to choking and death

Long-term effects: Ulcers (open sores) in the stomach lining; stomach cancer.

The use of alcohol impairs many body functions. **Describe how alcohol use affects brain function.**

THE USE OF ALCOHOL CAN CAUSE SERIOUS HEALTH PROBLEMS.

Reading Check

Identify Which parts of the body are affected by alcohol?

Alcohol and the Stomach

In the stomach, alcohol increases the flow of acid used for digestion. Some people become sick to their stomach. Most of the alcohol passes into the small intestine. Some, however, is absorbed into the bloodstream and causes the blood vessels to expand. From the bloodstream, alcohol passes into the liver.

Alcohol and the Heart

Alcohol affects the way the heart pumps blood through the body. Because it makes the blood vessels wider, the blood comes closer to the surface of the skin. This makes a person who is consuming alcohol feel warm, but his or her body temperature is actually dropping. Alcohol also slows down a person's heart rate.

Alcohol and the Liver and Kidneys

Short-term use of alcohol also affects the liver and kidneys. The liver acts like a filter, taking alcohol from the bloodstream and removing it from the body. However, the liver can filter only about half an ounce of alcohol from the bloodstream each hour. Any additional alcohol that is consumed stays in the bloodstream and affects the body. In addition, alcohol causes the kidneys to produce more urine. Extra urine production can lead to dehydration, or the loss of important body fluids. When people consume too much alcohol, they often feel more thirsty than usual the next day.

Alcohol Poisoning

If someone consumes a lot of alcohol very quickly, it can lead to alcohol poisoning. Binge drinking is a common cause of **alcohol poisoning**. Alcohol poisoning is a dangerous condition that results when a person drinks excessive amounts of alcohol over a short time period. A person who has too much too quickly may vomit, become unconscious, or have trouble breathing. Alcohol poisoning can result in death.

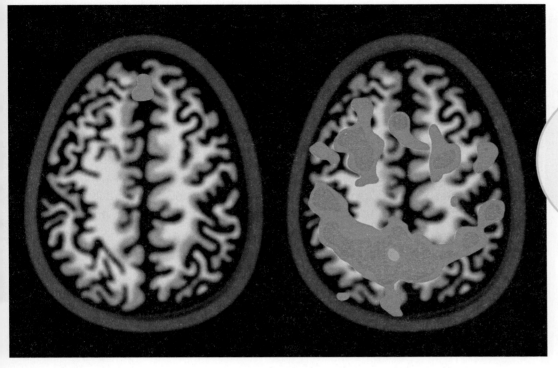

Alcohol affects the activity of the brain, as this CAT scan shows. **Name a possible effect of alcohol on brain activity.**

Long-Term Effects

MAIN IDEA Alcohol use affects all areas of a person's life.

Consuming alcohol regularly can lead to a number of serious health problems. Alcohol use can damage major organs and make existing health problems worse. It can also lead to learning and memory problems. Alcohol can have health effects on the following body systems.

Brain

Alcohol affects the parts of the brain which control memory and problem solving. Alcohol also destroys brain cells. Because brain cells do not grow back, this can be serious enough to limit everyday functions. Alcohol can also block messages sent to the brain. When this happens, people can have a hard time seeing, hearing, or moving.

Heart

Heavy drinking makes the heart weak and enlarged, which leads to high blood pressure. The risk of congestive heart failure and stroke also increases with excessive alcohol use.

Stomach

Alcohol causes your body to create more acid. Stomach acid usually helps with digestion. However, the extra acid created by alcohol consumption can eventually cause an ulcer to develop in the stomach lining. Drinking alcohol also makes the valve between your stomach and esophagus weak. This valve usually works to keep acid in the stomach. When it is weakened by alcohol use, acid comes up and causes heartburn.

Liver

Consuming alcohol regularly over a long period of time puts a serious strain on the liver. One potential risk is **fatty liver**, which is a condition in which fats build up in the liver and cannot be broken down. Heavy drinkers are particularly at risk of developing **cirrhosis**, or the scarring and destruction of liver tissue. This condition can be deadly. Cirrhosis creates scar tissue that prevents blood from flowing normally through the liver. If the liver is not working correctly, it cannot filter out wastes or remove other poisons from the blood. These poisons can eventually reach the brain and cause more damage. Damaged liver cells cannot process some vitamins. These include vitamin D and vitamin B. These low levels of vitamin cause nutrient deficiencies. In an athlete, this can reduce performance.

Other Dangers of Alcohol Use

MAIN IDEA Alcohol use can affect thoughts and behavior.

Because alcohol affects the brain, it also affects thoughts and behaviors. As a result, a person who consumes alcohol can cause arguments, physical fights, and vehicle accidents. The person may engage in risky behavior, such as using illegal drugs or engaging in sexual activity.

Alcohol and Driving

A person who uses alcohol experiences loss of coordination, concentration, and visual awareness. He or she also has slowed **reaction time**, or the ability of the body to respond quickly and appropriately to situations. Driving while intoxicated is extremely dangerous for the driver, his or her passengers, and others on the road. It is very important for your safety not to ride in a vehicle with a driver who has been using alcohol. If a person has been drinking, do your best to avoid letting them drive. You can always call someone else to come pick you up.

Reading Check

Compare and Contrast How are the short- and long-term effects of alcohol abuse and addiction similar and different?

Reading Check

Infer Why does slowed reaction time make driving while intoxicated so dangerous?

About one out of every three fatal car crashes is an alcohol-related crash. **Explain how you can reduce your risk of being involved in an alcohol-related crash.**

Beer
12 oz.

Wine
5 oz.

Liquor
1.5 oz.

Alcohol and Behavior

Using alcohol can also damage your mental/emotional and social health. Teens who use alcohol are more likely to do poorly in school. A teen who uses alcohol may start to lose interest in his or her favorite activities. In addition, that person may risk losing friends as a result of his or her alcohol use. Someone under the influence of alcohol might also engage in other risk behaviors such as tobacco use or sexual activity. Alcohol use can lead a person to make unhealthful decisions.

Alcohol and Pregnancy

If a pregnant woman consumes alcohol, it passes through her bloodstream to her baby. This can lead to what is known as **fetal alcohol syndrome (FAS)**. It's a group of alcohol-related birth defects that include both physical and mental problems. A baby born with fetal alcohol syndrome can have low birth weight and a smaller-than-normal brain. FAS can also cause serious heart and kidney problems. As they grow older, babies who were born with FAS may also develop speech problems and have learning disabilities.

Alcohol-impaired driving is a major cause of traffic accidents. **Describe the way police officers measure whether someone is legally intoxicated.**

Reading Check

Describe Tell about the effects of fetal alcohol syndrome.

Developing Good Character

Being a Responsible Friend One way of showing you are responsible is by looking out for the well-being of others. Don't let a friend get in a car with a driver who has been drinking. If your friend is using alcohol, urge that person to get help. Don't hesitate to talk to an adult if your friend is unwilling to seek help. This is not breaking your friend's trust. It is taking the first step in getting your friend the help he or she needs.

Lesson 2 Review

What I Learned

1. **DESCRIBE** What kinds of long-term damage can alcohol use cause?

2. **RECALL** Describe how alcohol affects the mind.

3. **VOCABULARY** Define *blood alcohol content*. Use the term in a sentence.

Thinking Critically

4. **APPLY** You are at a park with friends. When it is time to leave, a friend's brother offers you a ride. You smell alcohol on his breath. What should you do, and why?

5. **ANALYZE** Why is a person under the influence of alcohol more likely to engage in other high-risk behaviors?

6. **EXPLAIN** What advice do you think the coach of a college sports team would give her players about drinking alcohol during training and the night before matches? Why?

Applying Health Skills

7. **ACCESS INFORMATION** Some teens may believe myths about alcohol. With classmates, research several of these myths. Use your findings to create a poster showing the truth about these concepts.

Alcoholism and Alcohol Abuse

Before You Read

Quick Write Make a list of the reasons you can think of to avoid alcohol use.

Vocabulary

alcoholism
malnutrition
alcohol abuse
substance abuse
enablers
intervention
recovery
detoxification

BIG IDEA Alcohol is a highly addictive drug that can lead to disease and damage relationships.

Alcohol's Addictive Power

MAIN IDEA Alcohol is a powerful drug that can cause addiction.

Alcohol is a powerful drug. It is habit-forming, and can lead to addiction. Teens age 15 and younger are four times more likely to become addicted than older people. That's because the teen brain is not fully formed. Addictive substances have a more powerful effect on the teen brain. A person who uses alcohol in large amounts is even more likely to become addicted. An addiction to any drug can change a person's life. It takes the focus off healthful goals. It can also damage relationships with family and friends.

Alcohol use is a serious health issue. Estimates show that at least 17 million people in the U.S. have an alcohol problem. Alcohol addiction has negative impact on that person's entire life. A person who uses alcohol is unable to maintain relationships with others. That person cannot focus on maintaining healthy relationships. People with an alcohol addiction also have trouble taking care of their physical health. Many alcoholics do not eat healthful meals, so they do not get the nutrients they need. Also, alcohol addiction may lead to trouble with law enforcement. People with alcohol addictions may drive under the influence of alcohol. This can lead to accidents where the alcoholic or others are hurt or killed. The alcoholic may spend time in jail because of the addiction. Alcohol addiction affects all three sides of a user's health triangle—physical, mental/emotional, and social.

How can someone tell if a person has an alcohol problem? A person who is addicted to alcohol frequently uses it alone. He or she often uses alcohol to the point of becoming intoxicated. Using this drug typically becomes more important than anything else in a person's life.

Reading Check

Identify What makes alcohol such a powerful drug?

> ONE OF THE GREATEST DANGERS WITH ALCOHOL USE IS THAT IT IS HABIT FORMING.

The Disease of Alcoholism

MAIN IDEA The disease of alcoholism results from addiction and has physical, mental/emotional, and social consequences.

People who are addicted to alcohol suffer from **alcoholism**. It's a disease in which a person has a physical and psychological need for alcohol. They are called *alcoholics*. Alcoholics typically experience some or all of the following symptoms of alcoholism:

- **Craving** is a strong feeling of need to consume alcohol. This is likely related to alcohol's effects on the brain.
- **Loss of control** means the user is unable to limit his or her alcohol consumption.
- **Tolerance** is when your body needs more and more of a drug to get the same effect. If someone is an alcoholic, he or she will need to consume more and more alcohol in order to feel intoxicated.
- **Physical dependence** can lead to painful symptoms. If an alcoholic stops using alcohol, he or she may experience sweating, shaking, or anxiety.

Making excuses to drink is another common symptom. An alcoholic may be unable to limit how much he or she consumes at one time. Alcoholism may also cause a person to become irritable or violent. This can result in injury or abuse. Alcoholics may hurt themselves or others.

SUPPORT GROUPS WORK TO HELP PEOPLE BREAK THEIR PATTERNS OF ADDICTION.

An alcoholic typically experiences four symptoms of the addiction. **Name the four symptoms of alcoholism.**

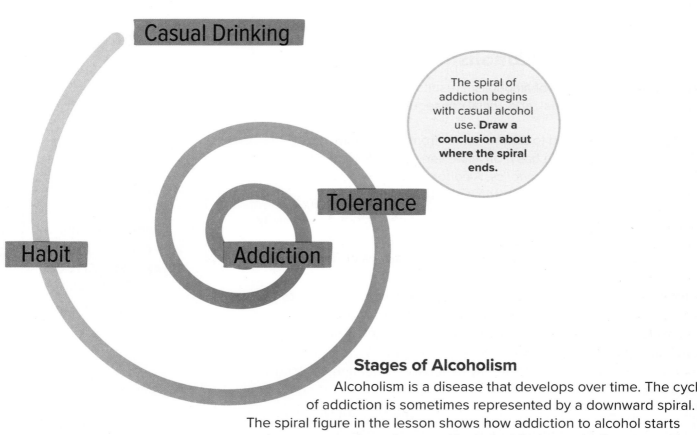

Casual Drinking

Tolerance

Habit

Addiction

The spiral of addiction begins with casual alcohol use. **Draw a conclusion about where the spiral ends.**

Stages of Alcoholism

Alcoholism is a disease that develops over time. The cycle of addiction is sometimes represented by a downward spiral. The spiral figure in the lesson shows how addiction to alcohol starts and progresses along the way. Alcohol addiction typically starts with casual drinking. After a person starts using alcohol casually, it can then become a habit. Once alcohol use is a habit, the body builds up tolerance, which spirals toward the addiction.

Each of the three stages of alcoholism may be either long or short. How long each stage goes on depends on the individual and on how old he or she is when the alcohol use begins. All alcoholics do not go through each stage in the same way.

- **Stage One—Abuse.** The user may have short-term memory loss and blackouts. He or she may also begin to lie or make excuses for drinking. It is also common for the person to begin saying or doing hurtful things to friends and family members.
- **Stage Two—Dependence.** The alcoholic loses control and cannot stop drinking. The person's body begins to depend on the drug. The user can become aggressive, avoid family and friends, or have physical problems. The user typically tries to hide his or her alcohol problem, but is unable to function well at home, school, or work.
- **Stage Three—Addiction.** The person may be intoxicated for long periods of time. The liver may be already damaged. Less alcohol may be needed to cause intoxication. Common at this stage are strange fears, hallucinations, and **malnutrition**. Malnutrition is a condition in which the body doesn't get the nutrients it needs to grow and function properly.

ALCOHOLISM IS A DISEASE THAT DEVELOPS OVER TIME.

How Alcoholism Affects Families

Alcoholism is a problem that affects more people than just the alcoholic. It can be a painful experience for family members as well. Children of alcoholics sometimes blame themselves, thinking they did something to drive a parent to alcohol. This is not the case. A child is never to blame for a parent's alcoholism.

Denial is also a problem for family and friends. Often, they do not want to admit that a loved one has an addiction. Family members may focus on helping the alcoholic and not take care of their own needs.

If the alcoholic is abusive, this can have a negative effect. Friends may try to help by making an alcoholic feel comfortable with his or her behavior. This only encourages the addiction. It can create an unhealthy pattern and keep the alcoholic from getting the help he or she needs.

How Alcoholism Affects Society

Teen alcohol use costs the U.S. more than $50 billion a year. The total cost of alcohol-related problems is estimated at $223.5 billion a year. That figure is higher than the total for smoking and other drug-related issues. The greatest impact is on health care, law enforcement, and the workplace. Doctors and nurses have to take care of people with alcohol problems. Police and the courts must deal with people who break alcohol laws. A business can lose money when an employee who uses alcohol does not work hard on the job.

Alcoholism can cause problems with relationships and within families. **Describe how denial about an alcohol problem can affect family members.**

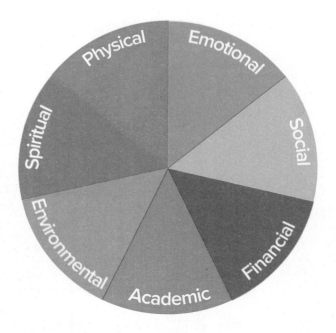

Reading Check

Determine Where does the cycle of addiction usually begin?

The Wellness Wheel

The effects of alcohol use can have a long-term impact on your wellness. Each area of a person's wellness wheel is affected by alcohol use and alcoholism. The person may be affected by personal alcohol use or alcohol use by someone within the home. The Wellness Wheel shows seven different aspects of health. They include physical, emotional, social, financial, academic, environmental, and spiritual. Some effects of alcohol use on each area are:

- **Physical** – Health problems associated with alcohol use can affect most body organs. They can also affect the health of a fetus and infant.
- **Emotional** – Addiction to alcohol can lead to emotional problems. The effects of alcohol use by family members may lead to neglect of children, depression, and harm to self-esteem.
- **Social** – Alcohol use can affect friendships. Friends who do not use alcohol may avoid someone who does. Addiction to alcohol will also affect friendships and other relationships.
- **Financial** – A person who uses alcohol or is addicted to it may use financial resources meant for other purposes to buy more alcohol. They may fail to pay bills. While influenced by alcohol, the person may become the target of theft.
- **Academic** – Alcohol use can impair academic performance. It impairs brain functioning.
- **Environmental** – Environmental impacts of alcohol use can relate to the physical environment in which the person lives. This impact can also be felt in families of an alcohol. The home may feel threatening and unsafe to family members of an alcoholic.
- **Spiritual** – Alcohol use and alcoholism can affect a person's self-image and self-esteem.

Reading Check

Identify Name a behavior that may occur when someone abuses alcohol.

Alcohol Abuse

MAIN IDEA Alcohol abuse is different than alcoholism.

Although the terms *alcoholism* and **alcohol abuse** are sometimes used in the same way, there is a difference. Alcohol abuse is using alcohol in ways that are unhealthy, illegal, or both. People who abuse alcohol are not physically dependent on the drug. Their bodies are not in extreme need for the drug.

Alcohol abuse is a type of **substance abuse**. It means using legal or harmful drugs, including any use of alcohol, while under the legal drinking age. Alcohol abuse has four main symptoms. These include:

- Failing to complete major work tasks or ignoring responsibilities at home or school.
- Drinking in situations that are dangerous. For example, driving when intoxicated or riding with someone who has been drinking can result in an accident and serious injury.
- Having ongoing financial or legal problems related to alcohol use.
- Continuing to drink even after a friend of family member has identified a problem.

Teen alcohol use costs the U.S. more than $50 billion a year. **Name two ways that alcohol use by teens impacts the financial health of local governments.**

Help for Alcohol Abuse

MAIN IDEA Family, friends, and organizations can all help someone with an alcohol problem.

People who have a problem with alcohol need help. They may not be able to admit they have a problem. Sometimes they are helped to continue the behavior by **enablers**. Enablers help an alcoholic continue drinking without facing any negative effects of their substance misuse.

One way to help an alcoholic understand and admit they have a problem is to have an **intervention**. This is a gathering in which family and friends get the problem drinker to agree to seek help. During an intervention, the problem drinker's family and friends talk about their concerns. They try to convince the abuser to stop using alcohol.

Before holding an intervention, meeting with a substance abuse counselor can provide guidance. The counselor can tell the family and friends what to say or not to say. The counselor can also plan for the problem drinker to get treatment.

Ways to Seek Help

Groups such as Alcoholics Anonymous (AA) can help people who are addicted to alcohol. The AA website is https://www.aa.org. A similar group that helps families and friends is Al-Anon https://al-anon.org.

Support groups work to help people break their patterns of addiction. They allow people to talk with others who are facing the same problem. You can search online to learn more about these groups. Another way to get help for people who abuse alcohol is through harm reduction. Harm reduction is a set of strategies to help reduce the costs of alcohol abuse. These strategies can include setting a weekly limit for alcohol intake. Other limits that can be set are not using alcohol when driving, when making important decisions, or when in a person is in an unsafe environment.

The Road to Recovery

Before an alcoholic can recover, they must agree to never to drink again. When that happens, **recovery** can begin. It's the process of learning to live an alcohol-free life. Recovery is usually long and difficult. It involves several steps that each person must follow.

- **Admission.** The person must first admit that they have an addiction and ask for help.
- **Detoxification.** This is the physical process of freeing the body of an addictive substance. It begins breaking the physical addiction. An alcoholic may go through withdrawal when he or she suddenly stops using alcohol.

Reading Check

Explain What makes recovery from alcoholism so difficult?

A recovering alcoholic must always fight the addiction. **Give examples of some healthful choices that would help someone remain alcohol free.**

- **Counseling.** Alcoholics need outside help from counselors and support groups to recover. Local organizations that can help provide a support group.
- **Resolution.** The alcoholic commits to accepting responsibility for their actions. After recovery, people who have had alcohol problems are called recovering alcoholics. A recovering alcoholic is someone who has an addiction to alcohol but chooses to live without alcohol.

Recovery is never final. After a person goes through recovery, that person must always fight addiction. Recovering alcoholics risk a relapse if they drink again.

Help for Families

MAIN IDEA Support groups can help families of alcoholics.

Living with someone who has an alcohol problem or is an alcoholic is difficult. Teens who live with an alcoholic might feel confused and sad. If the alcoholic becomes violent, they may fear getting hurt. They may also fear that a family member will be hurt.

Families and friends of alcoholics can also get help. Al-Anon teaches ways to live with an alcoholic. Alateen (https://al-anon.org/newcomers/teen-corner-alateen/) helps teens who live in a home where the adults are alcoholics. They may offer sessions for teens to share their concerns. They may help families find counseling and mental health resources. Al-Anon and Alateen also help educate the public.

Reading Check

Identify Name one example of a support group for families of alcoholics.

The support group was formed to help families of people who suffer from alcoholism. **What are some support groups for alcoholism in your area?**

Staying Alcohol Free

MAIN IDEA Choosing not to use alcohol is the best way to avoid its dangers.

As you have learned, the use of alcohol has serious physical, mental/emotional, and social consequences. Choosing to be alcohol free is the best way not to experience these dangers. You can try to avoid situations where people are drinking alcohol. If someone pressures you to drink alcohol, use refusal skills. Practice the S.T.O.P. strategy:

- **S**ay no in a firm voice.
- **T**ell why not.
- **O**ffer another idea.
- **P**romptly leave.

If you choose friends who are also alcohol free, you will have a support system. When you are around people who make healthful choices, it is easier for you to make healthful choices. Positive peer pressure can make it more likely that you and your friends will choose activities that do not involve alcohol.

Benefits of Staying Alcohol Free

Staying alcohol free is a choice to lead a healthy lifestyle. When you choose not to use alcohol, you are showing respect for yourself and your body. You are choosing to remain in control of who you are.

Another benefit is being able to focus on your future. An alcohol-free lifestyle allows you to care for your family and friends. Better relationships are a benefit of choosing not to use alcohol.

. . . .

Reading Check

Describe What are benefits of choosing an alcohol-free lifestyle?

. . . .

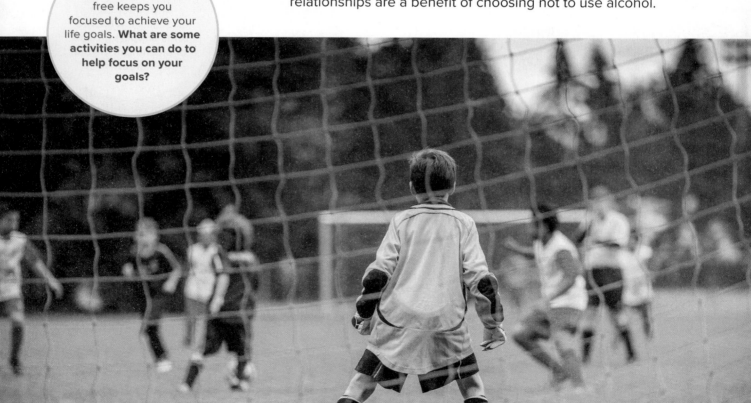

Staying alcohol free keeps you focused to achieve your life goals. **What are some activities you can do to help focus on your goals?**

Healthy Alternatives

When someone offers you alcohol, use your refusal skills as a healthful alternative. Refer to the S.T.O.P. strategy above. If you are offered alcohol, say no and explain why you have chosen not to drink. Offer a suggestion of an alcohol-free activity. If those steps do not work, promptly leave the area.

Finding another way to think or act will also help you avoid alcohol use. Instead of using alcohol, find a healthful way to spend your time. Join a club or sports group at school. Volunteer at a local organization, such as a food bank or animal shelter. Volunteering can give you a sense of purpose and can make you feel good about yourself. Another idea is to start a hobby or business with your friends. Alcohol use will never help you reach your goals, but positive activities such as these can help.

> STAYING ALCOHOL FREE IS A CHOICE TO LEAD A HEALTHY LIFESTYLE.

Reading Check

Explain What makes recovery from alcoholism so difficult?

Fitness Zone

Stress-relieving activities
Some people use alcohol to relieve stress. I think of other ways to deal with my stress. I can write a list of things I like to do that help me get rid of stress and stay healthy: run, play basketball, read a book, have a healthful snack with my mom, or ride my bike.

Lesson 3 Review

What I Learned

1. **VOCABULARY** Define *alcohol abuse*. Use the term in a complete sentence.

2. **EXPLAIN** What is the difference between alcoholism and alcohol abuse?

3. **DESCRIBE** How can a person get help for an alcohol problem?

Thinking Critically

4. **ANALYZE** Layla has started drinking alcohol. Now she is forgetting things and lying to others about her alcohol use. She also has been in more arguments with her friends and family. What stage of alcoholism is Layla likely experiencing?

5. **EVALUATE** Briefly explain why alcohol use is even more dangerous for teens.

6. **HYPOTHESIZE** How might you be affected if one of your close friends or family members developed an alcohol problem? Where could you find help? Explain your answer.

Applying Health Skills

7. **DECISION MAKING** A friend has been irritable and moody lately. He tells you that he really needs alcohol and asks you to help him get some. He thinks only alcohol will make him feel better. Use the decision-making steps to make a responsible choice.

Drugs

LESSONS

1 Drug Use and Abuse

2 Types of Drugs and Their Effects

3 Staying Drug Free

15

Drug Use and Abuse

Before You Read

Quick Write List two legal drugs or types of medicine. Then list two drugs you know are illegal or harmful.

Vocabulary

illegal drugs
drug misuse
drug abuse
drug-free zone
drug possession
drug trafficking
probation

Reading Check

Identify What are two categories of legal drugs?

BIG IDEA Using drugs affects your body, mind, emotions, and social life and can lead to consequences with the law.

What is a Drug?

MAIN IDEA Some drugs can help heal the body, but drugs can also be harmful to your health.

You have likely heard the word *drug* many times before. Some drugs are medicines, which may be able to save your life or the life of someone you love. Other drugs are illegal and dangerous.

Medicines prevent or cure illnesses or treat their symptoms. Some are available as over-the-counter (OTC) medicines, meaning you can find them on the shelf at a grocery store or pharmacy. Some OTC medicines can only be purchased by an adult, but most of them are available to anyone. Other medicines can only be legally obtained with a prescription, or written permission from a doctor.

When a drug is used for a purpose other than preventing or curing disease, it becomes an illegal drug. All types of drugs, including medicines, can be misused or abused. It is important to be careful when using any drug, including medicine prescribed by your doctor. Whether you use an OTC medicine or prescription drug, always closely read and follow the directions on the label. With a prescription medicine, your doctor will include instructions for how much to take, when to take it, and for how long. Some drugs are illegal because they are harmful to your health. However, even medicines that are legal can be dangerous if they are not used correctly. Medicines are only effective when they are used properly.

Why Do Teens Use Drugs?

MAIN IDEA Teens may choose to use drugs because of peer pressure, media, or social media influences.

Peer pressure is one of the leading reasons why teens use drugs. Wanting to fit in and be friends with peers who use drugs may try to influence other teens to use drugs. Using refusal skills can help teens say no to drug use. Teens are also influenced to use drugs through the media. Movies and television that show drug use rarely show the damage that it causes. Television shows and movies rarely show the progression from a casual user to an addiction or death.

As well as traditional media showing incomplete portrayals of drug abuse, social media sites fail to promote the risks associated with drug use. A study by the National Institute of Drug Abuse reported that one social media site includes pages with pro-marijuana messages. More than 70 percent of the followers of this page are under age 19. While some states in the U.S. have legalized marijuana use for adults, it is illegal for anyone under the age of 18 to use the drug. Sites that promote the legal use of drugs, such as marijuana and alcohol, are encouraged to also list the risks of using these substances.

Drug Misuse and Abuse

MAIN IDEA Any drug can be harmful to your health if abused or misused.

Using a medicine in a way that it was not intended makes the medicine an **illegal drug**. These are drugs that are made and used purely for their effects. Some drugs are considered illegal because they can be extremely dangerous or addictive. However, **drug misuse** is also dangerous and can even lead to abusing drugs. It is taking or using medicine in a way that is not intended. The Centers for Disease Control and Prevention (CDC) provides information on the harmful health effects of teens and substance abuse, including alcohol. A person who does any of the following is misusing drugs:

- Using the drug without following instructions on the label.
- Using a drug not prescribed for you.
- Allowing someone else to use a drug prescribed for you.
- Taking more of the drug than the doctor prescribed.
- Using the drug longer than advised by your doctor.

Drug use is associated with many negative consequences. **Explain how the media might influence teens to use drugs.**

Drug abuse is intentionally using drugs in a way that is unhealthful or illegal. It is dangerous to overall health. A person can abuse legal or illegal drugs. When a person uses legal drugs for nonmedical reasons, that person is also abusing drugs. Even prescription medicines can be dangerous if used improperly.

Misusing or abusing any drug can damage your body and lead to addiction, allergic reactions, illness, or even death. Addiction is a serious problem. Some drugs make it difficult to concentrate, or they may cause depression or anxiety. Your social health is also affected. Teens who abuse drugs may withdraw from family and friends and lose interest in school or other activities.

To stop using drugs, a person who is addicted must go through withdrawal. Withdrawal is a difficult and painful process. The person in withdrawal experiences a series of physical and mental/emotional symptoms when the addictive substances are no longer used.

Physical Consequences

Physical effects of drug use include sleeplessness, memory loss, irritability, nausea, heart failure, seizures, or stroke. If a drug user drives while taking drugs, that person is putting others at risk too. Many drug users develop an addiction. The symptoms of addiction can include:

- **Denial** occurs when a person cannot admit that he or she has a problem with drugs.
- **Tolerance** develops when a person uses a drug regularly. The user needs more and more of the drug to get the same effect.
- **Craving** is a primary symptom of addiction. A person will feel a strong need, desire, or urge to use drugs and will feel anxious if he or she cannot use them.
- **Loss of control** causes a person to take more drugs than he or she meant to take. Drug use may also happen at an unplanned time or place.
- **Physical dependence** makes it very difficult to quit using a drug. A person's body develops an actual physical need for drugs in order to function.

Drug misuse can be especially harmful to teens. Recall that the body is a system of interacting subsystems, including the reproductive system, the endocrine system, and the muscular system. Drugs can affect the long-term development of these body systems.

The nervous system is another body system negatively impacted by drug misuse. The teen brain is developing until about age 25 or 26. Drug misuse can permanently damage the brain. Some of this damage may appear as reduced motor skills, or the ability to move the muscles in normal ways. Simple tasks such as writing, speaking, or walking can be affected. This type of damage can often be permanent.

Knowing the facts about illegal drugs can convince you to avoid using them. **Name a way that using illegal drugs can damage the body.**

MISUSING OR ABUSING A DRUG CAN DAMAGE YOUR BODY.

Mental and Emotional Consequences

Drug abuse weakens a person's ability to think and learn, even though the person may not realize it at the time. Some drugs kill brain cells. The brain damage that results can interfere with the user's ability to think. Among teens, drug misuse and abuse is especially serious. The brains of teens are still developing, and misusing or abusing drugs can interfere with brain development.

Some teens may think that using drugs is a way to cope with problems. However, using drugs keeps you from learning to handle difficult emotions in healthful ways. Drug users often experience depression, anxiety, and confusion. These feelings can make an initial problem worse. People may also become dependent on drugs, creating more problems. Drug use also often leads to poor decisions and bad judgment. It can cause a person to engage in other risk behaviors.

Social Consequences

Drug abuse can change someone's personality, cause mood swings, or even lead to violence. Drug users often have low self-esteem and difficulty dealing with others, even those closest to them. A person addicted to drugs will start to think only of his or her need for the drug.

Teens who abuse drugs can also lose their friends. Some may end their friendships or lie to friends in order to cover up their addictions. After a while, obtaining and using the drug becomes more important than maintaining relationships with friends.

Teens who use drugs often miss school or do not learn well because they cannot pay attention. Teens may not be able to participate in school activities if they are caught using drugs. As a result, they lose the opportunity to learn new skills or have interesting new experiences. Teens who abuse drugs often hurt their chances of reaching their long-term goals, such as going to college or having a career.

Reading Check

Describe How can drug use affect a person's emotional health?

In females, drug use can negatively affect:

- height
- weight
- onset of first menstrual cycle
- regularity of periods
- breast development
- function of ovaries
- pregnancy
- the health of unborn babies

In males, drug use can negatively affect:

- height
- weight
- male hormone levels
- testicle size and function
- muscle mass and development
- the age at which the voice gets lower
- the age at which body and facial hair increases

The use of illegal drugs can have serious health consequences during the teen years. **Name two reasons why drug use is especially harmful to teens.**

Today, most schools have a **drug-free zone**. It's a 1,000-yard distance around a school where anyone caught with drugs can be arrested. Anyone caught with illegal substances within a drug-free zone can be arrested. Students caught using drugs or being under the influence of any drugs can be suspended or expelled from school.

Drug Use and the Law

MAIN IDEA Using drugs can lead to problems with the law.

Using drugs for any purpose other than the one the drug is intended is illegal. This includes taking drugs prescribed to other people, or giving prescription drugs to others. Alcohol use by teens is also illegal. Anyone who uses drugs illegally risks getting into legal trouble.

Using drugs can lead to legal problems for the user. Federal and state laws prevent addictive drugs from being used or sold. To learn the laws in your state, go to your state legislature's website. Use the search feature to find any laws related to drug use.

When a person is caught using illegal drugs, the legal consequences are very serious. A person who uses drugs illegally may be arrested for **drug possession**, which is when a person has or keeps illegal drugs. The person can also be arrested for **drug trafficking**, which is the buying or selling of drugs. Teens can spend time in a detention center if arrested or convicted of these crimes. They might also be sentenced to **probation** where they must regularly check in with a court officer. Probation is a set period of time during which a person who has been arrested must check in regularly with a court officer. Often, they and their parents may have to pay fines. Teens can also get a criminal record.

Drug use is also linked to gang activity. Gangs may acquire drugs and force members of the gang to transport the drugs from one location to another. A person who is caught with drugs may face serious legal consequences. Gangs may also require that gang members sell drugs which can also lead to legal consequences, such as jail.

The legal consequences of drug use can also impact a teen's social health. A criminal record can last forever. Such a background can restrict access to certain jobs. A criminal record can also affect relationships with family and friends.

> **USING DRUGS FOR ANY PURPOSE OTHER THAN THE ONE THE DRUG IS INTENDED IS ILLEGAL.**

Reading Check

Explain How can drug abuse lead to crime?

A person who has drugs in a drug-free zone at a school can be arrested. **Describe the difference between drug possession and drug trafficking.**

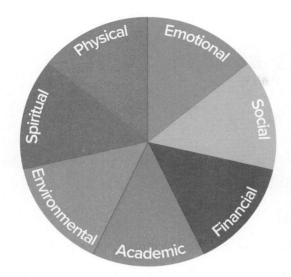

The Wellness Wheel

Different areas of wellness can change as you go through adolescence. Use the wellness wheel to list changes you may be experiencing related to drug use on the left side of the wheel. List strategies to use in response to each of these changes on the right side of the wheel. For example, you may experience a change in your social health if you find out a friend is using drugs and lying to you about it. On the left side of the wheel you might write "My friend said she had never tried any drugs. Then, I found out she had used inhalants at a party." What strategies could you list on the right side of the wellness wheel in response to this change?

. . . .

Developing Good Character

Citizenship Part of being a good citizen is helping to protect the health of others. The Partnership for Drug-Free Kids at drugfree.org or the National Institute on Drug Abuse (NIDA) advocate for the health of others.

Conduct research on organizations like these to find out how you can get involved. Present your findings to the class. Encourage your peers to get involved too.

. . . .

Lesson 1 Review

What I Learned

1. **VOCABULARY** Define *drug misuse*. Give at least one example in a complete sentence.

2. **NAME** What are three physical effects drugs can have on a person's body?

3. **DESCRIBE** How could using illegal drugs negatively affect your social health?

4. **EXPLAIN** How can drug misuse affect a teen's developing brain?

Thinking Critically

5. **ANALYZE** Why is a person who uses drugs more likely to be involved in a crime?

6. **APPLY** Less than seven percent of all middle-school aged children have tried marijuana. Fewer teens have tried other drugs, such as cocaine, methamphetamine, ecstasy, and hallucinogens. Write one paragraph explaining why teens do not use these substances.

Applying Health Skills

7. **ACCESS INFORMATION** Use online or library resources to investigate the legal consequences of drug abuse and abuse in your state. Create a brochure or electronic presentation to educate others about the legal consequences of drug use.

Types of Drugs and Their Effects

Before You Read

Quick Write List two illegal drugs you know about. Then briefly describe how these two drugs are harmful to a person's health.

Vocabulary

marijuana
THC
stimulant
club drugs
methamphetamine
amphetamines
CNS depressants
amnesia
hallucinogens
narcotics
opium
euphoria
inhalants
anabolic steroids

BIG IDEA All types of illegal drugs have both short- and long-term effects.

Marijuana

MAIN IDEA Marijuana is a drug that affects the body.

Marijuana is a drug that is usually smoked. It consists of dried leaves and flowers of the hemp plant, called cannabis sativa. It is also called pot or weed. Some states have legalized the use of recreational marijuana for people over the age of 21. It is illegal for anyone under age 21 to use or possess marijuana in the United States.

Some states allow marijuana to be used as part of a medical treatment plan. For people with diseases like cancer, marijuana can ease nausea. This allows the person to eat enough food to gain nutrients to fight the disease. The users of medical marijuana need a prescription from a doctor to obtain the drug.

Effects of Marijuana

Marijuana affects the brain. A chemical in it changes the way the brain processes what a person sees, feels, hears, and perceives. The main active chemical in marijuana is **THC**.

Marijuana use can cause a variety of reactions in people. Some users may feel a pleasant sensation, but others do not react well to this drug. Marijuana has both short-term and long-term effects on the human body. Some short-term effects are:

- Reduced reaction time.
- Reduced coordination.
- Impaired short-term memory.
- Increased fear, distrust, or panic.

Some long-term effects of marijuana use are:

- Increased heart rate.
- Increased risk of breathing problems.
- Disrupted brain development in people under age 26.

Marijuana is typically smoked. It can have the same effects on the heart and lungs as smoking. Also, the secondhand smoke from marijuana use can create a health risk for others who are not smoking the drug.

Marijuana can also affect a developing fetus. Some studies show that exposure to marijuana can increase the risk of a baby being born with low birth weight. It can also cause problems with a child's development after birth.

Health experts consider marijuana a gateway drug. This means that a frequent user of marijuana is more likely to try other drugs. It is thought that the THC in marijuana can be addictive. However, the majority of people who use marijuana do not try other drugs.

Stimulants, Depressants, and Club Drugs

MAIN IDEA Stimulants, depressants, and club drugs have many negative effects on the body.

A **stimulant** is a drug that speeds up the body's functions. It raises the heart rate, blood pressure, and metabolism. These drugs increase the activity of the central nervous system. A depressant affects the body in the opposite way. These drugs slow down the activity of the central nervous system. **Club drugs** are illegal drugs that are found mostly in nightclubs. They are very dangerous because the user doesn't know what's in the drug.

Reading Check

Explain How does marijuana affect the user?

Stimulants are drugs that increase the activity of the central nervous system. **Describe how depressants affect the central nervous system.**

Stimulants

Stimulants cause the heart to beat faster. Blood pressure and metabolism also rise. Someone who uses a stimulant will often move or speak more quickly than normal. That person may also feel excited or even anxious.

Illegal stimulants include cocaine, crack, and **methamphetamine**, or meth. It's a stimulant similar to **amphetamines**, which are strong stimulant drugs that speed up the nervous system. Some stimulants are legal and not necessarily harmful, such as caffeine found in coffee, tea, soda, and chocolate. The question of whether caffeine is harmful to human health is not clear. The American Heart Association reports that several studies have been done on this question. The results, however, are not clear.

Doctors sometimes prescribe amphetamines to their patients for certain medical conditions. However, stimulant abuse can be dangerous. Stimulants are an extremely addictive drug. The misuse of stimulants can lead to heart attack and death. Athletes that use amphetamines may not realize when they are injured. Continuing to play while injured can lead an even greater injury. In addition to the effects of stimulants on the body, the use of methamphetamine during pregnancy can harm a fetus. There is an increased risk of baby being born too early, or premature. Babies can also be born with low birth weight. They can be born with damage to the heart and brain.

Depressants

In contrast to stimulants, depressants slow down a person's motor skills and coordination. They can affect someone mentally and emotionally by giving a false sense of well-being through feelings of reduced anxiety or relaxation. However, when a depressant wears off, the user may experience mood swings and depression.

Most depressants come in tablet or capsule form. Depressants are legal when prescribed by a doctor to treat certain conditions. For example, doctors sometimes prescribe **CNS depressants**. These are substances that slow down the body's functions, including heart and breathing rates. They are used to treat people who suffer from anxiety or sleep disorders.

Reading Check

Describe What are some dangers of using depressants?

The biggest risk associated with stimulant abuse is damage to your heart, sometimes causing heart attacks or death. **What are some other harmful effects of stimulant abuse?**

Substance	Other Names	Forms	Methods of Use	Harmful Effects
Amphetamine	Crystal, ice, glass, crank, speed, uppers	Pills, powder, chunky crystals	Swallowed, snorted up the nose, smoked, injected	Uneven heartbeat, rise in blood pressure, physical collapse, stroke, heart attack, and death
Methamphetamine	Meth, crank, speed, ice	Pills, powder, crystals	Swallowed, snorted up the nose, smoked, injected	Memory loss, damage to heart and nervous system, seizures, and death
Cocaine	Coke, dust, snow, flake, blow, girl	White powder	Snorted up the nose, injected	Damage to nose lining and liver; heart attack, seizures, stroke, and death
Crack	Crack, freebase rocks, rock	Off-white rocks or chunks	Smoked, injected	Damage to lungs if smoked, seizures, heart attack, and death

Alcohol is also a depressant, and it is illegal for anyone under the age of 21 to use alcohol. Misuse and abuse of depressants, including alcohol, can lead to coma or even death. The risk is even higher when a person combines alcohol with a depressant drug.

Club Drugs

Club drugs are often used to make people feel more relaxed in a social setting. They are often made in home laboratories. Club drugs may be mixed with other drugs or harmful chemicals.

- **Ecstasy** is a synthetic chemical. It increases the heart rate and body temperature, which can damage a person's organs. A person using Ecstasy may experience tingly skin or clenched jaws. He or she can also feel anxious and paranoid.
- **Rohypnol** makes a person's blood pressure drop. The user feels dizzy and very sleepy. The drug also causes blackouts and **amnesia**, or a partial or total loss of memory. Rohypnol typically comes in pill form, although it can be crushed into a powder. It is also a drug that the user may not know that he or she has been given. If added to a drink, for example, it can make a person unconscious. As a result, this drug is unfortunately used to commit the crime of date rape.
- **Ketamine,** an anesthetic used in medical procedures, can be deadly if abused. It causes hallucinations, and people who use it often experience amnesia. An overdose of ketamine can cause a person to stop breathing.
- **GHB** is a depressant that is used to treat a neurological disorder called narcolepsy. It is sometimes used as a date rape drug. It causes drowsiness, nausea, vomiting, unconsciousness, slowed heart rate, slowed breathing, seizures, coma, and death.

. . . .

Reading Check

Compare and Contrast What is the main difference between the way stimulants and depressants affect the body?

. . . .

Hallucinogens

MAIN IDEA Hallucinogens are dangerous drugs that mainly affect the user's mind.

Hallucinogens are drugs that distort moods, thoughts, and senses. They interfere with thought processes and the ability to communicate. Hallucinogen users may become disoriented or confused.

Hallucinogens can cause many harmful effects, including death. **What are two examples of hallucinogens?**

Substance	Other Names	Forms	Methods of Use	Harmful Effects
PCP	Angel dust, supergrass, killer weed, rocket fuel	White powder, liquid	Applied to leafy materials and smoked	Loss of coordination; increased heart rate, blood pressure, and body temperature; convulsions; heart and lung failure; broken blood vessels; bizarre or violent behavior; temporary psychosis; false feeling of having superpowers
LSD	Acid, blotter, microdot, white lightning	Tablets; squares soaked on paper	Eaten or licked	Increased blood pressure, heart rate, and body temperature; chills, nausea, tremors and sleeplessness; unpredictable behavior; flashbacks; false feeling of having superpowers

. . . .

Reading Check

Determine Name three harmful effects of LSD.

. . . .

Strange behavior is common because the user can no longer tell what is real and what is not.

LSD (acid) and PCP (angel dust) are common hallucinogens. LSD is one of the strongest. It may come on tablet form or on absorbent paper. Someone who uses LSD may not know who or where they are. Harmful behaviors are common with LSD use. Users may even have terrifying flashbacks weeks or months after using the drug. The effects of PCP are similar to those of LSD.

Narcotics

MAIN IDEA Narcotics help to relieve the pain of severe disease, but are extremely addictive.

Narcotics are drugs that get rid of pain and dull the senses. They are highly addictive drugs. One type of narcotic are opioids. The CDC released data showing that between 1999 and 2017, deaths from prescription opioids, synthetic narcotics, such as fentanyl, and heroin have increased fivefold. This has raised concern about access to this class of drug and whether the drugs are misused and overused.

Historically, narcotics have been made from **opium**. It's a liquid from the poppy plant containing substances that numb the body. When used under a doctor's supervision, narcotics such as morphine and codeine are effective in treating extreme pain. However, laws control how all narcotics are sold and used because they are so addictive.

Heroin

Heroin is an illegal narcotic made from morphine. It is often inhaled or injected and sometimes smoked. It gives the user a sense of **euphoria**, or a feeling of well-being or elation. However, this feeling quickly wears off, and the user experiences symptoms of withdrawal. These symptoms may include nausea, cramps, and vomiting. People who use heroin risk unconsciousness and death. Since it is usually injected, heroin users also risk HIV or hepatitis infection from shared needles. Because it highly addictive, users commonly experience tolerance and dependence. If heroin is used during pregnancy, there is a risk that the baby will be born prematurely, or with a low birth weight. There is also a risk that the baby will be born dependent on heroin. This means the baby will have to be treated in a hospital until they are not dependent.

> PEOPLE WHO USE HEROIN RISK UNCONSCIOUSNESS AND DEATH.

Prescription Opiates

Prescription opioids are used to treat severe pain. Opioids are only available with a doctor's prescription. When used correctly, opioids can control pain. However, the drug is also highly addictive.

A person who is addicted to opioids needs more and more of the drug as time goes on. Addiction can lead a person to commit crimes to obtain the drug. Some may steal money and drugs from family members and friends. Some try to obtain prescriptions from multiple doctors. Others commit crimes to get money to pay for the drug.

. . . .

Reading Check

Identify Name two types of narcotics that may be used as medicine when prescribed by a health care professional.

. . . .

HEALTH EFFECTS OF NARCOTIC DRUG ABUSE

Can cause drowsiness, constipation, and depressed breathing.
Taking a large single dose could cause severe respiratory depression or death.
Can cause death if taken with certain medications or alcohol.
Can lead to physical dependence and tolerance. The body becomes used to the substance and higher doses are needed to feel the same initial effect.
Narcotics are highly addictive, often causing uncontrollable drug use in spite of negative consequences.
Withdrawal symptoms occur if use is reduced abruptly. Symptoms can include restlessness, muscle and bone pain, insomnia, diarrhea, vomiting, cold flashes with goose bumps, and involuntary leg movements.
Withdrawal from narcotics usually requires detoxification in a hospital. Although withdrawal is often a painful experience, it is not life-threatening.

When narcotics are abused, there is a risk of addiction and other health consequences. **Name some of these consequences.**

Effects of Opioid Use Opioid use affects physical, mental/emotional, and social health. Use of opioids can add stress within a family. It can lead to relationship problems with parents and other family members.

Opioid use includes short- and long-term physical effects. Short-term effects include feelings of euphoria, pain relief, drowsiness, and feelings sedated. Long-term effects include nausea and vomiting, bloating, constipation, liver damage, brain damage, and addiction.

Misuse of opioid use can lead to death. More than 47,000 people suffered fatal overdoses due to opioid use in 2017. It is estimated that between 26 to 36 million people throughout the world abuse opioids. In the U.S. 2.1 million people have opioid use disorder.

Think of all the ways you can have fun without using drugs. **What are some drug-free activities that you and your friends enjoy?**

One symptom of inhalant abuse is red or runny eyes. **What other parts of the body can inhalant abuse damage?**

Factors Leading to Opioid Abuse A report by the Partnership for Drug-Free Kids identified some factors that put teens at risk for opioid abuse. One factor is family problems. About 65 percent of young people who experienced physical and emotional abuse at home were more likely to become substance abusers.

Inhalants

MAIN IDEA Using inhalants can cause brain damage and even death.

Inhalants are the vapors of chemicals that are sniffed or inhaled to get a "high." They include some household products. Inhalants are substances that are only taken by inhaling. Toxic inhalants include hair spray, lighter fluid, air freshener, cleaning products, markers or pens, spray paint, and paint thinner. Inhalants are a central nervous system depressant. Inhalant use can cause nausea, dizziness, light-headedness, slurred speech, confusion, and loss of motor skills.

Inhalant abuse sends poisons straight to the brain. These chemicals can cause users to hallucinate. Inhalants cause permanent brain damage and affect a person's ability to walk, talk, or think. Other effects of inhalant abuse include damage to the liver, kidneys, and hearing. Inhalant abuse, even if it is only the first time, can make the user sick. Overdosing on inhalants can cause seizures and coma. In extreme circumstances, using inhalants can cause instant death. Experimenting with inhalants can cause death from choking, suffocation, or heart attack.

Warning Signs of Inhalant Abuse

A person who uses inhalants often shows symptoms of the abuse. Some common symptoms are listed below. If you notice these symptoms in someone you know, speak to a teacher or trusted adult about your concerns.

- Eyes that are red or runny.
- Sores or spots near the mouth.
- Breath that smells strange or like chemicals.
- Having ink or marker spots near the nose.

Steroids

MAIN IDEA Steroid use can cause serious health problems.

Some drugs mimic the behavior of chemicals made by the body. One example is **anabolic steroids**, which are substances that cause muscle tissue to develop at an abnormally high rate. Anabolic steroids act like the male hormone testosterone. They promote muscle growth. Doctors sometimes legally prescribe steroids to treat growth problems, lung diseases, and skin conditions.

Reading Check

Give Examples What are two symptoms of inhalant abuse?

Some athletes use steroids to try to increase their body weight, strength, or endurance. However, steroid use causes a variety of serious health issues. Effects on males are decreased sperm production, enlarged breasts, shrinking of the testicles, baldness, and testicular cancer. Effects on females include a deepening voice, decreased breast size, coarse skin, excess body hair, baldness. Both sexes may also become angry and aggressive, have delusions, have severe ache, oily scalp and skin.

Any nonmedical use of steroids is illegal. Athletes who use steroids can be dismissed from a team or an event. Illegal steroid users may also face fines and jail time. Athletes who are caught using steroids are often suspended or banned from their sport. Many have had their reputations damaged as a result.

The best way to improve your athletic performance is to practice. **How could using performance enhancing drugs affect your health?**

Lesson 2 Review

What I Learned

1. **VOCABULARY** Define *depressant*. Use the term in a complete sentence.

2. **EXPLAIN** What is the greatest long-term risk of marijuana use?

3. **NAME** What are three types of drugs that are legal only when prescribed by a doctor?

Thinking Critically

4. **APPLY** Suppose a friend told you inhalants were safe because they are items found in your own house. How would you respond? Is this valid health information?

5. **ANALYZE** What are some of the high-risk behaviors that could result from using hallucinogens or club drugs?

6. **APPLY** Suppose you read about a professional athlete who says they use marijuana during training. How could this affect their performance?

Applying Health Skills

7. **ADVOCACY** Write a script for a public service announcement for radio or television, explaining the short- and long-term effects of narcotics use.

Staying Drug-Free

Before You Read

Quick Write Write a short paragraph describing why you think drug use may be harmful for teens.

Vocabulary
drug free
drug rehabilitation

Reading Check

Explain Describe the influence that peers can have on a teen's decision to use alcohol or drugs.

BIG IDEA Many reasons and resources exist to help teens and their families stay drug free.

Why Do Some Teens Use Drugs?

MAIN IDEA Responding to peer pressure, the media, and personal problems can influence teens to try drugs.

Teen drug use is low. Less than seven percent of middle school students say they have tried marijuana. Teens who have tried drugs say they were influenced by media images or peer pressure. In the media, the consequences of drug use are rarely shown. These consequences include addiction, job loss, broken relationships with family and friends, breaking the law to get money for drugs, and being convicted for crimes.

The truth is that drug use can have serious effects on all sides of your health triangle. Physical consequences can include sleeplessness, addiction, heart failure, and stroke. Mental/emotional consequences can include irritability, trouble thinking or concentrating, depression, anxiety, and impaired brain development. Social consequences can include troubled relationships with family and friends and withdrawal from fun activities. Drug use has also been linked to gang activity and violence.

Stages of Drug Dependence

MAIN IDEA There are predictable stages that people who become addicted to drugs go through.

People who become addicted to drugs usually go through stages as they continue to use drugs. Being aware of these stages can help you recognize if someone has a problem and get help before they become addicted.

- **Stage One – Experimentation.** Experimentation is the voluntary use of drugs. Some people may try drugs out of curiosity. Some people may try drugs to cope with stress or other problems.
- **Stage Two – Regular use.** Regular use means that someone is using drugs on a continued basis. There may be a pattern to the drug use. People may use drugs on the weekends with friends or after a stressful day at work.

- **Stage Three – Problem Use.** During the problem use stage, a person experiences problems as a result of their drug use. These problems could include problems at work or school and changes in relationships with friends. Legal problems could include getting speeding tickets or driving while under the influence of drugs.

- **Stage Four – Dependence.** A person in the dependence stage develops an increased tolerance to the effects of the drugs. They have to use more of the drugs to get the same effect. Social, economic, and legal problems continue and may increase during the dependence stage.

- **Stage Five – Addiction.** A person who is addicted to drugs can no longer control their use. They will experience cravings as the effects of the drug wear off. People in the addiction stage experience job loss and major changes to their relationships. They can become homeless.

Ways to Stay Drug Free

MAIN IDEA Staying drug free has many benefits.

The choice to avoid illegal drugs and the improper use of legal drugs may be the most healthful decision you can make. A person who is drug-free is showing good character. This person has decided not to use illegal drugs. Being drug-free can also refer to a place where illegal drugs are not used. It means you have taken charge of your life and your health. It is important to make decisions that promote both a healthy body and mind. Staying **drug free** can be difficult, but it has many benefits. Staying drug free is a characteristic of a person not taking illegal drugs or of a place where no illegal drugs are used.

Reading Check

Compare Describe the differences between the experimentation stage and the problem use stage of drug dependence.

Most teens enjoy a drug-free life. **Describe what it means to be drug-free.**

Have fun at drug-free and alcohol-free events. Avoid environments where alcohol and other drugs are present. Use positive peer pressure to help others avoid these environments.

Improve your talents or skills. Choose an activity you like and practice it until you become an expert. Become a great skateboarder, a computer whiz, or the best artist at school.

Be part of a group. Join a sports team, a club, or a community group.

Start your own business. Make yourself available for babysitting, yard work, or other jobs. Let friends and neighbors know.

"Drug Free Zones" have been established near schools. **Describe how drug-free zones work.**

Reasons *to* Avoid Drugs

Making wise choices about drug use will have a positive effect on your health triangle. To help teens avoid drug use, many schools have established drug-free zones near schools. The consequences of having or using drugs in a drug-free zone are enhanced. For example, jail sentences are increased.

- **Physical health.** You show that you care for yourself and your health. You will not suffer the physical consequences of drug abuse on the body.
- **Control over your actions.** You are better able to stay in control and act more responsibly.
- **Obeying the law.** You show respect for the law and are a good citizen.
- **Protecting your future.** You are able to set goals and work toward them. You are able to concentrate better and will do better in school.
- **Healthier relationships.** You are able to enjoy other interests with family and friends.
- **Self-respect.** You want to avoid harm to your body and mind. You can be confident about your decisions. You know you have made healthful choices.

Alternatives to Drug Use

If someone offers you drugs or alcohol, what would you do? You could suggest an alternative. Offering positive alternatives can help relieve some of the pressure you may be feeling. It also gives you the chance to be a positive influence on your peers. Some possible alternatives are:

- Start a physical activity program.
- Volunteer in your community.
- Join a school club.
- Participate in a drug-free event.
- Create art or write your thoughts in a journal.
- Add balance to your life, such as balancing physical activity with rest.
- Find friends who are also drug-free.

Saying No to Drugs

Developing skills to refuse drugs is very important. The best way to avoid being pressured to use illegal substances is to use refusal skills. Refusal skills can help you say no to other unhealthy behaviors as well. You can resist negative peer pressure without feeling guilty or uncomfortable. Saying no in a clear and confident way lets others know you respect yourself and your health. If you feel pressure to experiment with drugs, remember the S.T.O.P. strategy:

- **S**ay no in a firm voice.
- **T**ell why not.
- **O**ffer positive alternative ideas or activities.
- **P**romptly leave.

This strategy is helpful when you are faced with a difficult situation. However, you can also take steps to help you avoid even having to use your refusal skills. You can choose to make friends with people who have also chosen to avoid drugs. Friends who are committed to being drug free will support your decision and help you avoid situations where drugs or alcohol may be present. You can also look for healthful ways to deal with whatever issues you may face. If you feel lonely or depressed, or if you need help solving a personal problem, talk to a parent or another adult you trust. Making wise choices about how you spend your time, who your friends are, and how you deal with your feelings can all have a positive effect on your ability to be drug free.

Friends can be an important influence during teen years. **How can friends help you stay drug free?**

Reading Check

Identify What are two positive alternatives to drug use?

Help for Drug Users and their Families

MAIN IDEA Resources are available to help drug users and their families face substance abuse.

Stopping drug abuse after it has started is much harder than resisting drugs in the first place. Some effects of drug abuse are permanent. However, drug addiction is treatable. Many resources are available to help drug users overcome the pattern of addiction. One website that provides resources about prevention, intervention, and treatment for drug use is the Substance Abuse and Mental Health Services Administration (SAMHSA) at https://store.samhsa.gov.

Drug Treatment Options

Drug addiction is a disease. People who are addicted cannot function without the drug. Treatment requires changes in behavior. People who are addicted to drugs must first admit that they have a problem. Then they need to seek help to recover.

Many times, the first step in this process will be detoxification. It is the physical process of removing drugs from the body. Detoxification is usually done in a hospital or drug treatment facility.

After detoxification, the person might choose to enter **drug rehabilitation**. It's a process where the person relearns how to live without the abused drug. It is also called treatment. While in treatment, the person stays in a specialized facility. In treatment, the person relearns how to live without the drug. The person may return home and continue counseling and other treatment at an outpatient facility.

Support groups bring together people who are facing similar problems. They work together with the goal of staying drug free. Support groups give people in recovery a place to learn from others who are facing the same challenges. Support groups include Narcotics Anonymous and Cocaine Anonymous.

Some people find the support and help they need to stay drug free through counseling. Counseling provides an opportunity to openly share thoughts and feelings with a trained expert. It can help addicts deal with their psychological dependency on drugs. Counseling may involve only the addict or the person's entire family.

Harm reduction strategies are also used. These reduce the costs of drug abuse. One harm reduction strategy for drug abusers is to reduce the spread of HIV/AIDS through needle-exchange programs. Another is to reduce the cravings caused by heroin and opioid withdrawal by creating methadone maintenance programs.

> STOPPING DRUG ABUSE AFTER IT HAS STARTED IS MUCH HARDER THAN RESISTING DRUGS IN THE FIRST PLACE.

Reading Check

Recall What are some drug treatment options?

Help for Families

When someone is addicted to drugs, that person's family also needs help. One of the many resources for families is Nar-Anon. Like Al-Anon, Nar-Anon helps family members learn how to deal with the problems caused by drug addiction.

Support from friends can help you stay substance free. **What resources are available in your community for teens with substance abuse problems?**

Developing Good Character

Citizenship You can demonstrate good citizenship by encouraging others to stay drug free. Find out about programs in your school or community that educate teens on the dangers of drug use and tell your classmates about them. Identify what methods they use to reach teens and find out how students can get involved.

What programs would you be interested in participating in? Why?

Lesson 3 Review

What I Learned

1. **VOCABULARY** Define *drug rehabilitation*. Use it in a complete sentence.

2. **LIST** What are two reasons why some teens might choose to use drugs?

3. **DESCRIBE** What are some of the ways support groups help people become drug free?

4. **DESCRIBE** What are the stages of drug dependence and what changes may a person experience during each stage?

Thinking Critically

5. **EVALUATE** How can suggesting a positive alternative to alcohol or drug use help you stay substance free? Explain your answer.

6. **APPLY** What do you think is the most important reason for you to stay drug free? Explain your reasoning in a short paragraph.

Applying Health Skills

7. **REFUSAL SKILLS** Think about ways to say no to harmful behaviors. Team up with a classmate. Role-play a situation where you use these strategies to say no to illegal drugs.

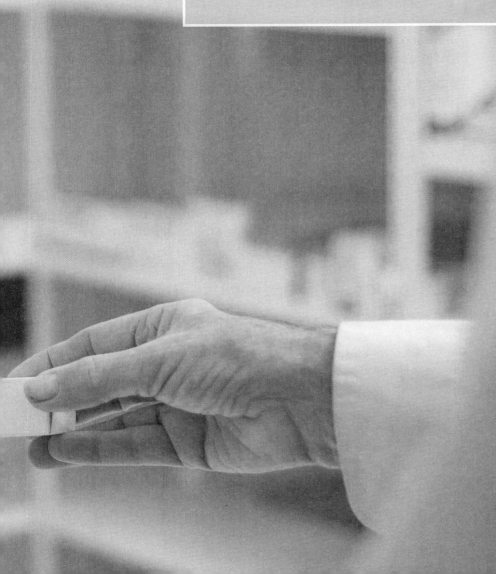

Using Medicines Wisely

LESSONS

1 Types of Medicines

2 Using Medicines Correctly

16

Types of Medicines

Before You Read

Quick Write Describe a time when you used a medicine. State what the medicine was intended to treat and how you used it.

Vocabulary

medicine
antibiotics
over-the-counter (OTC) medicine
prescription medicine
pharmacist

Reading Check

Define What are medicines?

BIG IDEA Using medicines wisely is a sign of good personal and consumer health.

What Are Medicines?

MAIN IDEA A medicine is a drug that can prevent or cure an illness or ease its symptoms.

If you have a cold or feel as if you are getting a fever, you might go to the drugstore to buy some **medicine** that can help you feel better. A medicine is a drug that prevents or cures an illness or eases its symptoms. You might also choose to use a medicine to help relieve the aches and pains caused by an injury. When sickness or injury occurs, medicines can often help a person feel better or recover from the illness.

In earlier times, many medicines were taken from plant leaves. People might eat the leaves or drink tea brewed from them. Today most medicines are made in laboratories. Many come in the form of pills or liquids and are swallowed. Medicines may also be injected into the bloodstream using needles, inhaled into the lungs, or rubbed into the skin.

Medicines in the United States are carefully controlled by the Food and Drug Administration (FDA). This agency is part of the federal government's Department of Health and Human Services. The FDA sets standards for medicine safety and effectiveness. The agency tests and approves medicines before they can be sold.

Various types of medicines are used in various ways. Some medicines protect you from getting certain diseases. Some cure diseases or kill germs. Still some medicines are used to manage chronic, or ongoing, conditions such as asthma. Other medicines help relieve symptoms of illness or treat minor injuries.

The Purpose of Medicines

MAIN IDEA Different medicines serve different purposes in the body.

Medicines are used for four purposes. They are to prevent disease, fight germs, relieve pain, or manage disease. Vaccines are medicines that prevent disease. Some vaccines are created from a weakened version of the disease. This creates an immune response in the body. The immune response is triggered when the person is exposed to the disease. Antibiotics are an example of a drug that fight germs. Taking an antibiotic kills germs that are invading the body.

Other drugs that contain ingredients such as ibuprofen relieve pain. An example is taking a drug to relieve a headache. Other drugs manage diseases that cannot be cured. One type of drug is taken each day for life to manage thyroid diseases. This type of medicine makes the symptoms of the disease more manageable.

Preventing Disease

A vaccine protects against diseases that can spread from person to person. When you are vaccinated, your body can make substances called antibodies that will attack or kill off the germs that cause a disease. Some vaccines provide protection for many years. Others, such as the flu vaccine, protect for only about a year.

Fighting Germs

Antibiotics are medicines that reduce or kill harmful bacteria in the body. They may be prescribed to treat an infection. However, improper use can make bacteria resistant. That means certain antibiotics may no longer stop an infection. An antibiotic cannot help the body fight an illness such as the common cold. Other drugs are used to fight viruses and fungi. Several common types of antibiotics are listed in the figure in the lesson.

Reading Check

Explain Name three different kinds of medicine, and tell what each does.

Type of Medicine	Some Examples	Disease or Problem
Vaccines	• MMR vaccine • Varicella vaccine • HPV vaccine • Pertussis vaccine	• Measles, mumps, and rubella • Chicken pox • Human Papillomavirus • Whooping Cough
Antibiotics	• Penicillin • Cephalosporin • Tetracycline • Macrolides	• Strep throat, pneumonia, STDs • Meningitis, skin rash • Urinary tract infections, Rocky Mountain spotted fever • Given to patients allergic to penicillin
Pain Relievers	• Aspirin, acetaminophen, ibuprofen, codeine	• General pain relief

Different medicines do different things in the body. **Name the four purposes of medicines.**

Relieving Pain

You're probably familiar with medicines that relieve pain and to reduce swelling. Pain relief medicines to help headaches, toothaches, and other minor ailments are commonly used. These are called **over-the-counter (OTC) medicines.** They are medicines that you can purchase without a doctor's prescription. OTC medicines are safe to use if the directions are followed. These medicines can cause harm if they are not used according to directions. They include aspirin, ibuprofen, and acetaminophen. OTC medicines also include vitamin and mineral supplements.

The use of some medicines is controlled by your doctor. These medicines are called **prescription medicines**. They can only be obtained by getting a doctor's written prescription. Prescription medicines can only be sold through a pharmacy. The distribution of prescription medicines has more risk than OTC medicines. That's why prescription medicines are controlled. When the doctor prescribes a medicine, they intend for that person to take the medicine. The doctor has selected the dosage for the medicine based on needs of the person who is intended to use it. Sharing prescription medicines can be risky. For example, the person using the shared medicine may be allergic to it. Also, the doctor has prescribed a certain dosage for the patient. The full dosage must be taken to cure the disease. Sharing medicines can prolong illness.

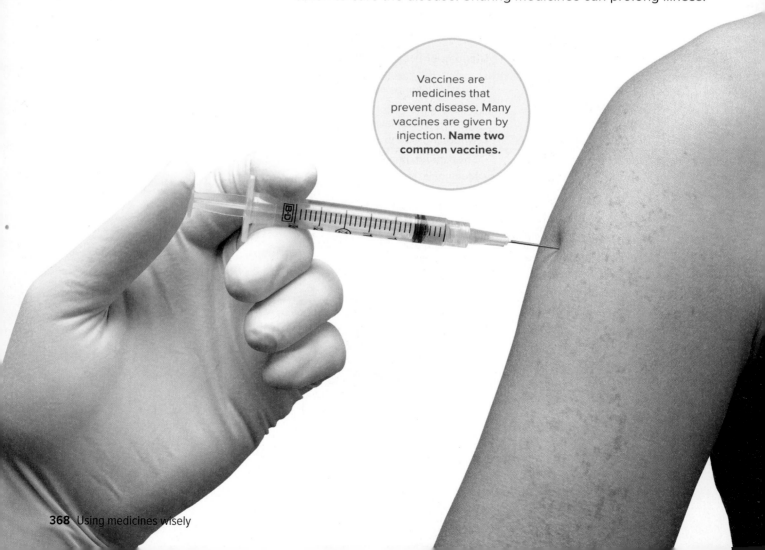

Vaccines are medicines that prevent disease. Many vaccines are given by injection. **Name two common vaccines.**

Managing Disease

Some medicines help people manage chronic diseases or conditions, such as allergies, asthma, diabetes, or mental illnesses such as anxiety and depression. People with diabetes take insulin to help control their blood sugar. People with allergies can take antihistamines to treat swelling and other allergy symptoms.

Often, medicine is taken by swallowing. However, medicine can be given in a number of different ways. The various methods of delivering medicine to the body include:

- **Swallowing, or ingestion.** A pill, tablet, capsule, or liquid moves through the stomach into the bloodstream and then through the body. Most pain relievers are taken this way.
- **Injection, or shot.** Injected medicines begin to work more quickly because they directly enter the bloodstream. These are administered by a needle that pierces the skin.
- **Inhalation.** Medicine can be inhaled, or breathed in, as a mist or fine powder. People with asthma may use an inhaler. Cold or sinus medication can also be inhaled through the nose.
- **Topical application.** Creams and ointments can be rubbed directly onto the skin. Patches containing medicine may also be applied to the skin.

Reading Check

Contrast What is the main difference between prescription and over-the-counter (OTC) medicines?

Some medicines are sold over-the-counter while others can only be sold with a doctor's prescription. **Explain why some medicines require a prescription.**

Myth vs. Fact

Myth Only babies and toddlers need vaccines.

Fact Several important vaccines are required for those between the ages of 10 and 18. Most preteens and teens will need to receive a vaccine or booster shot for diphtheria and tetanus, chicken pox, hepatitis B, measles, mumps, and rubella, and meningococcus.

• • • • • • • • • • •

Prescription and Non-Prescription Medicines

MAIN IDEA Some medicines require a doctor's permission, while others are available without a prescription.

As you have learned, prescription medicine can be obtained only with a written order from a doctor. Written permission is required because prescription medicines carry more risks. You do not need a doctor's permission to buy an over-the-counter medicine. However, both prescription and over-the-counter medicines must be used very carefully.

If you are prescribed medicine, your doctor will write out instructions that explain how much to take, when to take it, and for how long. Prescriptions must be filled by a pharmacist who is trained to prepare and distribute medicines. Your doctor's instructions will appear on the prescription medicine label. The FDA requires medicine labels to have specific information. Always read the label closely before taking any medicine and follow the directions.

You can find over-the-counter (OTC) medicines in groceries and drugstores. They are considered safe to use without a doctor's permission. However, always be careful when you use OTC medicines. Follow the directions, because even OTC medicines can be harmful if not used correctly. If you have any questions about a medicine, ask a doctor or **pharmacist**. A pharmacist is a person trained to prepare and distribute medicines.

Your doctor or pharmacist can help you understand a prescription medicine label. **What types of information can you find on a prescription medicine label?**

Pharmacy Identification

Name of Pharmacist

Prescription Number

Date prescription was filled

Name and address of patient

Directions for use

Name of prescribing doctor

Name of the medicine

Strength and/or amount per container

Number of refills allowed

Expiration date

Special Instruction

McGrath Pharmacy
123 Main St. Miller, NJ 09009
(609) 555-1122
Pharmacist: T. Lewis

RX #125690
Date Filled: 4/09/14
Dr. Tobe Friedland

Candace Sanchez
1578 Lakeside Lane
Miller, NJ

Take one capsule every six hours, one hour before a meal

Erythromycin
250 mg capsules
Quantity: 60 capsules
Refills: 0
Drug Expires: 04/09/15

Delayed release capsules — do not crush or break

Take medication on an empty stomach.

Finish all medication unless otherwise directed by a doctor

Various types of medicines are used to treat or prevent various illnesses. **Which types of medicines have you been given in the past?**

Lesson 1 Review

What I Learned

1. **VOCABULARY** Define the term *antibiotics*. Use it in an original sentence.

2. **NAME** What is the type of medicine that prevents a disease from developing?

3. **LIST** What are the four main purposes of medicines?

Thinking Critically

4. **ANALYZE** A friend of yours on the football team wants an energy burst before a game. He wants to take a handful of vitamins. When you express concern, he says, "They're over-the-counter vitamins." Respond to this comment.

5. **EVALUATE** Why are some drugs distributed over-the-counter and other drugs require a prescription from a doctor?

Applying Health Skills

6. **PRACTICING HEALTHFUL BEHAVIORS** Lanie's doctor gave her a six-day prescription of an antibiotic for her sore throat. After only three days, all of her symptoms are gone. Should Lanie continue taking the antibiotic? Explain why or why not.

Using Medicines Correctly

Before You Read

Quick Write Write about a time when you had to take medicine and how it affected you.

Vocabulary
side effect
medicine misuse
medicine abuse
overdose

BIG IDEA Medicines can contribute to good health when used properly.

How Medicines Enter the Body

MAIN IDEA A medicine may be swallowed, injected, inhaled, or applied to the skin.

Different medicines are used in different ways. For example, a mild sunburn or itchy mosquito bite can be treated with a cream or lotion. This method is referred to as topically, or applied to the skin. Other ways that medicines can be taken include by swallowing, injection, or inhalation. The effects of medicines are different depending on how they are taken.

Swallowing, or ingestion, is the most common way to take medicines. Pills, tablets, capsules, and liquids are taken orally, or by mouth. The medicine moves into the stomach and small intestine. From the digestive system, the medicine passes into the bloodstream and circulates throughout the body. Cold medicines and pain relievers are often delivered this way.

Injection, or a shot, is another way that medicines can enter the body. A needle injects the medicine directly into the bloodstream. Injected medicines begin to work more quickly than other types.

Inhalation is yet another delivery system. Medicine can be inhaled as a mist or fine powder. People with asthma often use inhalers. You can breathe in, or inhale, cold or sinus medication through your nostrils.

Medicines are also given topically, or applied to the skin. You can apply creams and ointments this way. Skin patches that release medicine over time are another type of topical medicine.

Problems with Medicines

MAIN IDEA Medicines affect different people in different ways.

Because every person's body is unique, medicines affect people in different ways. Combining medicines may also affect the way they work. Some medicines do not interact well with others and can cause harmful reactions. Some people are allergic to certain medicines and cannot take them at all. Your age, weight, and overall health can determine how a medicine affects you.

Reading Check

Identify What is the most common way medicines are taken?

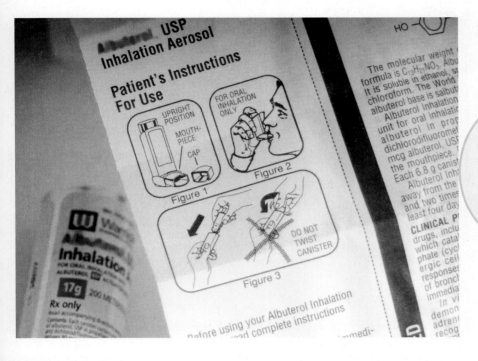

Inhalers deliver medicine to the body. People with asthma carry inhalers to ease breathing. **Name two other ways that medicines enter the body.**

Side Effects

Medicines can help you, but if you do not use them properly, they can also hurt you. A **side effect** is a reaction to a medicine other than the one intended. Sometimes, a side effect is simply unpleasant. For example, a medicine might make you feel sick to your stomach, sleepy, dizzy, or cause a headache. A more serious side effect is an allergic reaction, which requires immediate medical attention.

To avoid most side effects, follow the instructions you get from the doctor or pharmacist on how to take a medicine. Always tell your doctor of other medicines you are taking. Include any vitamins you take as well as over-the-counter and prescription drugs.

Before taking a medicine, read the label. The label gives instructions for use, such as how much medicine to take and how often to take it. The label may also tell you to avoid the use of other drugs while taking the medicine. Medicine labels also tell you about any activities that could be dangerous to do while on the medicine. Some medicines cause sleepiness and warn about driving or using machinery. Ask the doctor or pharmacist if there's something on the label you do not understand.

Drug Interactions

Taking two or more medicines at once can be dangerous. They may cause unexpected drug interactions. For example, one drug may become more effective or less effective, or the combination may produce different, more dangerous side effects that neither drug would have caused when taken by itself. Taking medicines with certain foods can also cause interaction problems. Always let your doctor and pharmacist know what other medicines you are taking before starting a new medicine.

> IF NOT USED PROPERLY, MEDICINES CAN BE AS HARMFUL AS ILLEGAL DRUGS.

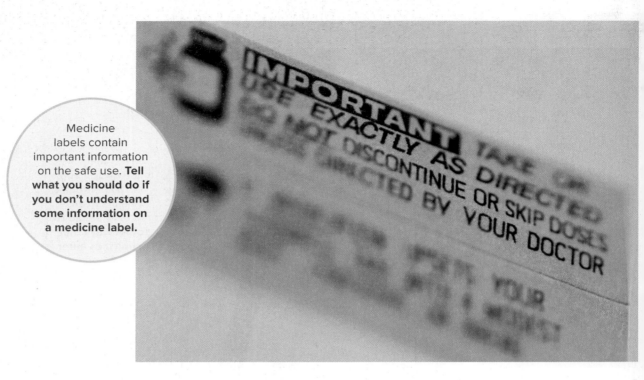

Medicine labels contain important information on the safe use. **Tell what you should do if you don't understand some information on a medicine label.**

Tolerance

When someone uses a particular medicine for a long period of time, the person's body may develop a tolerance. The concept of tolerance can apply to medicine as well as alcohol and other drugs. In some cases, tolerance may cause a medicine to lose its effectiveness over time.

A person may go through withdrawal when he or she stops using the medicine. These symptoms will gradually ease over time. If you experience withdrawal after using a medicine, talk to your doctor. You may need to be prescribed a different medication. Symptoms of medicine withdrawal can include:

- Nervousness.
- Insomnia.
- Severe headaches.
- Vomiting.
- Chills.
- Cramps.

Antibiotic Resistance

The more an antibiotic is used, the less effective it becomes. This is especially true when an antibiotic is overused. Why? Antibiotics kill harmful illness-causing bacteria in the body. With frequent exposure, however, bacteria often build up a resistance to antibiotics. Bacteria adapt to, or overcome, the medicine. Bacteria can also develop a resistance when antibiotics are not taken as prescribed. For example, if the prescription label says to take the medicine for 14 days, and you stop after 7 days, the bacteria may still be in your body and could make you sick again. Medicines should always be used wisely, and they should be used only as directed.

Reading Check

Identify Name two risks of medicines.

Improper Use of Medicines

MAIN IDEA Misusing medicines can be as harmful as using illegal drugs.

Medicines are types of drugs. Medicines are intended to be helpful, not harmful. They can prevent and cure diseases, fight germs, and relieve pain. If they are not used properly, however, medicines can be as harmful as illegal drugs. They can cause addiction, injury, and even death. If you use medicines improperly as a teen, it can result in serious health problems not only now but also later in life.

Prescription medicines and over-the-counter medicines carry warning labels. These labels list possible side effects. To avoid the side effects, follow a medicine's instructions. This includes taking the correct dose. Using too much medicine or using it too often can damage body systems. For example, using too much of a medicine could cause liver or kidney failure. Some medicines can harm unborn babies, newborns, or young children. A female who is pregnant or plans to become pregnant should talk to a doctor before she takes any medicine.

Medicine Misuse

Medicines have many benefits when they are used correctly. Research studies show that most teens—96 percent—do use medicines correctly.

Medicine misuse can be dangerous. It is taking medicine in a way that is not intended. This is why medicines need to be taken with great care. Medicine misuse may prevent you from getting full benefits of a medicine. It can seriously harm your health. Examples of medicine misuse include:

- Taking more medicine than a doctor instructs.
- Not following the directions on the label.
- Taking a medicine that is past its expiration date.
- Taking a medicine that was prescribed for another illness.
- Taking a medicine that was prescribed to another person.

Reading Check

Recall How can medicine misuse affect a person's health?

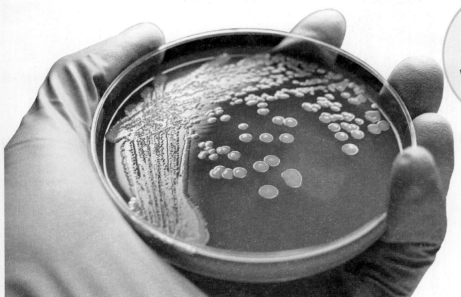

Antibiotic resistance is a growing problem in the U.S. **Name two ways that antibiotic resistance occurs.**

Medicine Abuse

Medicine abuse is intentionally using medicines in ways that are unhealthful and illegal. It's a form of drug abuse. Some teens believe that prescription and OTC medicines are safer than illegal drugs. However, medicines are safe only if used properly. People may abuse medicines for several of the following reasons:

- **To lose weight.** A healthy diet and exercise are the safest ways to maintain a healthy weight.
- **To stay awake.** Getting plenty of sleep and learning strategies to manage your time wisely will help you study effectively.
- **To get "high".** A dangerous trend is the practice of having "pill parties," in which partygoers mix whatever OTC and prescription medicines may be available. Using medicine that is not prescribed for you is both illegal and dangerous and could even cause death.

Taking a medicine prescribed to someone else is medicine abuse. It is also illegal and unsafe. A doctor orders a medicine for a specific person to treat a specific illness. Even if you think you have the same illness, your body may need a different medication or dose.

A danger of both medicine abuse and medicine misuse is the risk of drug **overdose**. An overdose is taking more of a drug than the body can tolerate. Misusing medicines can also lead to addiction or death. The best way to make sure you are using a medicine safely is to follow the instructions on the label.

When using medicines, it is important to keep out or reach of young children. **Why is it important to be responsible when using medicines?**

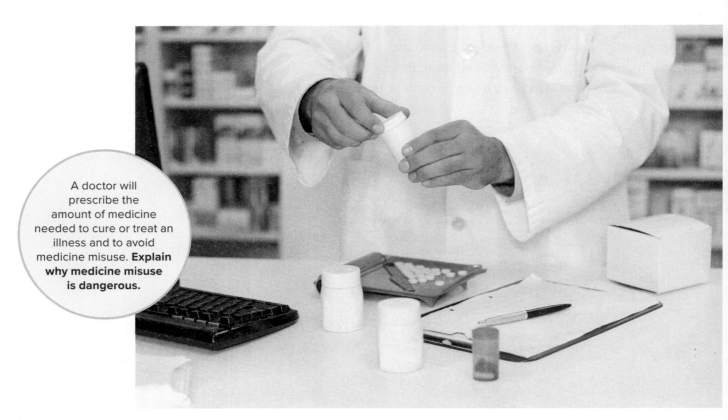

A doctor will prescribe the amount of medicine needed to cure or treat an illness and to avoid medicine misuse. **Explain why medicine misuse is dangerous.**

How to Use Medicines Safely

MAIN IDEA Medicines are helpful only when they are used properly.

Medicines can be helpful, but they can also cause serious harm. This is why they should be taken with great care. To avoid misuse, follow these guidelines:

- Follow the instructions on the label.
- Take the correct dosage for the recommended length of time. If you experience side effects from a prescription drug, contact your doctor before you stop using the medicine.
- Do not take medicines after their expiration date.
- Do not give prescription medicines to someone else. Do not use a medicine prescribed for an earlier illness without asking your doctor.
- Contact your doctor if you do not understand the label instructions, if you experience any unusual or unpleasant side effects, or if you accidentally take too much medicine.
- Store all medicines safely—in a cool, dry place, in their original containers, and out of reach of children.

Reading Check

Give examples What are three ways to use medicines safely?

Fitness Zone

When I get regular exercise and eat healthfully, I don't get sick very often. As a result, I don't need to take medicine very often and risk possible bad side effects. Physical activity helps me stay healthy.

How can a physical fitness routine contribute to your good health?

Some medicines must be taken every day or at certain times of the day. **List two ways that medicines can be misused.**

Lesson 2 Review

What I Learned

1. **VOCABULARY** Define the term *side effect*. Use it in an original sentence.

2. **DESCRIBE** What factors determine a medicine's effect on the body?

3. **COMPARE AND CONTRAST** How does medicine abuse differ from medicine misuse?

Thinking Critically

4. **HYPOTHESIZE** Milla's doctor has prescribed a medicine to treat a case of poison ivy. She also regularly takes medicine because she has trouble concentrating. Should Milla tell her doctor what medicine she is already taking? Why or why not?

5. **EVALUATE** Tasha's friend has offered her some of her prescription medication. Tasha asks you whether she should take it. What advice would you give her and why?

Applying Health Skills

6. **ACCESSING INFORMATION** Go online to research a popular drug that you have seen advertisements for. What condition does the drug treat? What are its side effects?

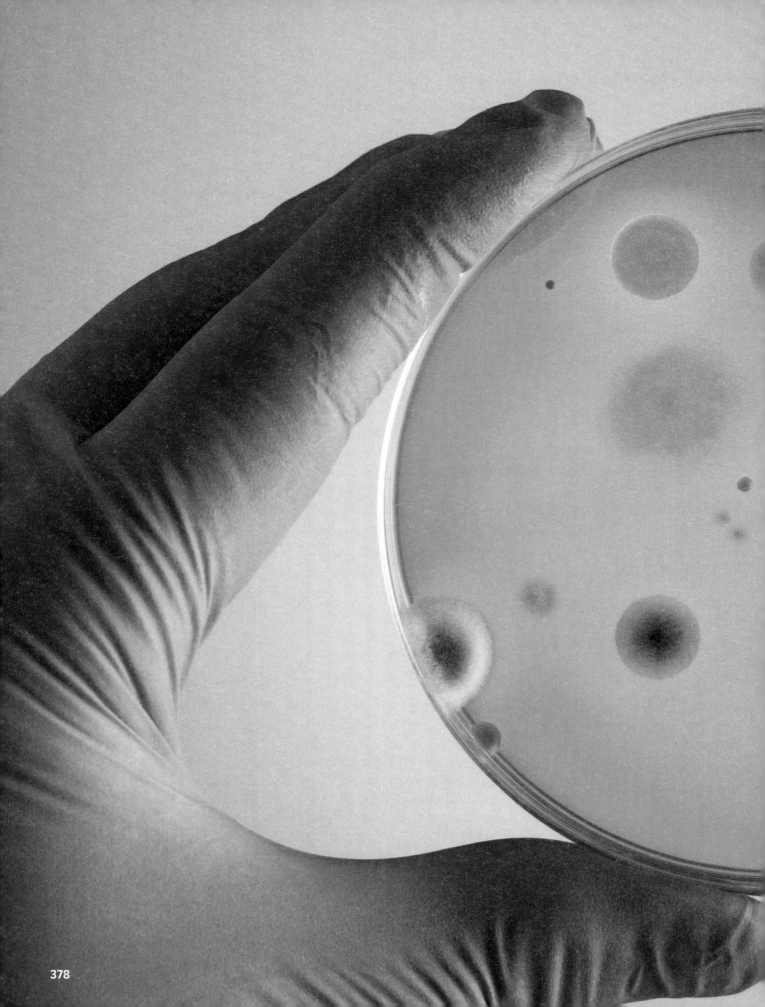

Communicable Diseases

LESSONS

- **1** Preventing the Spread of Disease
- **2** Communicable Diseases
- **3** Sexually Transmitted Diseases
- **4** HIV/AIDS

17

Preventing the Spread of Disease

Before You Read

Quick Write How do you think people catch colds? Explain your answer.

Vocabulary

disease
communicable disease
pathogens
viruses
bacteria
fungi
protozoa
vector
host
contagious

BIG IDEA Good personal hygiene and other healthful behaviors can help protect you from communicable diseases.

Germs and Disease

MAIN IDEA Communicable diseases are caused by germs.

Having a cold makes a person feel miserable. A cold is a temporary illness. After a few days, most people feel better. Their cold is gone. It hasn't always been this easy to recover from an illness or disease. Scientists study diseases and illnesses and try to come up with treatments. In some cases, they have come up with medicines that prevent diseases.

A cold is one type of **disease**. A disease is any condition that interferes with the proper functioning of the body or mind. A disease that can spread to a person from another person, an animal, or object is called a **communicable disease**. Communicable diseases are caused by **pathogens**, or germs that cause disease. You probably call them germs. Germs are all around us. Some germs are not harmful. When a harmful germ, or pathogen, enters your body, it can multiply and cause an infection.

Types of Pathogens

Pathogens are not all alike. However, all pathogens can cause disease. There are four main types of pathogens: viruses (VY•ruh•suhz), bacteria (bak•TIR•ee•uh), fungi (FUHN•jy), and protozoa (proh•tuh•ZOH•uh).

- **Viruses** are the smallest and simplest pathogens. They are so tiny they can only be seen with a special microscope. Viruses are not alive. They are usually made up of genetic material and protein. Some strains, or types, of viruses are not harmful. Other types can cause diseases such as the common cold, flu, upper respiratory infections, and measles. Most infections caused by viruses cannot be cured with antibiotics. The most harmful viruses can cause serious diseases. AIDS and hepatitis are serious diseases caused by viruses. AIDS is an immune system disorder. Hepatitis is a disease of the liver.

- **Bacteria** are simple one-celled organisms. They are everywhere. Some bacteria are helpful to humans. Some are harmful. For example, the bacteria in your digestive tract help you break down the food you eat. Other bacteria live on your skin and help prevent harmful bacteria from infecting you. Harmful bacteria can cause diseases such as pneumonia and strep throat. Most infections that are caused by bacteria can be treated with antibiotics.

- **Fungi** are organisms that are more complex than bacteria, but cannot make their own food. They must break down other organisms and absorb their nutrients. Most fungi are harmless to humans. Mushrooms, for example, are fungi. Some mushrooms are safe for humans to eat. Molds and yeast are also fungi. Fungi like to live in warm and moist places. They can cause a variety of diseases. Athlete's foot and ringworm are caused by fungi. Fungi can also cause serious lung infections.

- **Protozoa** are one-celled organisms but are more complex than bacteria. Some protozoa, called parasites, attach themselves to healthy cells. They rob the cell of nutrients, but they don't kill the cell. Some protozoa cause serious illnesses in humans. One of the most well-known and deadly diseases caused by protozoa is malaria. Malaria is found in tropical regions. It is spread to humans by infected mosquitoes. If an infected mosquito bites a person, the protozoa are transferred into the body through the skin. The figure located in the lesson shows some kinds of pathogens and lists the diseases they cause.

How Pathogens Spread

How can you keep from being infected by a communicable disease? Well, to understand how to avoid becoming infected, it's important to understand how pathogens are spread. There are four main ways pathogens are spread.

> **HARMFUL BACTERIA CAN CAUSE DISEASES SUCH AS PNEUMONIA AND STREP THROAT.**

. . . .

Reading Check

Compare How are viral infections different from bacterial infections?

. . . .

> Direct contact is a common way that pathogens are spread. **Describe how you can prevent the spread of pathogens through direct contact.**

Direct Contact

One of the most common ways pathogens are spread is by physically touching another person. For example, if someone with a cold sneezes into his right hand and then you shake hands with him; you will likely get cold pathogens from him. Shaking hands with him passes along the germs that are on his skin. Washing your hands often with soap can help stop the spread of pathogens.

Some pathogens are spread through contact with infected blood. People can come into contact with infected blood by injecting drugs using a needle that someone else has used. Dirty needles used for tattooing or piercing can also spread pathogens. It is also possible for the blood of an infected person to infect someone else if the blood comes into contact with broken skin. Some pathogens are spread through sexual contact. Abstinence is the best way to prevent sexually transmitted diseases.

Indirect contact

If someone sneezes or coughs, pathogens are spread through the air. It is important to use tissues to cover your nose when you sneeze and your mouth when you cough. Pathogens are also spread when people touch or share items that are contaminated with a pathogen. Some items that can spread pathogens are cutting boards and countertops. Drinking glasses, eating utensils, toothbrushes, and razors are other examples. These items should never be shared.

Contact with animals or insects

Animals and insects can spread pathogens. An organism, such as an insect, that transmits pathogens is a **vector**. If an animal is sick with rabies, for example, that animal can spread rabies if it bites someone or another animal. Insects spread many diseases. For example, infected deer ticks can spread Lyme disease. Infected mosquitoes can spread malaria or West Nile virus. Using insect repellent when you go outdoors can help protect you from insect bites.

Contaminated Food and Water

Bacteria that cause food poisoning can be spread through contaminated food or water. To prevent foodborne illness, cook meat fully. If food is undercooked or improperly stored, pathogens can grow. The pathogens can cause illness.

Illnesses caused by contaminated foods are called foodborne illnesses. To help prevent foodborne illnesses, make sure meat is fully cooked. Undercooked meat may contain bacteria that will make you ill. Remember to properly store foods that can spoil. This includes dairy products and meat. You should also wash fruits and vegetables. Many food packages say the product has been washed. However, it never hurts to wash it yourself. Meat, poultry, fish, and eggs need to be handled carefully. Bacteria from raw meats and eggs can spread on your food preparation surface. Keeping your food preparation area clean is very important.

Infected mosquitoes can spread diseases like West Nile virus. **Describe how you can protect yourself from mosquito bites.**

You should always wash knives and surfaces that meat, poultry, fish, and eggs have touched. Use warm, soapy water to clean food preparation surfaces.

The Epidemiological Triangle

Scientists use the epidemiological triangle to learn more about a communicable disease and how it spreads. The top of the triangle is labelled *agent*. It is what causes the disease, which is a pathogen in the case of communicable disease. The bottom left corner is labelled *host*. The **host** is the organism that is either carrying the disease or is sick. The bottom right corner is labelled *environment*. The environment is conditions outside of the host that cause or allow the disease to be transmitted.

Some types of food poisoning are caused by the bacterium *E. coli*. On the epidemiologic triangle, *E. coli* would be the agent. The host for *E. coli* would be humans. The environment for *E. coli* is meat or water contaminated with animal feces. If meat that contains *E. coli* is not fully cooked, it can make people sick.

Stopping the Spread of Pathogens

MAIN IDEA Practicing certain healthy behaviors can stop the spread of pathogens.

Although you can't completely stop the spread of pathogens, you can protect yourself and others from the spread of pathogens. Good personal hygiene is one of the best ways to help stop the spread of pathogens. You can also help stop their spread by eating nutritious foods and getting enough sleep. Engaging in regular, physical activity will also help your body fight pathogens. Keep your environment, the space around you, clean. This will also keep down the number of pathogens. Stopping the spread of pathogens will help to protect your health and the health of others.

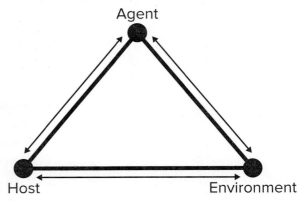

The Epidemiologic Triangle

Scientists use the epidemiologic triangle to learn how communicable diseases are spread. **Describe the difference between the *host* and the *agent*.**

Pathogens	Diseases
Viruses	Colds, chicken pox, influenza, measles, mononucleosis, mumps, hepatitis, herpes, HPV, HIV, yellow fever, polio, rabies, viral pneumonia
Bacteria	Pinkeye, whooping cough, strep throat, tuberculosis, Lyme disease, most foodborne illnesses, diphtheria, bacterial pneumonia, cholera, gonorrhea
Fungi	Athlete's foot, ringworm
Protozoa	Dysentery, malaria, trichomoniasis

Communicable diseases are caused by pathogens. **Name the two diseases listed on the chart that are caused by fungi.**

Reading Check

Analyze What personal practices can prevent the spread of communicable diseases?

Developing Good Character

Keep It to Yourself When you have a cold, take action to prevent spreading your cold to others. Be careful to cover your mouth and nose when you cough or sneeze. Keep as much distance between yourself and others as you can. Avoid sharing cups, utensils, or other personal items.

What character traits are you demonstrating when you take steps to prevent spreading communicable diseases?

Protecting Yourself

So, what can you do to protect yourself from pathogens? In addition to the tips mentioned above, follow these guidelines to help keep yourself from getting sick.

- Keep your distance from people who are sick with a communicable disease. Be especially careful if they are still **contagious**, or able to spread to others by direct or indirect contact.
- Do not share eating or drinking utensils or toothbrushes or other personal hygiene items.
- Wash your hands thoroughly in warm soapy water before preparing or eating food. It is important to wash after using the bathroom, playing with pets, and visiting a sick person. It is also necessary to wash after touching trash or garbage or other sources of pathogens.
- Keep your hands and fingers away from your mouth, nose, and eyes. Don't bite your fingernails.
- Handle and prepare food safely. This is especially important for meats, fish, and poultry. Eggs should be handled carefully, too. Meats, fish, poultry, and eggs should be cooked thoroughly.
- Wash vegetables and fruits before eating.
- Wash counters thoroughly with paper towels or a clean sponge or cloth. If you use a sponge or cloth, clean them thoroughly and frequently. Using sponges and cloths repeatedly can spread more germs than they remove.
- Keep your environment clean. Empty the trash often. Keep the trash cans clean. Clean up after your pets.

Another way to maintain good health is to practice healthy lifestyles. If you are exposed to pathogens, a healthy lifestyle can help you avoid getting sick. These healthy lifestyle practices include the following.

- Eat a balanced diet.
- Bathe or shower regularly using soap. Be sure to wash your hair using shampoo.
- Avoid tobacco products, alcohol, and other drugs.
- Get 8-9 hours of sleep every day.
- Rest when you are sick.
- Check with parents or guardians to make sure your immunizations are up to date.

Hand washing is the most important way to prevent the spread of pathogens. **Name two other healthy lifestyle behaviors that can help you prevent pathogens.**

- Manage your stress well. Learn ways to relieve your stress. Too much stress can weaken your immune system and make you more vulnerable to communicable diseases.
- Get regular physical checkups. Talk with your doctor if you have questions.

Protecting Others

You might be carrying pathogens and not realize it. You may have a communicable illness that hasn't shown symptoms yet. If you come into contact with others, you can easily spread the pathogens. Help protect the people you come into contact with. Think ahead and follow these healthful behaviors.

- If you are sick, stay home and away from others. Tell your parent or guardian and get medical help if you need it. If you become ill at school, let the school medical personnel know as soon as possible. Getting medical help early can keep the illness from getting too bad. It can also help keep it from spreading.
- When you sneeze or cough, cover your mouth with a tissue. Only use the tissue once. If you don't have a tissue, what should you do? Sneeze or cough into the crook of your elbow. This helps to keep pathogens from spreading. Wash your hands immediately after you sneeze or cough.
- If you have a prescription to take medication, follow the directions exactly. Take all of the medicine you are supposed to take. If you stop taking the medicine before you are supposed to, you might become sick again.

Maintaining cleanliness can prevent the spread of pathogens that cause illness. **Describe how taking out the trash might prevent the spread of pathogens.**

Lesson 1 Review

What I Learned

1. **EXPLAIN** What is a communicable disease?

2. **IDENTIFY** What are four ways pathogens can be spread?

3. **DESCRIBE** How can staying home when you are sick help keep others healthy?

Thinking Critically

4. **ANALYZE** How can hand washing help keep a community free from communicable diseases?

5. **APPLY** Imagine that you woke up with a sore throat and headache. Your team is playing in the community soccer finals today and you're the starting goalie. What should you do?

Applying Health Skills

6. **ADVOCACY** Use the knowledge you have gained from this lesson to create one or more brochures to put in your school health office. The brochures should contain information that explains to students how they can help keep themselves and others safe from the spread of pathogens. List at least five things students can do to keep themselves safe and five things students can do to keep others safe from the spread of pathogens.

Communicable Diseases

Before You Read

Quick Write Write down the names of three common diseases. What are the symptoms of each?

Vocabulary

infection
influenza
contagious period
mononucleosis
hepatitis
hepatitis b
tuberculosis (TB)
pneumonia
strep throat

Reading Check

Explain Why should you stay home for 24 hours after your cold symptoms appear?

BIG IDEA Common communicable diseases include colds, the flu, strep throat, pneumonia, mononucleosis, hepatitis, and tuberculosis

The Common Cold

MAIN IDEA Just about everyone has a cold once in a while.

The cold is the most common communicable disease. Colds are responsible for more school absences than any other illness. Cold symptoms include runny nose, headache, sore throat, coughing, sneezing, and mild fever.

You might be wondering, if colds are so common, why they can't come up with a vaccine to guard against colds. Well, there are hundreds of different viruses that are responsible for the common cold. A virus is the smallest and simplest pathogen. Because there are so many, scientists can't develop vaccines for all of them. Besides, scientists believe there are probably twice as many viruses that cause the cold than they already know about. The common cold is here to stay.

So, what can you do if you have a cold? The first thing you should do is protect others. When you sneeze, cover your nose and mouth to prevent the **infection** from spreading. Infection is a condition that happens when pathogens enter the body, multiply, and cause harm. When you sneeze, thousands of infected droplets containing the cold virus are released into the air. They can be spread through direct or indirect contact to others.

To recover from a cold, rest in bed. During the first 24 hours your cold is contagious. So, it is best if you stay at home and don't infect others. Your parents might also give you over-the-counter medicines to help relieve your cold symptoms. You should also drink plenty of fluids when you have a cold. If your cold symptoms get worse, or your sore throat lasts for several days, you should see a doctor.

> WHEN YOU SNEEZE, COVER YOUR NOSE AND MOUTH TO PREVENT INFECTION FROM SPREADING.

The Flu

MAIN IDEA The flu is a common communicable disease.

Your joints and muscles ache. You have fever, chills, fatigue, and a headache. **Influenza** (in•floo•EN•zuh), or the flu. It's a highly communicable viral disease characterized by fever, chills, fatigue, headache, muscle aches, and respiratory symptoms. You'll notice the symptoms of the flu are different from the common cold. Flu symptoms usually affect you more quickly and more seriously than cold symptoms do. The flu can be spread through both direct and indirect contact. Resting and drinking lots of fluids can help you recover faster from the flu. Some types of flu are serious and require a doctor's care.

December through March is considered to be "the flu season." This is because most cases of the flu are reported during these months. Flu viruses are not the same as the viruses that cause colds. Every year, certain strains of the flu virus are stronger and spread faster than the year before. Scientists meet every year to determine which strains will spread fastest during the next flu season. Then they develop vaccines for those strains of flu.

Some types of flu can be dangerous. Just after World War I, a flu outbreak killed about 20 million people around the world. Over 600,000 of those people were in the United States. Today scientists are doing their best to keep that kind of flu outbreak from happening again.

Other Communicable Diseases

MAIN IDEA All communicable diseases have a contagious period.

Communicable diseases have a **contagious period**, or the length of time that a particular disease can be spread from person to person. Often, the contagious period begins before the person starts to show symptoms. The contagious periods for several communicable diseases are listed in the chart in the lesson. Chicken pox, measles, and mumps all have specific contagious periods.

- Chicken pox is contagious for about a week before symptoms appear. Typical symptoms for chicken pox include an itchy, bumpy rash, fever, and aching muscles. The itchy bumps will blister and then dry up. When they are dry, chicken pox are no longer contagious. A vaccine for chicken pox became available in 1995. Since then, the disease is much less common.
- Measles has symptoms that include a rash, fever, and head and body aches. Measles are contagious a few days before symptoms appear. The contagious period lasts about five days after the symptoms appear. Measles is a dangerous disease. Around the world, over 1 million children die each year from measles. Vaccines are available in the United States and some other countries. Because of the vaccine, measles is less common than it once was.

Reading Check

Compare How are the flu and the common cold similar?

Some people should get flu shots before each flu season including people 65 and older and anyone with a weakened immune system. **Explain why the flu vaccine changes every year.**

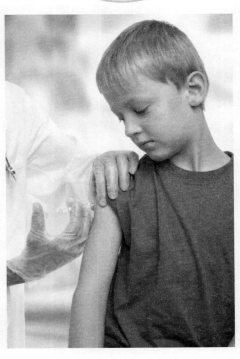

- Mumps causes a fever, headache, and swollen salivary glands. Mumps are contagious about a week before symptoms appear and for about nine days after that. More than 90 percent of the children in the United States are vaccinated against mumps. Fewer people get the disease now.

COVID-19

The coronavirus disease 2019, also called COVID-19, is a viral disease characterized by fever, cough, and shortness of breath. Other symptoms include a sore throat, a runny nose, fatigue, and muscle aches. The virus was first discovered in December 2019 in Wuhan, Hubei Province, China. In March 2020, the World Health Organization declared that the COVID-19 outbreak was a pandemic, which means that it had spread globally.

The virus spreads through infected droplets in the air. These droplets are released when people who have the virus cough or exhale. The droplets also land on surfaces. When people touch surfaces with virus particles and then touch their eyes, nose, or mouth, they can transmit the virus to their respiratory system. People may develop symptoms for up to 14 days after being exposed to the virus. During that time, scientists think that people are contagious. Some people may not have any symptoms, but are still contagious.

In spring 2020, scientists began working to develop a vaccine for COVID-19. They were also testing medications to help fight the disease.

Mononucleosis

Mononucleosis (MAH•noh•nook•klee•OH•sis), or mono, is a viral disease characterized by a severe sore throat and swelling of the lymph glands in the neck and around the throat. Mono most commonly infects teens and young adults. People in these age groups tend do more dating and more kissing. This makes them more vulnerable to the disease. It is spread through contact with the saliva of an infected person. It can also be spread by contaminated drinking glasses and eating utensils.

Besides a sore throat and swollen lymph glands in the neck, symptoms also include fatigue, loss of appetite, and headache. Severe cases may include an enlarged spleen and an infected liver.

People diagnosed with mono are advised to take it easy. Rest is the best treatment for mono. The good news is that once a person is fully recovered from mono, they cannot get it again.

Reading Check

Explain How have the vaccinations for chicken pox, measles, and mumps affected people?

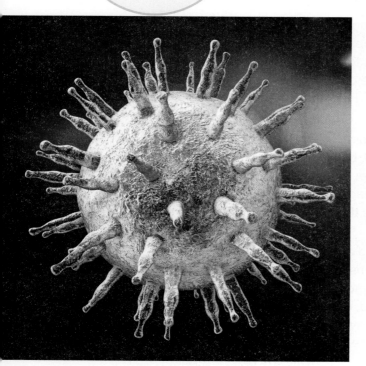

This is an illustration of the Epstein-Barr virus, the cause of infectious mononucleosis. **What is the best way to treat mono?**

Hepatitis

Hepatitis (hep•uh•TY•tis) is a viral disease characterized by an inflammation of the liver and yellowing of the skin and the whites of the eyes. Other symptoms include fatigue, weakness, loss of appetite, fever, headaches, and sore throat.

There are three common strains of hepatitis: A, B, and C. A different virus causes each strain. Hepatitis A is commonly found in areas that have poor sanitation. It spreads when infected human wastes contaminate the food and water sources. If someone consumes contaminated water or food, they can become infected. An open wound that is cleaned with contaminated water can also cause the infection to spread. There is no treatment for hepatitis A. However, it will usually clear up in a few months. Rest and healthful food choices help in the recovery from hepatitis A.

Hepatitis B is a disease caused by the hepatitis B virus that affects the liver. Hepatitis B and C can be more dangerous. These strains can cause permanent damage to the liver. They can lead to cirrhosis and liver cancer. These types of hepatitis are usually spread through contact with infected blood or other body fluids. They can spread when drug users share needles. They can also spread through sexual contact. There are vaccinations for hepatitis A and B. There are medications that can help treat hepatitis B and C.

Disease	Symptoms	Contagious Period	Vaccine
Chicken pox	Itchy rash, fever, muscle aches	One to five days before symptoms appear to when spots crust over	Yes
Pneumonia	High fever, chest pain, cough	Varies	For some types
Rubella	Swollen lymph nodes, rash, fever	Seven days before rash starts to five days after	Yes
Measles	Fever, runny nose, cough, rash	Three to four days before rash starts to four days after	Yes
Mumps	Fever, headache, swollen areas in neck and under jaw	Seven days before symptoms to nine days after	Yes
Whooping cough	Fever, runny nose, dry cough (with a whooping sound)	From inflammation of mucous membranes to four weeks after	Yes
Tuberculosis	Fever, fatigue, weight loss, coughing blood	Varies	Yes
COVID-19	Fever, cough, shortness of breath	At least 14 days	Yes

Communicable diseases can be spread in many ways and cause illness for varying lengths of time. **Identify which diseases have similar symptoms and name the symptoms.**

Tuberculosis

Tuberculosis (TB) (too•ber•kyuh•LOH•sis) is a bacterial infection that usually affects the lungs. Symptoms include cough, fatigue, night sweats, fever, and weight loss. It is spread through the air when an infected person coughs or sneezes. The cough or sneeze carries tiny droplets that are then inhaled by another person.

A person can have tuberculosis, or TB, and not even know it. He or she may not show any symptoms. They may not be sick, but they can still spread the disease. Doctors and other health care providers often test people to be sure they do not carry TB.

Tuberculosis can be treated with medications. However, the treatment period is longer than for most other bacterial infections. Treatment for TB can take up to nine months. Your age, overall health, and type of medication used determine how long it might take. If TB is not treated properly, it can be fatal.

Until 1985, TB was rare in developed countries. The spread of HIV, the virus that causes AIDS, has contributed to the increase in TB. People with HIV have a weakened immune system. This keeps them from fighting off bacterial infections such as TB. In countries where TB is a problem, babies are given a vaccination for TB. The vaccination is not 100 percent effective. So, scientists are working to develop a better vaccine.

Pneumonia

Pneumonia is a serious inflammation of the lungs. It is usually caused by bacteria or viruses. Bacteria or viruses live in your sinuses, nose, mouth, and the environment. When these spread to your lungs, pneumonia can occur. Symptoms include fever, cough, chills, and nausea. Pneumonia can also cause vomiting, chest pains, and difficulty breathing. Pneumonia is spread through direct or indirect contact with an infected person. You can catch it from people whether or not they show symptoms of illness.

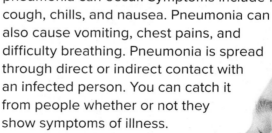

> **A PERSON CAN HAVE TUBERCULOSIS, OR TB, AND NOT EVEN KNOW IT.**

Eating healthful foods provides your body with the nutrients needed to fight illnesses. **Name two other behaviors that can help you fight pathogens.**

Pneumonia caused by bacteria can be treated with antibiotics. Pneumonia caused by viruses can be treated with antivirals. People with pneumonia need to rest and drink plenty of fluids. People who have other illnesses such as diabetes or HIV/AIDS are at greater risk to catch pneumonia.

There are ways to lower your risk for getting pneumonia. First, pneumonia can be prevented with vaccines. Next, practicing good hygiene can reduce your risk for pneumonia. Wash your hands often and thoroughly. Keep hard surfaces such as counters and desks that you touch often clean. Cough or sneeze into a tissue or into your elbow or sleeve. Reduce your exposure to cigarette smoke. All of these can help reduce your risk for pneumonia.

Strep Throat

Have you ever had a sore throat? Sore throats can be very painful and uncomfortable. Most sore throats are caused by a virus. **Strep throat** is a sore throat caused by the streptococcal bacteria. It can be treated with antibiotics.

Strep throat is spread through direct or indirect contact. When someone infected with strep throat breathes or coughs, they release droplets into the air. If you happen to breathe in some of those droplets, you will likely get strep throat.

How do you know if you have strep throat? If you have a red and painful throat, fever, and swollen lymph nodes in your neck, you may have strep throat. Your symptoms might also include headache, nausea, and vomiting. If you have these symptoms, get medical help. If strep throat is left untreated, it can lead to more serious illnesses. It can lead to rheumatic fever, which can damage the heart. It can also lead to nephritis, which can damage the kidneys. Treatment with antibiotics can prevent these illnesses.

Reading Check

Compare How are the symptoms for pneumonia and strep throat similar?

Fitness Zone

Staying Healthy I've always heard that exercising regularly can help you stay healthy. I never really believed it, though. A couple of years ago, I started getting sick all the time. It seemed like I would just get over a cold and I'd get another one, or I'd get the flu. After several trips to the doctor, she suggested that I eat more healthful foods and exercise more. I tried it. And you know what? It worked. I feel a lot better. I have a group of friends that I ride bikes with. We also play soccer and shoot hoops. We don't eat as much fast food as we used to. We have a lot of fun together. But the best part is that I haven't been sick in quite a while. I guess it's true — exercise and eating right really can help you stay healthy.

Lesson 2 Review

What I Learned

1. **DEFINE** What is the *contagious period* of a disease?

2. **EXPLAIN** How might a person become infected with hepatitis A?

3. **DESCRIBE** Why is it important to get treatment for diseases such as strep throat?

Thinking Critically

4. **ANALYZE** Japan is a small country with lots of people. If someone is ill and must go out, she or he will wear a surgical mask. Why do you think they do this?

5. **EVALUATE** You read on the Internet that a scientist has come up with the cure for the common cold. How can you know if his claim is valid?

Applying Health Skills

6. **PRACTICING HEALTHFUL BEHAVIORS** Create a poster that provides students with tips on how to stay healthy and how to keep others healthy. Ask for permission to post your finished product in a school hallway.

Sexually Transmitted Diseases

Before You Read

Quick Write Write a paragraph about why it is important for teens to avoid sexual activity.

Vocabulary

sexually transmitted diseases (STDs)
chlamydia
genital herpes
genital warts
trichomoniasis
gonorrhea
syphilis
bacterial vaginosis
pelvic inflammatory disease

Reading Check

Define What is a sexually transmitted disease?

BIG IDEA Sexually transmitted diseases are infections spread through sexual activity.

What Are Sexually Transmitted Diseases?

MAIN IDEA Sexually transmitted diseases are a growing problem in the United States.

Sexually transmitted diseases (STDs) are infections that are spread from person to person through sexual contact. They are also called sexually transmitted infections (STIs). The pathogens that cause STDs are transferred from person to person through sexual contact. A person who has sexual relations with someone who has an STD can be infected with the disease. Many times, a person with an STD may not even be aware that they have a disease. Some STDs are silent, meaning that they have mild or no symptoms.

STDs are a major health problem in the United States. The Centers for Disease Control and Prevention (CDC) tracks health issues in the United States. In 2017, the CDC reported that 2.3 million new cases of three STDs were diagnosed in the U.S. This is a 31 percent increase over previous years. Treating STDs costs the U.S. about $16 billion each year. If untreated, many STDs can cause infertility.

The CDC estimates that young people make up about 27 percent of the sexually active population in the United States. This age group experiences about 50 percent of all STDs. The message is that if you're young, practice abstinence to avoid STDs.

> MANY TIMES, A PERSON WITH AN STD MIGHT NOT EVEN BE WARE THAT THEY HAVE A DISEASE.

Common STDs

MAIN IDEA Sexually transmitted diseases include a wide range of diseases.

STDs are passed from one partner to another through sexual activity. All STDs affect both males and females. STDs can cause serious health problems if they are not treated. If you or someone you know suspect you have an STD, seek medical attention. Some common STDs are described here.

Chlamydia (kluh•MI•dee•uh) is a bacterial STD that may affect the reproductive organs, urethra, and anus. Chlamydia is the most often reported bacterial STD in the United States. It is estimated that there are 2.8 million cases in the U.S. each year.

Chlamydia is transmitted from person to person through sexual relations. The greater the number of sexual partners a person has, the greater the risk for the infection. Chlamydia can also be passed from an infected mother to her baby during childbirth.

Chlamydia is often called a "silent" disease because most infected people don't show any symptoms. A person can have it and not know about it. If symptoms do occur, they include genital discharge and pain when urinating. Untreated chlamydia can cause pelvic inflammatory disease in females. This can cause infertility. Chlamydia can also cause infertility in males. If left untreated, it can affect the reproductive organs, urethra, and anus.

Chlamydia can be cured with antibiotics. However, reinfection is common if sexual partners have not been treated. Abstinence from sexual activity is the only sure way to prevent chlamydia.

Genital herpes (HER•peez) is a viral STD that produces painful blisters on the genital area. It causes painful blisters on the genital area. It is transmitted by skin-to-skin contact. Sexual relations do not have to take place for herpes to be transmitted.

People can have herpes for many years and not know it. There may not be any obvious symptoms. Sometimes, though, there are outbreaks of painful sores and blisters. Even when the blisters go away, the virus is still in the body. A person with herpes can infect another person even if the infected person has no blisters or sores. Symptoms also include pain in the lower genital region and genital discharge. Severe symptoms include fever and swollen glands.

There is no treatment that can cure herpes. There are medications that help to reduce the number of outbreaks. They also might prevent outbreaks. Other medications might reduce the possibility of transmission to sexual partners. Abstaining from sexual contact is the only sure way to prevent transmission of genital herpes.

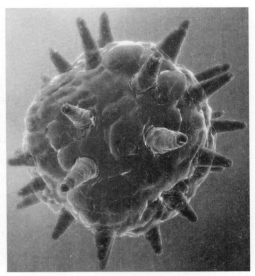

Viruses, bacteria, and parasites cause STDs. **List the names of the viral STDs.**

Reading Check

Identify Name two STDs that are caused by viruses.

Genital warts are growths or bumps in the genital region caused by certain types of human papillomavirus or HPV. HPV is the most common sexually transmitted infection in the United States. About 6 million people become infected each year.

There are more than 40 types of HPV. They infect the genital areas of both males and females. HPV is passed through genital contact. Sexual intercourse does not have to occur for the virus to be transmitted. It can be passed to another person whether or not symptoms are present.

HPV is another "silent" disease. Most people don't develop symptoms or health problems. In 90 percent of cases, the immune system clears the body of the virus in about two years. In some cases, the body does not clear the infections. Then genital warts may appear. Warts may appear as a small bump or several bumps. On rare occasions, warts appear in the throat. Cervical cancer and cancers of other reproductive organs can also be caused by HPV.

There are vaccines that can protect against the most common types of HPV. When taken by girls and women ages 11 through 26, the vaccines can protect against genital warts and most cervical cancers. When males ages 9 through 26 take the vaccines, they may be protected against genital warts and anal cancers.

Condoms and dental dams may also lower the risk of transmitting HPV. They may also lower the risk of genital warts and cervical cancer. However, condoms do not cover all of the areas that can be infected. So, condoms do not fully protect against HPV. As with all other STDs, abstinence is the only sure way to prevent infection with HPV.

Pubic lice are also called crabs. They are a parasite that affect the pubic or genital area. Pubic lice are transferred from one person to another through sexual contact. They can also be spread through contact with clothing, bed linens, and towels if these items were used by someone with pubic lice.

Pubic lice are usually transferred from one person to another through sexual contact. They are very contagious.

Pubic lice cause itching around the genitals. You might also notice eggs or nits attached at the base of a shaft of hair. These are small and might be difficult to see without a magnifying glass.

There are a number of over-the-counter medications to kill pubic lice. These include lotions and shampoos. When using these products, it is important to follow the directions exactly. It is also important to kill

Pubic lice are a parasite that affect the genital area. **Explain how pubic lice are treated.**

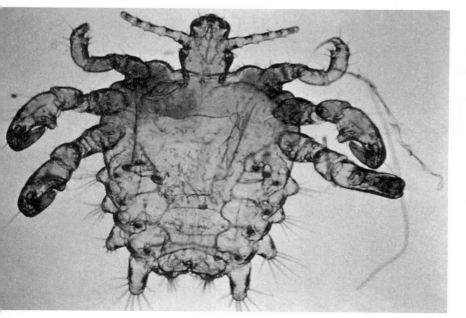

lice remaining on clothing, bedding, or towels to prevent re-infestation. Items should be washed in hot water and dried in a hot dryer cycle. Items that cannot be washed should be dry cleaned. If items cannot be cleaned by any of these methods, they should be sealed in a plastic bag for two weeks to kill the lice.

Trichomoniasis (RREE.koh.moh.NI.ah.sis) or "trich" is an STD caused by a protozoa. About 3.7 million people in the U.S. get the disease each year. The parasite is passed from person-to-person during sexual activity.

Trichomoniasis is a silent STD. Most people have no symptoms. If symptoms are present, they include vaginal discharge and discomfort during urination. Irritation or itching in the genital area can also occur.

Having trichomoniasis increases the chance of getting other STDs. Antibiotics are used to treat the disease. Both partners should be treated. About 20 percent of people who are cured are re-infected within three months because they have sexual relations with a person who is untreated.

Gonorrhea (gahn.uh.REE.uh) is a bacterial STD. It affects the mucous membranes, especially in the genital area. About 700,000 people in the U.S. get gonorrhea each year. Teens and young adults are at the most risk.

Gonorrhea is spread through sexual activity with someone who has the disease. A person who is treated can become re-infected. Mothers can pass the infection to their infant during childbirth.

The symptoms of gonorrhea are different for males and females. Some males and females may not have symptoms. If a male has symptoms, they can include a thick, yellowish discharge from the penis and a burning sensation while urinating. Symptoms in females can be mistaken for a bladder or vaginal infection.

Gonorrhea can cause permanent health problems. It can spread to joints and the heart. Females who are not treated can become infertile. Gonorrhea can be treated with antibiotics.

Syphilis (SIH.fuh.luhs) is a bacterial STD. It affects many parts of the body. It is spread during sexual relations if there is contact with a syphilis sore. A pregnant female can spread syphilis to her fetus. The disease cannot be spread through contact with toilet seats, bathtubs, shared clothing, swimming pools, or hot tubs.

Syphilis affects the body in stages. During the first stage painless sores appear at the place where the infection occurs. If untreated, symptoms in the next stage include a body rash. Other symptoms are fever, swollen lymph glands, sore throat, patchy hair loss, headaches, weight loss, muscle aches, and fatigue. In the final stage, the infection moves through the body. It damages the brain, nerves, eyes, heart, blood vessels, liver, bones, and joints. Penicillin is used to cure syphilis if the person has been infected less than a year. If the infection is not treated during this time, multiple injections of penicillin may be needed.

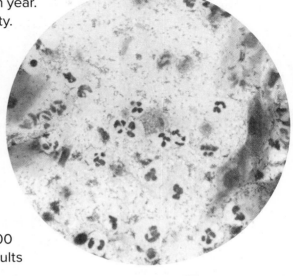

Trichomoniasis is a silent STD. **Describe why an STD is labeled as a silent STD.**

Bacterial vaginosis (BV) is a condition that happens when there is too much of certain bacteria in the vagina, changing the normal balance of bacteria. It is the most common STD affecting females between the ages of 15 to 44. Scientists don't know how BV is spread. It occurs in females who are sexually active. Females who are not sexually active rarely get BV. Having BV increases the risk of getting other STDs. Symptoms of BV include:

- a white or gray vaginal discharge,
- pain, itching, or burning in the vagina,
- a strong odor, and
- a burning feeling while urinating.

BV is treated with antibiotics. To avoid BV, practice abstinence from sexual activity. Douching should also be avoided. It can spread the bacteria that causes BV.

Pelvic inflammatory disease (PID) is an infection of the female's reproductive organs. It can be treated with antibiotics. The symptoms of PID may be mild. Some female have no symptoms. If they do occur, symptoms can include:

- pain in the lower abdomen
- fever
- unusual discharge and bad odor from the vagina
- pain or bleeding during sex
- burning sensation during urination
- bleeding between periods.

PID can damage reproductive organs if it is untreated. Having the disease can result in the inability to get pregnant. A female with untreated PID may also have pelvic/abdominal pain throughout her life.

Preventing and Treating STDs

MAIN IDEA Abstinence is the only sure way to prevent STDs.

The media is full of messages showing sexual activity as exciting. Those messages never mention the risk associated with sexual activity. Those risks include the spread of STDs or an unplanned pregnancy. Ads also don't show the social and emotional problems that can result from becoming involved in a serious relationship while a young teen.

Acquiring an STD can affect the rest of a person's life. Some cannot be cured. Having genital herpes cannot be cured. A person with this disease needs to be cautious throughout their life to avoid spreading the disease to others.

The only 100 percent effective way to avoid STDs is to abstain from sexual activity. Deciding to say no to sexual activity may be a difficult decision to make. Some teens face pressure from friends to become sexually active. Choosing friends with similar values can make saying no easier.

When you know the facts about STDs, you have the power to avoid them and prevent them from spreading. **Describe the best way to avoid getting STDs?**

Reading Check

Apply What might you say if a date wants to engage in sexual activity?

If abstinence is not an option, a condom or dental dam should be used. Condoms and dental dams can reduce the spread of STDs if used correctly. Most STDs are transmitted through skin-to-skin contact or through the exchange of body fluids. If used correctly, condoms and dental dams help prevent skin-to-skin contact and the exchange of body fluids.

A person who suspects that he or she may have an STD should seek medical help right away. If left untreated, STDs can cause permanent damage to the reproductive system and other serious problems. If telling parents isn't safe, a trusted adult can help understand what options are available. Free clinics are open in many cities that can treat STDs free-of-charge.

Group activities can help you avoid situations where you may feel pressure to engage in sexual activity. **Name two activities that you and your friends like to do together.**

Lesson 3 Review

What I Learned

1. **DEFINE** What is an *STD*?

2. **EXPLAIN** What are the consequences of syphilis if left untreated?

3. **APPLY** What is the best way to avoid becoming infected with STDs?

Thinking Critically

4. **APPLY** A teen thinks he or she has an STD. Why is it important for the teen to seek medical help?

5. **EVALUATE** Why might drinking alcohol increase your risk of getting an STD?

Applying Health Skills

6. **COMMUNICATION SKILLS** Using the S.T.O.P. strategy can help you refuse to participate in unhealthy activities. The steps to the S.T.O.P. strategy are: Say no in a firm voice. Tell why not. Offer other ideas. Promptly leave. Use these steps to develop a list of responses to pressure to engage in sexual activity.

HIV/AIDS

Before You Read

Quick Write. Write a response to this question: What would happen if the cells that control your immune responses began to be destroyed?

Vocabulary

HIV (human immunodeficiency virus)
AIDS (acquired immunodeficiency syndrome)
opportunistic infections
carrier

BIG IDEA HIV causes AIDS, which is a deadly disease that interferes with the body's immune system.

What Are HIV and AIDS?

MAIN IDEA The human immunodeficiency virus (HIV) causes the acquired immunodeficiency syndrome (AIDS), a serious disease that attacks the body's immune system.

HIV (human immunodeficiency virus) is the virus that causes AIDS. HIV and AIDS are diseases that attack the immune system. HIV attacks lymphocytes called T cells. When it attacks a T cell, it replaces the cell's genetic information with its own genetic information. Then it begins to multiply. As more T cells are taken over, the immune system weakens. Eventually, the T cell count drops so low that the immune system can no longer protect the body. When this happens, AIDS develops. The art in the lesson describes how HIV attacks the immune system.

AIDS (acquired immunodeficiency syndrome) is a deadly disease that interferes with the body's natural ability to fight infection. It continues to weaken the immune system. When AIDS is present in the body, the likelihood of developing an **opportunistic infection** increases. Opportunistic infections are diseases that attack a person with a weakened immune system and rarely occur in a healthy person. For example, many people with AIDS develop a type of pneumonia that can cause death. Drugs can help delay the onset of AIDS. However, there is no cure. People with AIDS will eventually die from diseases that a healthy immune system could have resisted.

In the United States, about 1.1 million people are living with HIV. About 160,000 of those people are unaware that they have HIV. This group accounts for about 40 percent of all new HIV infections.

How HIV is Spread

You cannot get HIV through casual contact with a person who has the virus. HIV circulates through the body in infected blood and other body fluids. Body fluids that can transmit HIV include fluid from the vagina, breast milk, and semen. Semen is the fluid that carries sperm. There are several ways these fluids spread from one person to another.

Reading Check

Identify What are five ways HIV cannot be spread?

- **Engaging in sexual intercourse with an infected person.** The most common way HIV spreads is through sexual intercourse. Having any form of sexual intercourse with an infected person can spread HIV. During sexual activity, semen and vaginal fluids are transferred from one person to the other. HIV can be transferred in these fluids.

The virus circulates in the body even before it destroys the immune system. People can have the virus and not know it. This makes it extremely risky to have sexual intercourse outside of a monogamous relationship. People with multiple partners are at the greatest risk of getting the virus. Only one incident of sexual activity with an infected person can spread the virus.

- **Using a contaminated needle.** Drug users can get HIV from a needle that has already been used by an infected person. A tiny drop of blood left on a needle is all it takes to transmit HIV to another person. Needles used for tattooing and piercing can also be contaminated. They, too, can transmit the virus. People who use needles to take medication for diabetes or other illnesses should do so under the care of a medical professional.
- **Other ways HIV is spread.** HIV can be spread from a pregnant woman with HIV to her developing baby. It can be transmitted to the child during birth or during breast feeding. Certain drugs can reduce the rate of transmission of HIV to the unborn child during pregnancy.

In the years before it was known that HIV caused AIDS, some people were infected during blood transfusions through a contaminated blood supply. However, since 1985, all blood is screened for HIV. The blood supply in the United States is considered to be extremely safe from the virus.

In the U.S. about 1.1 million people are living with HIV. **Identify the approximate number of people who are unaware that they have the virus.**

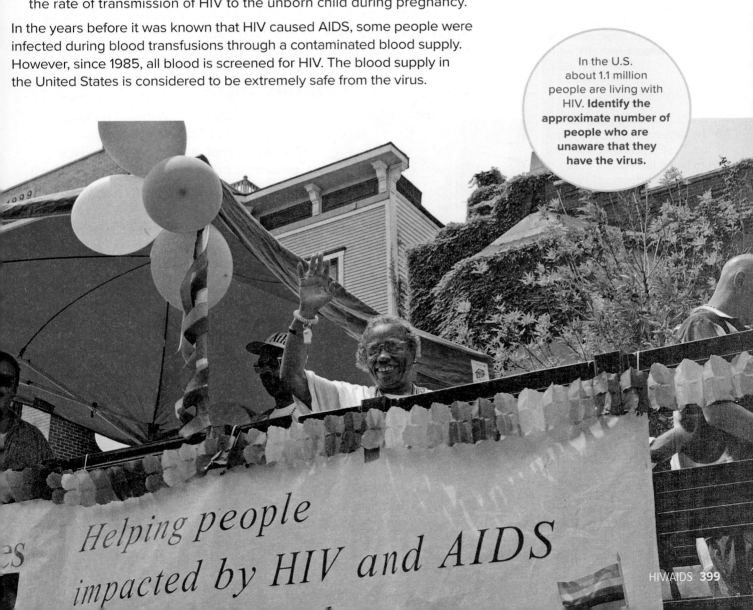

Helping people impacted by HIV and AIDS

How HIV is NOT Spread

Remember that HIV can only be spread through specific infected body fluids. The HIV virus cannot survive for long outside these specific body fluids. HIV *cannot* be spread by:

- Breathing the air near an infected person.
- Being bitten by a mosquito that has bitten an infected person.
- Touching the tears of an infected person.
- Touching the sweat of an infected person.
- Touching objects touched by someone with HIV.
- Shaking hands with an infected person.
- Hugging an infected person.
- Swimming in a pool with an infected person.
- Sharing utensils with an infected person.
- Donating blood.
- Using the same shower, bathtub, or toilet as an infected person.
- Sharing sports equipment with an infected person.

It is not necessary to avoid people with HIV or AIDS. You cannot get sick just by being around them. They need your friendship and respect just like other people.

> **YOU CANNOT GET SICK BY JUST BEING AROUND PEOPLE WITH HIV OR AIDS.**

> HIV damages the immune system. **Name the type of infections that people with HIV and AIDS may experience.**

1 The virus enters the body. When that T cell is activated, it will start producing more of the virus instead of performing T cell functions. Eventually, the host cells are destroyed.

3 More T cells are infected and destroyed. Without T cells, the body loses its ability to fight diseases and infections.

2 These viruses infect other T cells and multiply. T cells are destroyed, and more copies of the virus are released.

Key
- HIV
- T cell
- Pathogens

Fighting AIDS

MAIN IDEA AIDS is a worldwide problem.

The World Health Organization (WHO) estimated that in a recent year that almost 38 million people worldwide were infected with HIV. In that same year 770,000 people died from AIDS. Recent estimates by the Centers for Disease Control and Prevention (CDC) state that 12 million people in the United States are infected with HIV. Young people ages 13-29 account for 39 percent of the new HIV infections in recent years. CDC also believes that 20 percent of the people with HIV do not know they have it. CDC is the government agency that protects and promotes the health of the people of the United States through research and education.

Knowledge and abstinence until marriage are the best weapons against HIV and AIDS. Around the world, scientists, doctors, and others are working to educate people about the disease. When people know how it is spread, they can take measures to avoid becoming infected.

Detecting HIV

A person can be a **carrier** of HIV without having AIDS. A carrier is a person who is infected with a virus and who can pass it on to others. It's important that people learn their status because about 1 in 7 people in the U.S. have HIV and are unaware that they have the disease. The CDC's guidelines for testing are:

1. Every person between the ages of 13 and 64 be tested once for HIV. This test should be done as part of a routine medical exam.

2. Every person being treated for an STD should also be tested for HIV. The person should be told why they are being tested for HIV. Testing should only take place if the person agrees to it.

Laboratory testing is the only way to show whether the virus antibodies are present. However, if a person was recently infected, the virus antibodies may not show up in a lab test. The CDC recommends that people who are in a high-risk category should be retested in three months.

The CDC also provides information on how and where to get tested. Resources for testing include your personal doctor, medical clinics, substance abuse programs, community health centers, and hospitals. To learn where you can get tested in your community, use these resources:

- Call 1-800-CDC-INFO (232-4636)
- Visit gettested.cdc.gov
- Text your zip code to KNOW IT (566948)

Laboratory tests can detect the presence of HIV in the blood. **Tell who should be tested for HIV.**

Reading Check

Explain What treatment is available for people with HIV?

Treating HIV and AIDS

A diagnosis of HIV changes a person's life. When it is confirmed, the CDC suggests that the person begin counseling. It should include emotional support, the need for behavioral changes, the need to stop using drugs or alcohol, and the need for ongoing medical care.

Today, HIV treatment slows the progress of the disease. The most common treatment is called antiretroviral therapy, or ART. Patients on ART take a combination of drugs. These drugs prevent HIV from overwhelming the immune system.

For ART to work best, the patient must follow a strict schedule for taking medications. Patients are cautioned not to skip taking a drug. They are also told to take their drugs on the strict schedule provided by the doctor. In some cases, ART has made HIV undetectable. However, resistance to some ART drugs can develop. Patients are told to report all of their symptoms and side effects to their doctor.

Fighting HIV infection is costly and difficult. So, scientists and educators work hard to keep people from getting HIV in the first place. Knowledge and abstinence are still the best weapons in the fight against HIV and AIDS.

Stopping the Spread of HIV

Work to develop drugs that eliminate the disease continue but progress has been made. Scientists are working to develop a vaccine to prevent the spread of HIV. Some drugs that have been developed to fight the infection have reduced the level of HIV in the blood of some carriers.

At this time, a vaccine is not available to prevent the spread of HIV. Until a vaccine is available, the best way to prevent HIV is to avoid high-risk behaviors. There are three main ways to avoid HIV and AIDS:

- **Practice abstinence.** Abstinence is consciously choosing to not participate in high-risk behaviors. Avoiding sexual activities until marriage is one type of abstinence. People who participate in sexual activity expose themselves to risks. They may not know whether their partner has an STD. They may not know whether their partner has HIV or AIDS. The more sexual partners a person has, the greater their risk of getting an STD such as HIV.

Counseling of a newly diagnosed person can help prevent the spread of HIV and AIDS. **Name three things that counseling a person with HIV should achieve.**

- **Avoid drugs and alcohol.** Using drugs and alcohol can make a person lose their ability to make good decisions. When their judgment is impaired, they are more likely to engage in risky behaviors. The risky behaviors might include sexual activity.
- **Avoid sharing needles.** Shared needles can carry enough blood to inject HIV into your blood stream. This includes needles used for injecting drugs or for tattoos or piercing. Sharing needles exposes you to any diseases the previous user has.

The only sure way to avoid getting HIV is to avoid contact with sources of the virus. Abstaining from sexual activity and not injecting drugs or sharing needles are ways to avoid contact with the virus. Avoid peer pressure to engage in these activities. Talk to a parent or trusted adult if you are being pressured. Find new friends if you need to. Find friends who think the way you do, who share your values, and who respect your decisions. If abstinence is not an option, a condom or dental dam should be used. Condoms and dental dams can reduce the spread of HIV if used correctly. Make responsible choices. It could save your life.

Developing Good Character

Caring Practicing abstinence from sexual activity until marriage and avoiding illegal drug use, especially the use of injectable drugs, will help protect you from HIV infection.
How does avoiding these risk behaviors show that you care about yourself and others?

The AIDS Quilt program serves as a memorial for the lives lost to AIDS. **Hypothesize another way that the AIDS Quilt program helps prevent the spread of HIV.**

Lesson 4 Review

What I Learned

1. **LIST** What are three ways HIV is spread?

2. **IDENTIFY** What are four ways you cannot become infected with HIV?

3. **DEFINE** Provide definitions for HIV and AIDS.

Thinking Critically

4. **ANALYZE** What is the relationship between HIV and AIDS?

5. **APPLY** Why is it important to see a health care provider if someone thinks they may have an STD?

Applying Health Skills

6. **ADVOCACY** Create a brochure or pamphlet about HIV and AIDS. Include the basic facts about the diseases and explain how to avoid getting the disease. Include ways it can be transmitted and ways it cannot be transmitted. Provide copies for your school health office.

Noncommunicable Diseases

LESSONS

1 Causes of Noncommunicable Diseases

2 Cancer

3 Heart and Circulatory Problems

4 Diabetes and Arthritis

5 Allergies and Asthma

18

Causes of Noncommunicable Diseases

Before You Read

Quick Write Make a list of as many factors as you can think of that can cause noncommunicable diseases.

Vocabulary
noncommunicable disease
chronic diseases
degenerative diseases
congenital disorders
birth defects
genetic disorders

BIG IDEA Noncommunicable diseases can result from hereditary or lifestyle choices, or may have an unknown cause.

What Are Noncommunicable Diseases?

MAIN IDEA Noncommunicable diseases are diseases that cannot be spread from person to person.

When Hannah is at her friend Danielle's house, her eyes get red and itchy. She starts to sneeze. Why? Hannah is allergic to Danielle's dog. An allergy is one type of a **noncommunicable disease**, or a disease that cannot be spread from person to person. You cannot catch an allergy or another noncommunicable disease, such as diabetes, from someone who has this disease.

Some noncommunicable diseases are chronic diseases. Asthma is one type of **chronic disease**. These are diseases that are present either continuously or off and on over a long time. Some of the most common noncommunicable diseases are chronic diseases. Other noncommunicable diseases include heart disease, cancer, allergies, diabetes, and arthritis. Some of these diseases are called **degenerative diseases**, which are diseases that cause further breakdown in body cells, tissues, and organs as they progress. Multiple sclerosis (MS) is a type of degenerative disease. MS damages nerve cells. This can cause nerve signals to slow down or stop. Over time, there is an increased loss of nerve function.

What Causes Noncommunicable Diseases?

MAIN IDEA Risk factors for noncommunicable diseases include heredity, lifestyle choices, and environmental factors.

Noncommunicable diseases can have many causes. Some may be present at birth. Others may be caused by lifestyle choices or a person's environment. Some types, such as rheumatoid arthritis and Alzheimer's disease, have no known cause.

Reading Check

Compare and Contrast What are the similarities and differences between chronic diseases and degenerative diseases?

> YOU CANNOT CATCH A NONCOMMUNICABLE DISEASE FROM SOMEONE WHO HAS THIS DISEASE.

Diseases Present at Birth

Some babies are born with **congenital disorders**, or disorders that are present when a baby is born. Babies may be born with physical or mental disabilities caused by **birth defects** or **genetic disorders**. Birth defects are an abnormality present at birth that causes physical or mental disability or death. Genetic disorders are disorders that are caused partly or completely by a defect in genes. Many birth defects have no known cause. Two types of congenital disorders are cystic fibrosis and sickle-cell anemia. These diseases are caused by heredity. Traits are passed from parents to their children. Tay-Sachs disease is another type of an inherited disease. It is a disorder that destroys the central nervous system.

Congenital heart disease is another type of congenital disorder. This is a defect in a heart valve or one of the big blood vessels leading out of the heart. The cause of congenital heart disease is often unknown. In some cases, it may be inherited. In some cases, it may result from medications or infections the mother may have taken during her pregnancy.

A mother's lifestyle choices can also lead to birth defects. For example, a pregnant female who drinks alcohol may give birth to a child with fetal alcohol syndrome (FAS). Symptoms of FAS include heart defects, poor coordination, and problems with speech, thinking, or social skills.

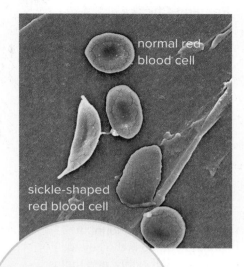

normal red blood cell

sickle-shaped red blood cell

Sickle-cell anemia is a hereditary disease that is passed on from parents to their children. **Name another inherited disorder and how it affects the body.**

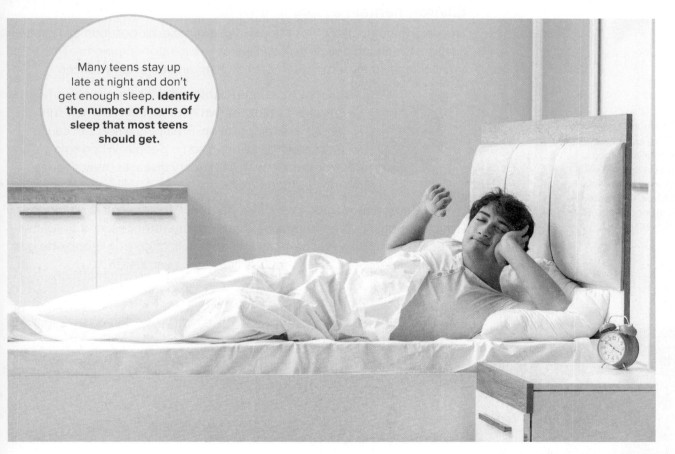

Many teens stay up late at night and don't get enough sleep. **Identify the number of hours of sleep that most teens should get.**

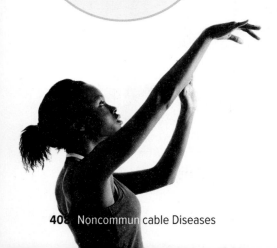

Reading Check

Analyze What personal health care practices can prevent the spread of noncommunicable diseases?

Cultural Literacy

What Hand Washing Does

When you wash your hands with soap and water, the rubbing of the hands scrapes off dirt, oils, and other particles, including pathogens. The amount of rubbing you do is the key to washing thoroughly, and that is why rubbing the hands for 30 seconds or more is essential.

Environmental factors can make outdoor activities difficult for people with respiratory diseases. **List the four health problems that can occur due to air pollution.**

Lifestyle Choices and Disease

It is hard to know who will develop a certain disease. However, some risk factors increase a person's chance of developing a disease. Heredity, age, gender, and ethnic group are factors over which people have no control.

People can, however, control one major group of risk factors — their lifestyle choices. When people make unhealthy lifestyle choices, they increase their risk of disease. Drinking too much alcohol can cause cirrhosis of the liver. Smoking or chewing tobacco can cause cancer and other diseases. Heart disease can result from a lack of physical activity, being overweight, or eating foods high in fat. To decrease your risk of disease, practice the following healthful lifestyle behaviors:

- **Eat healthful foods.** Eat plenty of whole grains, fruits, and vegetables. Go easy on foods high in fat, sugar, or salt.
- **Stay physically active.** Teens should be physically active at least 60 minutes every day. Regular physical activity strengthens the heart, lungs, muscles, and bones and helps the body systems work better.
- **Maintain a healthy weight.** Keep your weight within the recommended range for your gender, age, height, and body frame.
- **Get enough sleep.** Teens need at least eight hours of sleep a night.
- **Manage stress.** Use time management and other healthful strategies to reduce stress.
- **Avoid tobacco, alcohol, and other drugs.** These substances harm many parts of the body.

Environmental Factors and Disease

The environment affects your health. For example, air pollution can cause disease. Breathing polluted air can worsen respiratory problems. These can include asthma, emphysema, bronchitis, and even lung cancer.

You may have seen photos of cities covered by smog. Smog is a yellow-brown haze that forms when sunlight reacts with air pollution. Breathing smog can cause respiratory diseases in some people. When smog is heavy, people with respiratory diseases, may need to limit their outdoor activities.

Carbon monoxide is another harmful substance. It is a colorless, odorless gas. Fumes from car exhaust, some furnaces, and fireplaces emit carbon monoxide when used. High levels of carbon monoxide gas can be dangerous. It cause serious illness or even death.

THE ENVIRONMENT AFFECTS YOUR HEALTH.

Exposure to tobacco smoke, or secondhand smoke, can also lead to health problems. Children exposed to secondhand smoke have an increased risk of asthma attacks, respiratory infections, and ear infections. In nonsmoking adults, exposure to secondhand smoke can cause heart disease and lung cancer.

Air pollution is an environmental factor that can cause respiratory illness. **Name two other environmental factors that can cause respiratory disease.**

Lesson 1 Review

What I Learned

1. **VOCABULARY** Define *noncommunicable disease* and use it in a sentence that shows its meaning.

2. **GIVE EXAMPLES** What are examples of two congenital disorders?

3. **IDENTIFY** What are three risk factors that can cause noncommunicable diseases?

Thinking Critically

4. **EVALUATE** How can lifestyle choices affect a person's health?

5. **SUGGEST** How can communities lower the risk of diseases caused by the environment?

Applying Health Skills

6. **ACCESSING INFORMATION** Using print or online resources, research some common noncommunicable diseases. Choose one example and create a fact sheet describing what causes the disease and how it affects the body. Include information on how this disease is treated or managed. Share your findings with the class.

Cancer

. . . .

Before You Read

Quick Write Write a couple of sentences that explain why people need to wear sunscreen when they go outdoors.

Vocabulary

cancer
tumor
benign
malignant
carcinogen
biopsy
radiation therapy
chemotherapy
remission
recurrence

. . . .

BIG IDEA Cancer is a noncommunicable disease that occurs when abnormal cells multiply out of control.

What is Cancer?

MAIN IDEA Cancer is characterized by the rapid and uncontrolled growth of abnormal cells.

Cancer is a disease that occurs when abnormal cells multiply out of control. It's a noncommunicable disease that can affect people of any age. Any tissue in the body can become cancerous. Many cancers can be treated successfully. However, cancer is the second leading cause of death in the United States. Only heart disease kills more Americans each year.

How does cancer develop? The adult human body contains more than 100 trillion cells. These cells constantly divide to make more cells so the body can grow and repair itself. Most of the body's cells are normal at any given time. However, even in healthy bodies, some cells become abnormal. Your body's immune system usually destroys these cells. However, some abnormal cells can survive and begin to divide. Some of these abnormal cells grow in a clump called a **tumor** (TOO•mer). A tumor is a group of abnormal cells that form a mass.

Tumors are either **benign** (bi•NYN) or **malignant** (muh•LIG•nuht). Benign tumors are non-cancerous and do not spread. Malignant tumors are cancerous and can multiply out of control and sometimes spread to other parts of the body.

Types of Cancer

Any tissue in the body can become cancerous. Some types of cancer are more common than others. Skin cancer is the leading type of cancer. More than a million new cases of skin cancer are reported every year in the United States. They make up about half of all the cancer cases reported. Most cases of skin cancer are highly curable if detected early and treated appropriately.

Lung cancer is the deadliest form of cancer. It kills an estimated 160,000 people per year. More deaths result from lung cancer than do from colon cancer, prostate cancer, and ovarian cancer combined. The good news is that this number has been falling in recent years.

Causes and Risk Factors

Some types of cancers develop for unknown reasons. However, doctors have identified certain risk factors for some types of cancer. Risk factors can include inherited traits, age, lifestyle choices, and environmental factors. For example, a high-fat, low-fiber diet may be a risk factor for developing colon and rectal cancers.

Some types of cancer have well-known causes. For example, asbestos is a mineral that once was used in construction and manufacturing. Breathing asbestos dust can cause lung cancer. Asbestos is a **carcinogen** (kar•SI•nuh•juhn), which is a substance that can cause cancer. Other common carcinogens are the chemicals in tobacco. These chemicals are linked to lung and mouth cancers. The American Cancer Society has named about 100 substances that are known carcinogens for humans.

Not all carcinogens are chemicals. For instance, ultraviolet light from the sun can cause skin cancer. Exposure to radiation in large doses, including X-rays, can also cause cancer. Infection from certain viruses has been linked to specific types of cancer. For example, a chronic infection of hepatitis B virus is known to cause liver cancer.

Reading Check

Identify What are two carcinogens and the cancers they cause?

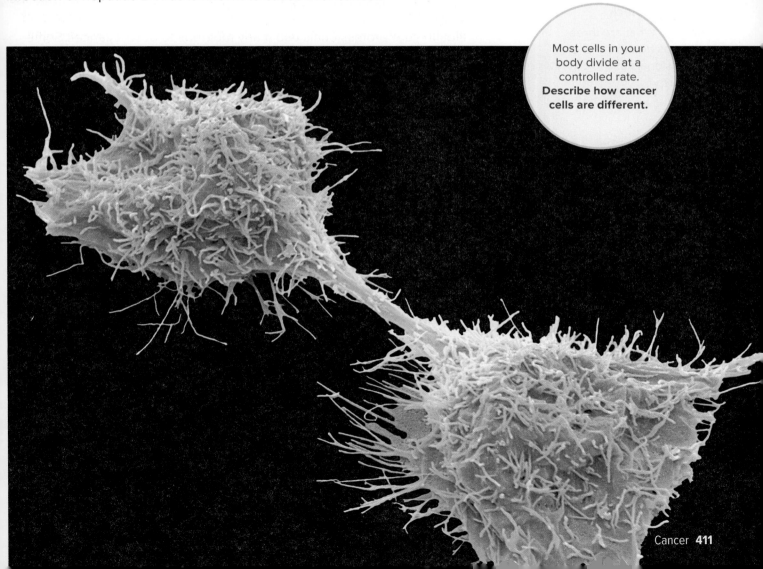

Most cells in your body divide at a controlled rate. **Describe how cancer cells are different.**

Type of Cancer	Important Facts
Skin cancer	is the most common kind of cancer. Excessive exposure to direct sunlight is the major cause of skin cancer.
Breast cancer	most often occurs in females over 50. It can occur in younger females and occasionally in males.
Reproductive organ cancer	can occur in the testicles and prostate gland in males. It can occur in the ovaries, cervix, and uterus in females.
Lung cancer	is the leading cause of cancer deaths in the United States. Smoking is the biggest risk factor for both males and females.
Colon and rectal cancer	develop in the digestive tract. Early detection and better screening have greatly reduced the number of cases of colon and rectal cancers.
Leukemia	is a cancer of the white blood cells that starts in the bone marrow. An increase in cancerous white blood cells interferes with the healthy white blood cells' immune response.
Lymphoma	is a cancer that starts in the lymphatic system. It weakens the immune system and increases the risk of developing infections.

The table lists some of the most common types of cancer. **Explain why people with lymphoma may struggle with other diseases as well.**

Reading Check

Identify What are three methods health care professionals use to detect cancer?

Diagnosing Cancer

MAIN IDEA Health care professionals use a variety of methods to detect and diagnose cancer.

Health care professionals use many methods to detect cancer. Some are very simple. For example, a doctor might spot a group of skin cells that don't look normal. He or she might feel a lump where the tissue should be soft.

Health care professionals also use more involved methods. They can use X-rays and other scanning equipment to look for unusual cell formations. If tissue shows a suspicious lump or formation, it usually undergoes a **biopsy**. A biopsy is the removal of a sample of tissue from a person for examination. The tissue from a biopsy goes to a lab for careful examination to see if the cells are cancerous. If they are, technicians will do other tests to learn more about the cancer. Together, a team of health care providers and the patient can decide on a plan for treatment.

One way to increase the chances of successful treatment is to detect cancer as early as possible. In order to detect cancer early, it is important to watch for warning signs. Some warning signs are listed in chart in the lesson. A person who shows a warning sign of cancer should be examined by a physician as soon as possible.

> ONE WAY TO INCREASE TH CHANCES OF SUCCESSFUL TREATMENT IS TO DETECT CANCER AS EARLY AS POSSIBLE.

Treating Cancer

MAIN IDEA Three common cancer treatments are surgery, radiation therapy, and chemotherapy.

The best way to treat cancer depends on many factors. These include the type of cancer, the stage of the disease, and the age and general health of the patient. The most common cancer treatments include surgery, radiation therapy, and chemotherapy. Cancer patients usually receive at least two of these treatments and sometimes all three.

Surgery removes cancer cells from the body. It is used to treat some cancers, including breast, lung, and colon cancers. Surgery is most effective when the cancer is isolated in one part of the body. **Radiation therapy** is a treatment that uses X-rays or other forms of radiation to kill cancer cells. It works best to kill cancer cells limited to a single area and to kill those that may still remain after surgery. More than half of all people with cancer are treated with radiation therapy. Doctors use **chemotherapy**, which is the use of powerful medicines to destroy cancer cells. It fights cancers that have spread beyond one location or occur throughout the body, such as leukemia.

Reading Check

Explain Which treatment is commonly used for a cancer that has spread?

THE MOST COMMON CANCER TREATMENTS ARE SURGERY, RADIATION, AND CHEMOTHERAPY.

Sometimes, x-rays are used to look for signs of cancer. **Explain how x-rays and other scans can be used to detect cancers.**

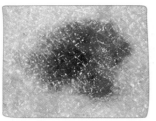

Asymmetry
One side of a mole looks different from the other side.

Border Irregularity
The edges are jagged or blurred.

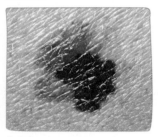

Color
The color is not uniform, or the same, throughout. If a mole is tan and brown, black, or red and white, have it checked.

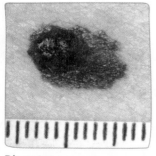

Diameter
The diameter is greater than 6 millimeters, about the size of a pencil eraser. A growth that has expanded to this size over time should be checked.

Side Effects of Treatment

Although cancer treatments are improving, all of them have side effects. Side effects of radiation therapy and chemotherapy include nausea, fatigue, and temporary hair loss. Side effects differ from person to person. They depend on the patient's age, the type of treatment, and the location of the cancer in the body. When cancer treatment is successful, the cancer is in **remission**. This is a period during which cancer signs and symptoms disappear. Sometimes, a **recurrence** happens. This is the return of cancer after remission.

Reducing the Risk of Cancer

MAIN IDEA The risk of cancer can be reduced by practicing certain healthful behaviors.

Anyone can get cancer, but you can protect yourself from some types of the disease. For example, staying tobacco free can greatly reduce your risk of developing lung cancer. Here are some tips on how to reduce your cancer risk.

- **Avoid tobacco and alcohol.** Cigarette smoking is the single major cause of cancer deaths in the United States. Excessive alcohol use increases the risk of several types of cancer. Two examples include liver cancer and cancer of the esophagus.
- **Eat well and exercise.** Many cancers, such as colon and rectal cancers, may be linked to diet. Eating well and staying fit can help you avoid these cancers.
- **Limit sun exposure.** Skin cancer is the most common form of cancer in the U.S. UV rays from the sun can cause cancer. To protect yourself from UV rays, avoid being in the sun between 10:00 a.m. and 4:00 p.m. That's when the sun's rays are the strongest. If you need to be outdoors during this time, apply a sunscreen with an SPF of at least 15 before you go outdoors. Also, wear a hat that shades your neck and the tops of your ears. You should also avoid tanning beds. They produce UV rays that can lead to skin cancer and damage the immune system.

Checking moles for irregularities can reveal skin cancers early. **Name the four factors to look for when checking a mole.**

- **Perform self-examinations.** Females should perform a breast self-exam once a month. Males should perform a testicular self-exam once a month. Ask your health care provider about the correct way to perform these exams. Directions on performing self-exams are included in the reproductive health lessons. If you notice any unusual lumps, see a health care provider right away. Also, check all moles and other skin growths frequently. See a health care provider immediately if you notice any changes in them.

- **Know the seven warning signs of cancer.** The American Cancer Society has identified seven possible signs of cancer. The first letter of each sign spells the word CAUTION. You play the most important role in early cancer detection. If you notice any of the warning signs, tell a parent, guardian, or health care professional right away.

Change in bowel or bladder habits

A sore that does not heal

Unusual bleeding or discharge

Thickening or lump in a breast or elsewhere

Indigestion or difficulty swallowing

Obvious change in a wart or mole

Nagging cough or hoarseness

Knowing the warning signs of cancer can help detect it in it's early stages. **Explain why early detection of cancer is important.**

Myth vs. Fact

Myth Indoor tanning beds and booths are not associated with skin cancer.

Fact Tanning beds expose skin to UVA radiation; the same type of energy emitted by the sun. Indoor tanning has been linked to at least two types of skin cancer, including melanoma, the deadliest type of skin cancer. Cancers of the eye are also linked to the use of tanning beds.

Reading Check

Give Examples List two ways in which you can take an active role in protecting yourself against cancer.

Lesson 2 Review

What I Learned

1. **VOCABULARY** Define *cancer*.

2. **NAME** What are three methods for diagnosing cancer?

3. **EXPLAIN** What are the differences among the three most common treatments for cancer?

Thinking Critically

4. **SYNTHESIZE** Moles get larger when the skin cells in the mole divide. Why do you think moles larger than 6 millimeters might be a warning sign of cancer?

5. **EVALUATE** Why is it important to diagnose cancer as early as possible?

Applying Health Skills

6. **ADVOCACY** During the summer, Zachary works as a lifeguard at the beach. He has noticed that a lot of teens don't wear sunscreen. He wants to create a pamphlet that talks about skin cancer and how it can be prevented. Create a similar pamphlet and hand it out to anyone who enjoys outdoor activities.

Heart and Circulatory Problems

Before You Read

Quick Write Explain three ways to keep your heart healthy.

Vocabulary

cardiovascular disease
arteriosclerosis
atherosclerosis
heart attack
arrhythmia
hypertension
stroke
angioplasty
pacemaker

Reading Check

Explain What factors contribute to heart disease?

BIG IDEA Heart disease is any condition that reduces the strength or function of the heart.

What is Heart Disease?

MAIN IDEA Heart disease can be caused by a variety of factors.

The heart is a muscle. Heart disease, or **cardiovascular disease**, is a disease of the heart and blood vessels. Common forms of heart disease include high blood pressure and hardening of the arteries.

The CDC reports more than 27 million people in the United States have heart disease. More than half a million people per year die from this condition. Heart disease leads to 25 percent of all deaths in the United States. Sometimes heart disease is due to heredity. However, most heart disease is related to lifestyle factors. Using tobacco, eating foods high in saturated fat, and lack of physical activity, can increase the risk of developing heart disease. Teens who make healthy lifestyle choices are less likely to develop heart disease as adults.

Coronary Heart Disease

Body cells must have a constant supply of fresh oxygen to survive. The body's tissues and organs depend on the flow of blood through the arteries to deliver oxygen to the cells. When the arteries are clear and healthy, the blood flows through them freely. When the arteries are damaged or blocked, the blood does not flow as well.

Two disorders can result when blood does not flow properly through the arteries. One is **arteriosclerosis** which is a group of disorders that cause a thickening and hardening of the arteries. The arteries become rigid, which means that less blood is able to flow through them. Another is **atherosclerosis**, which is a condition that occurs when fatty substances build up on the inner lining of the arteries. This buildup shrinks the space through which blood can travel. The illustration on the next page shows how blood flow to the heart is decreased if the arteries are blocked. If the space in the arteries narrows, the heart may not get enough oxygen.

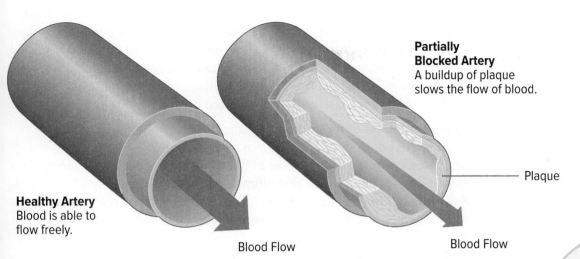

Partially Blocked Artery
A buildup of plaque slows the flow of blood.

— Plaque

Healthy Artery
Blood is able to flow freely.

Blood Flow

Blood Flow

Heart Attack

If the heart does not get enough oxygen, a **heart attack** is likely. A heart attack is a serious condition that occurs when the blood supply to the heart slows or stops and the heart muscle is damaged. The illustration of the heart shows a diseased or blocked artery. Explain what can happen if blood flow to the heart is cut off. If the blood to the heart is cut off for more than a few minutes, the heart muscle cells are damaged and die.

For males, symptoms of a heart attack include pain or pressure in the chest, or pain in the upper body including the arms, left shoulder, jaw, back, or abdomen. Males may also be short of breath, experience dizziness and have cold skin. They may vomit, feel tightness in the chest, or pass out.

Blood is able to flow freely through an unblocked artery. **Identify the reason that arteries become blocked.**

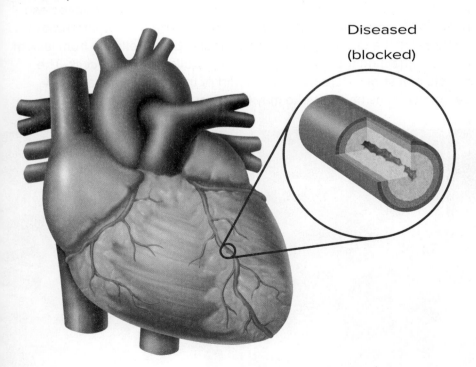

Diseased (blocked)

A heart attack happens when blood flow is cut off to the heart. **Describe how cardiac arrest is different.**

Females often have different symptoms of a heart attack than males. They may experience symptoms such as pain in the center of the chest, upper back, shoulder, arms, or jaw. They may also feel lightheaded. Prior to having a heart attack, females also report feeling more tired than usual, feeling anxious, and having indigestion. Only 30 percent of females feel discomfort in their chest before a heart attack. They may also report feeling symptoms up to one month before having a heart attack.

A heart attack is different from cardiac arrest. When cardiac arrest occurs, the heart stops beating in a normal way. The heartbeat becomes irregular and the flow of blood through the arteries stops. This condition is caused by an **arrhythmia**, or irregular heartbeat. Treatment depends on the type of arrhythmia. For some, doctors will prescribe drugs. Some may require an implanted device to regulate the heartbeat. These devices are pacemakers and implantable defibrillators. Sometimes surgery can fix the condition that led to the arrhythmia.

Other Cardiovascular Problems

Blood pushes against the walls of the blood vessels as it flows through them. One of the most common forms of heart disease is high blood pressure. High blood pressure occurs when a person's blood pressure is higher than normal. He or she has high blood pressure, or **hypertension**. It's a condition in which the pressure of the blood on the walls of the blood vessels stays at a level that is higher than normal. Hypertension can lead to a heart attack or a stroke. People with hypertension can manage the condition by following a healthful eating plan, exercising regularly, avoiding stress, and taking medicine if needed.

Like the heart, the brain needs plenty of oxygen and nutrients to function. It is possible for an artery to leak or develop a thick mass called a clot. When this happens, a **stroke** occurs. Stroke is a serious condition that occurs when an artery of the brain breaks or becomes blocked. During a stroke, blood flow to part of the brain is interrupted. That part of the brain is damaged as a result. Brain cells die from lack of oxygen, affecting the part of the body controlled by those cells. The effects of a stroke depend on what part of the brain is damaged. A person who has had a stroke may have trouble moving or speaking.

Reading Check

Identify What are two causes of strokes?

To reduce the risk of heart disease later in life, adopt healthy lifestyle habits now. **Name two healthy lifestyle habits that you've learned.**

Treating Heart Disease

MAIN IDEA Health care professionals can treat heart disease with angioplasty, medication, pacemakers, surgery, or transplants.

Health care providers can usually treat heart disease with the following methods:

- **Angioplasty** (AN•je•uh•plas•tee) is a surgical procedure in which an instrument with a tiny, balloon, drill bit, or laser attached is inserted into a blocked artery to clear a blockage. It can be performed in several ways. Using the balloon method, doctors inflate the tiny balloon until it pushes the blockage up and against the artery wall. Lasers or drill bits cut or burn away the blockage.
- **Medications** can break up blood clots that may block arteries. Aspirin may be prescribed to prevent platelets from clumping together to form blood clots. Other medications may be prescribed to help lower cholesterol levels.
- **Pacemakers** are devices placed inside the chest. A pacemaker is a small device that sends steady electrical impulses to the heart to make it beat regularly.
- **Implantable defibrillators are also placed inside the chest.** This device helps the heartbeat return to a normal rhythm.
- **Bypass surgery** creates new pathways for the blood. Surgeons use a healthy blood vessel from another part of the body, such as the leg. This surgery creates a new route for blood to flow around the blocked artery.
- **Heart transplants** completely replace a damaged heart with a healthy heart from someone who has just died. Heart transplants are complex. They are done only when the heart is severely damaged and no other treatment will work.

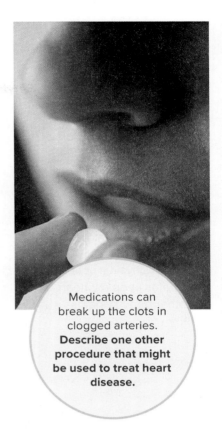

Medications can break up the clots in clogged arteries. **Describe one other procedure that might be used to treat heart disease.**

Reading Check

List What are two ways to treat heart disease?

Plaque can prevent the blood from flowing freely through an artery. **Name two ways that blocked arteries can be treated.**

Preventing Heart Disease

MAIN IDEA Strategies for preventing heart disease include getting regular physical activity, eating healthful foods, and staying tobacco free.

Symptoms of heart disease usually do not appear until adulthood. However, heart disease can begin developing in childhood. Making healthy choices today can lower your risk of developing heart disease when you are older. Strategies for keeping your heart healthy include:

- **Eat healthful foods and limit foods that are high in cholesterol, saturated fats, and trans fats.** Choose plenty of fresh fruits and vegetables, whole grains, and lean sources of proteins. Foods high in cholesterol, saturated fats, and trans fats are linked to heart disease.
- **Participate in regular physical activity.** The heart is a muscle and all muscles need regular physical activity to stay strong. Activities that benefit the heart raise the heart rate into the target heart zone.

SYMPTOMS OF HEART DISEASE USUALLY DO NOT APPEAR UNTIL ADULTHOOD.

Regular physical activity strengthens the heart. **Describe what impact physical activity may have on hypertension.**

- **Maintain a healthy weight.** Your heart works best if your weight is within a healthy range. Talk to a healthcare provider about the range that is best for you.
- **Manage stress.** Learning to relax will help you keep your blood pressure within a healthy range.
- **Stay tobacco free.** Chemicals in tobacco can cause heart disease, heart attacks, hypertension, and strokes. Staying tobacco free will help you avoid all of these problems.
- **Stay alcohol free.** Alcohol has been linked to high blood pressure and heart failure. Staying alcohol free helps you avoid these problems.

Eating a healthful diet most of the time can reduce the risk of hypertension. **List two other ways that this risk can be reduced.**

Lesson 3 Review

What I Learned

1. **VOCABULARY** Define *arteriosclerosis* and *atherosclerosis*.

2. **EXPLAIN** What is *angioplasty*?

3. **DESCRIBE** How do the symptoms of heart attacks in females differ from those in males?

Thinking Critically

4. **ANALYZE** Why is it a good idea to have your blood pressure checked regularly by a health care professional?

5. **SYNTHESIZE** Why do you think that medical professionals focus on preventing heart disease even though there are so many treatments for it?

Applying Health Skills

6. **PRACTICING HEALTHFUL BEHAVIORS** Both of Madison's grandmothers died of heart disease. Madison wants to reduce her own risk of heart disease. Make a list of positive health behaviors Madison can do to reduce her risk of heart disease.

Diabetes and Arthritis

Before You Read

Quick Write Name two factors you already know about diabetes.

Vocabulary

diabetes
insulin
type 1 diabetes
type 2 diabetes
arthritis
osteoarthritis
rheumatoid arthritis

BIG IDEA Diabetes and arthritis can be treated medically and managed by making healthful lifestyle choices.

What is Diabetes?

MAIN IDEA Diabetes affects children and adults and can be managed with medical treatments and healthful behaviors.

Diabetes mellitus (dy•uh•BEE•teez MEH•luh•tuhs), or **diabetes**, affects more than 30 million adults and children in the United States. It's a disease that prevents the body from converting food into energy. When you eat, your body breaks down the food to get the energy it contains. To do so, it turns food into a form of sugar called glucose. After your body digests food, glucose levels in the bloodstream rise. Some of the glucose begins to enter cells with the help of a hormone in your body called **insulin** (IN•suh•lin). It's a protein made in the pancreas that regulates the level of glucose in the blood. Your cells then use the glucose for energy to carry out life functions.

Some people who have diabetes do not have enough natural insulin. As a result, glucose cannot get into cells. Other people make enough insulin, but the insulin does not do its job properly. In both cases, the glucose remains in the blood. This leads to symptoms of feeling tired, losing weight, excessive thirst, and increased urination. If left unmanaged, diabetes can cause other long-term health problems such as kidney disorders, blindness, and heart disease.

Types of Diabetes

Diabetes consists of two main types: type 1 and type 2. **Type 1 diabetes** is a condition in which the immune system attacks insulin-producing cells in the pancreas. When the cells that produce insulin are killed, the body cannot control how much glucose is in the bloodstream. Type 1 diabetes often starts in childhood, but it may also begin in adulthood. About 5 percent of diabetics have type 1 diabetes.

Type 2 diabetes is a condition in which the body cannot effectively use the insulin it produces. Between 90 and 95 percent of diabetics have type 2 diabetes. In the past, type 2 diabetes typically began in adulthood. Type 2 diabetes is closely linked to a lack of physical activity and being overweight. Type 2 diabetes is also becoming more common among children and teens.

Reading Check

Identify What is the most common form of diabetes?

Diabetics must check their insulin levels throughout the day. **Name two methods for delivering insulin to the body of someone with diabetes.**

Managing Diabetes

People with diabetes must learn how to manage the disease. People with type 1 diabetes usually need to have injections of insulin, or they receive insulin from an insulin pump attached to their bodies. People with type 1 diabetes need to carefully manage their health.

People who have type 2 diabetes may also need insulin or other medications. Many of them, however, can control their disease by practicing healthful habits. They can eat nutritious foods, watch their weight, and be active.

People with diabetes must manage their health carefully. **List two symptoms of diabetes.**

Healthful Eating Plan	A healthful eating plan can help keep blood glucose levels within a normal range.
Weight Management	Regular physical activity helps people with diabetes maintain a healthy weight.
Insulin	People with type 1 diabetes and some people with type 2 diabetes receive insulin through a syringe or a pump.
Medical Care	People with diabetes need to be under the care of a medical professional.

What is Arthritis?

MAIN IDEA Arthritis is a disease marked by painful swelling and stiffness of the joints.

Arthritis (ar•THRY•tus) is a disease of the joints marked by painful swelling and stiffness. It affects about 50 million people in the United States. Although arthritis strikes older adults, children can also develop it. Two main types of arthritis include **osteoarthritis** (ahs•tee•oh•ar•THRY•tus) which is a chronic disease that is common in older adults and results from a breakdown in cartilage in the joints. The second type is **rheumatoid arthritis**. It's a chronic disease characterized by pain, inflammation, swelling, and stiffness of the joints. When rheumatoid arthritis affects a young person, it is called juvenile rheumatoid arthritis (JRA).

Types of Arthritis

Osteoarthritis is the most common form of arthritis. Osteoarthritis develops as a result of wear and tear on the joints, such as those of the knees and hips. The hard, slippery tissue in the joints between the bones is called cartilage. When cartilage in a joint wears down, the bones in the joints rub against each other. This rubbing causes pain, swelling, and morning stiffness. Risk factors for osteoarthritis include age, heredity, and being overweight. Maintaining a healthy weight and increasing physical activity can reduce stress on joints and help lower the risk of osteoarthritis.

Rheumatoid arthritis is usually more serious than osteoarthritis. People develop rheumatoid arthritis when their immune systems attack healthy joint tissue. These attacks damage joint tissue and cause painful swelling. The disease can affect any joint, including those in the hands, elbows, shoulders, hips, and feet. Symptoms include soreness, joint stiffness and pain, aches, and fatigue.

Juvenile rheumatoid arthritis (JRA) is the most common form of arthritis in young people. JRA appears most often in young people between the ages of 6 months and 16 years. Early symptoms include swelling and pain in the joints. The skin covering the joints may be red and warm to the touch. Children with JRA also typically get rashes and high fevers. Many children with JRA continue to have arthritis as adults. Some children with JRA, however, get better after puberty.

Managing Arthritis

No cure exists for arthritis. People with the disease can learn to manage it. They usually work with health care professionals to develop a plan to reduce the symptoms of arthritis. Many plans involve a combination of the following:

- **Physical activity and rest.** People with arthritis suffer less if they balance rest with low-impact physical activity. Rest helps handle fatigue that comes with the disease. Physical activity reduces swelling in the joints. It also allows joints to bend more easily.

Reading Check

Compare and Contrast How do the symptoms of osteoarthritis differ from those of rheumatoid arthritis?

Reading Check

Explain How do physical activity and rest help people who have arthritis?

- **Healthy eating and maintaining a healthy weight.** Eating healthy foods helps maintain overall health and keeps weight under control. Maintaining a healthy weight reduces stress on arthritic joints in the knees and feet.
- **Joint protection.** People can wear braces and splints to support arthritic joints. This equipment wraps around the joint and holds it steady.
- **Heat and cold treatments.** Hot baths ease the pain of some kinds of arthritis. Cold treatments can help reduce the swelling.
- **Medication.** Medicine can help slow the progress of some kinds of arthritis. Over-the-counter (OTC) medicines and prescription medicines can also help ease the pain and swelling of arthritic joints.
- **Massage.** A trained massage therapist can help some arthritis patients by gently massaging affected areas. This helps to relax the joints and increase blood flow to sore areas.
- **Surgery and joint replacement.** In extreme cases, surgeons can repair a joint or correct its position. They may even replace the damaged joint with an artificial one.

What Teens Want to Know

How can I prevent getting osteoarthritis? More than half of Americans will have some evidence of osteoarthritis by age 65. Heredity plays a part in whether a person gets osteoarthritis, but teens can take steps to prevent having the disease. One way to prevent joint injuries is by using protective gear when playing sports. Another preventive measure is to maintain a healthy weight. Excess weight puts stress on joints.

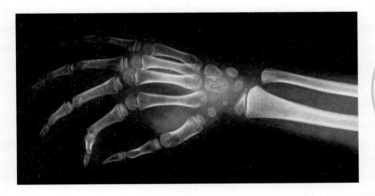

Arthritis can cause joint pain and difficulty with movement. **Name three ways to ease the symptoms of arthritis.**

Lesson 4 Review

What I Learned

1. **VOCABULARY** Define the terms *diabetes* and *arthritis*.

2. **DESCRIBE** What practices can some people with type 2 diabetes use to manage their disease without medication?

3. **DESCRIBE** What happens to a person's joints when osteoarthritis develops?

Thinking Critically

4. **EVALUATE** How are type 1 diabetes and rheumatoid arthritis similar?

5. **SYNTHESIZE** Based on what you know about arthritis, how can you help someone with arthritis manage the disease?

Applying Health Skills

6. **PRACTICING HEALTHFUL BEHAVIORS** Participating in regular physical activity can help reduce your risk of developing type 2 diabetes. If you enjoy what you do, you are more likely to participate in the activity on a regular basis. List ten physical activities you enjoy or might enjoy doing. Make a check mark beside three activities you will participate in during the next month. Make a commitment to do each of these activities at least once a week.

Allergies and Asthma

Before You Read

Quick Write List three allergy symptoms. How can they be managed?

Vocabulary

allergy
allergens
pollen
histamines
hives
antihistamines
asthma
bronchodilator

BIG IDEA Allergies and asthma are two kinds of noncommunicable diseases that can be managed with medicine and avoiding allergens.

What Are Allergies?

MAIN IDEA An allergy is an extreme sensitivity to a substance or allergen, such as pollen, insect bites or stings, or food.

Your immune system keeps you healthy. It helps your body fight off foreign substances. However, some people's immune systems react to fairly harmless substances. These reactions are allergic responses. Between 40 million and 50 million Americans have some type of **allergy**, or an extreme sensitivity to a substance.

A person can be allergic to a number of substances in the environment. **Allergens** are substances that cause allergic responses. For example, people who are allergic to ragweed are allergic to the tiny **pollen** grains from the ragweed plant. Pollen is a powdery substance released by the flowers of some plants. When pollen comes into contact with a person who is allergic to it, the pollen causes an allergic reaction.

When an allergen enters or comes in contact with a person's body, the immune system reacts as though it were harmful. When people are allergic to cats, for example, their allergic response is usually caused by dander on the cat's skin or fur. The photos in the lesson show several allergens that people can be exposed to each day.

Allergic Reactions

Many different substances can cause allergic reactions. Some are easy to avoid while others are not. The four categories of substances that cause allergic reactions are pollen from plants, insect bites and stings, plants such as poison ivy, oak, or sumac, and food allergies. When you are allergic to something, your immune system reacts quickly. It thinks that your body is under attack. Your immune system makes antibodies. Antibodies protect your body from the allergen. Antibodies are a special kind of protein that locks onto cells. Antibodies cause certain cells in the body to release **histamines** (HIS•tuh•meenz). Histamines are chemicals that the immune cells release to draw more blood and lymph to the area affected by the allergen. Histamines cause the symptoms of the allergic reaction. When you are exposed to the same allergen again, the same antibody response will occur. You'll have an allergic reaction every time you come into contact with that allergen.

Allergic reactions can range from minor irritations to severe problems. For example, some people may have sneezing or a runny nose. Although these symptoms are uncomfortable, they are harmless. Some people may get **hives**, or raised bumps on the skin that are very itchy, as part of an allergic reaction. Most allergic reactions happen within seconds or minutes of the time the allergen enters the body.

Some people are at risk for severe allergic reactions. Some reactions may include throat swelling or extreme difficulty breathing. In severe cases, some people could die if they are not treated immediately. They may need to carry medicine called epinephrine (eh•pin•EFF•rihn). Epinephrine slows down or stops the allergic reaction. It prevents the body from releasing histamines. Learning how to control your allergies will help you avoid or reduce the symptoms of allergic reactions.

ALLERGIC REACTIONS CAN RANGE FROM MINOR IRRITATIONS TO SEVERE PROBLEMS.

Reading Check

List What are three ways that the body responds to allergens?

Sneezing is one response to contact with an allergen. **Name the drug that can be used when severe allergic reactions occur.**

Managing Allergies

Although no cure exists for allergies, they can be managed in three basic ways.

- **Avoid the allergen.** For example, if you know you are allergic to poison ivy, learn what it looks like and stay away from it. Wear long sleeves and pants if you go into the woods. If you have a food allergy, check the ingredient labels on food products. When you go to restaurants, ask about the ingredients of menu items you want to order. Nut allergies are dangerous because they often cause severe reactions. People with allergies to peanuts or other nuts need to be especially careful about what they eat or come into contact with.

- **Take medication.** Some allergens such as dust and pollen are nearly impossible to avoid. People with these allergies often take medicines to help reduce the symptoms. **Antihistamines** are medicines that reduce the production of histamines.

- **Get injections.** Sometimes a long-term series of injections can help people overcome allergies. The injections contain a tiny amount of the allergen. Over time, the immune system is less sensitive to the allergen.

Reading Check

Describe What are three ways to manage allergies?

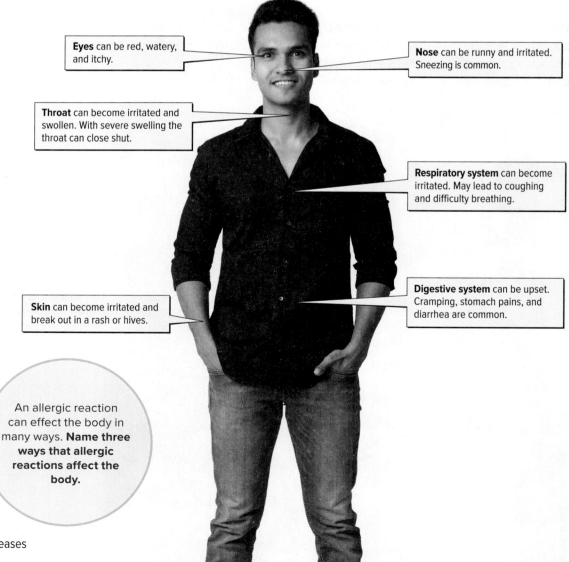

Eyes can be red, watery, and itchy.

Nose can be runny and irritated. Sneezing is common.

Throat can become irritated and swollen. With severe swelling the throat can close shut.

Respiratory system can become irritated. May lead to coughing and difficulty breathing.

Digestive system can be upset. Cramping, stomach pains, and diarrhea are common.

Skin can become irritated and break out in a rash or hives.

An allergic reaction can effect the body in many ways. **Name three ways that allergic reactions affect the body.**

What Is Asthma?

MAIN IDEA Asthma occurs when the airways in the lungs narrow, making it difficult to breathe.

Asthma is a chronic inflammatory disorder of the airways that causes air passages to become narrow or blocked, making breathing difficult. In the United States, about 24 million people are reported to have asthma. More than 6 million of these asthma sufferers are people under the age of 18. Worldwide, about 300,000 million people have asthma. Many substances and conditions can cause an asthma attack. Common triggers include:

- Allergens such as mold, dust, pollen, and pets.
- Physical activity.
- Air pollutants such as paint and gas fumes, cigarette smoke, industrial smoke, and smog.
- Infections of the respiratory system, such as colds and the flu.
- Weather changes, especially dramatic changes.
- Rapid breathing which often happens under stress, when laughing, or when crying.

What are the symptoms of an asthma attack? A person may wheeze, cough, or feel short of breath during an attack. Symptoms can also include tightness or fullness in the chest. The illustration shows how an asthma attack affects a person's airway.

> IN THE UNITED STATES, ABOUT 24 MILLION PEOPLE ARE REPORTED TO HAVE ASTHMA.

> When an asthma attack occurs, breathing becomes difficult. **Describe what happens to the airways during an asthma attack.**

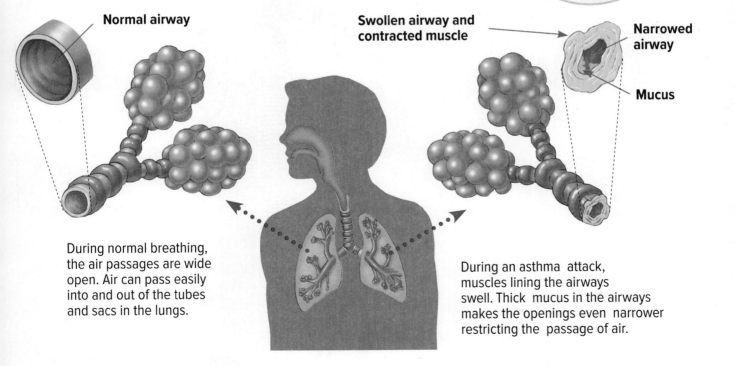

Normal airway

Swollen airway and contracted muscle

Narrowed airway

Mucus

During normal breathing, the air passages are wide open. Air can pass easily into and out of the tubes and sacs in the lungs.

During an asthma attack, muscles lining the airways swell. Thick mucus in the airways makes the openings even narrower restricting the passage of air.

Managing Asthma

Since there is no cure for asthma, people with the disease must learn to manage it and lead active lives. Managing asthma is often a team effort. Parents or trusted guardians, health care professionals, and friends can help people with asthma stay healthy. Strategies that people with asthma can use to help avoid asthma attacks include:

- **Monitoring the condition.** People with asthma must pay attention to the early signs of an attack. That way, they can act quickly if they sense an attack coming. They also can track their long-term lung capacity, or ability to take in air. An instrument called an airflow meter measures lung capacity. When used regularly, an airflow meter helps people know when their airways are narrowing.
- **Managing the environment.** For example, if dust and mold trigger your asthma, reduce them in your environment. It helps to keep floors, bedding, and pets clean.

Reading Check

Describe How can a person with asthma manage his or her condition?

Keeping the home and other environments clean can help reduce asthma attacks. **Explain how keeping the environment clean helps people with asthma breathe easier.**

- **Managing stress.** Stress is a major cause of asthma attacks. Panicking during an attack can make it even worse. Relaxing and staying calm can help those with asthma avoid attacks. Relaxing will help even during an attack.
- **Taking medication.** Two kinds of medicines can treat asthma: relievers and controllers. Relievers, such as **bronchodilators** (brahng·koh·DY·lay·turhz) are a medication that relaxes the muscles around the air passages. It helps to reduce symptoms during an asthma attack. People usually use an inhaler to take a bronchodilator. This small device sends medicine directly to the respiratory system. Controller medicines are taken daily, and help prevent attacks by making airways less sensitive to asthma triggers.

Lesson 5 Review

What I Learned

1. **VOCABULARY** Define *allergy* and *asthma*.

2. **NAME** What are four common types of allergens?

3. **DESCRIBE** What are the symptoms of an asthma attack?

Thinking Critically

4. **HYPOTHESIZE** Why do you think it's important for people with food allergies to be careful when eating out?

5. **SYNTHESIZE** If your friend is having an asthma attack, what are two ways you could help?

Applying Health Skills

6. **COMMUNICATION SKILLS** Samantha tried out for the basketball team and made it! She has asthma, however, and is too embarrassed to tell anyone about it. Write a short letter to Samantha and tell her why you think she should tell her coach and teammates that she has asthma. How can she manage her asthma during practice and games?

Safety

LESSONS

19

Building Safe Habits

Before You Read

Quick Write List three examples of accidental injuries that people experience.

Vocabulary

accident
accidental injuries
safety conscious
hazard
accident chain

Reading Check

Explain What is the difference between an accident and an accidental injury?

Developing Good Character

Responsibility When you put your belongings in their proper place, they're not in the way, so they're less likely to cause accidents. Putting away clothes and equipment also helps cut down on clutter.

BIG IDEA Being safety conscious means being aware that safety is important and acting safely.

Safety First

MAIN IDEA Accidents and accidental injuries can affect people of all ages.

"Buckle up!" "Look both ways before you cross the street!" You've probably been hearing warnings like these for as long as you can remember. You might have helped teach these safe habits to a younger brother or sister. **Accidents** are any event that was not intended to happen. When you stay safe and avoid accidents, you help yourself and those around you stay healthy.

You might not think serious accidents can happen to you. However, the Centers for Disease Control and Prevention (CDC) reports that **accidental injuries** are the leading cause of death in teens. Accidental injuries are injuries resulting from an accident. They are the fifth leading cause of death in the United States across all age groups. The leading causes of nonfatal accidental injuries in the home are falls. Other types of injuries are being struck or cut by something, being bitten or stung, overexertion, and poisoning.

How Accidental Injuries Happen

MAIN IDEA Accident chains include a situation, an unsafe habit, an unsafe action, and the resulting injury.

The first step in staying safe is being **safety conscious**. This means being aware that safety is important and acting in a safe manner. It is easier to prevent injuries than to treat them. To prevent injuries, develop safety habits and avoid **hazards**, or the potential source of danger. Some strategies that should become safety habits are:

- Avoid risky behaviors.
- Resist negative peer pressure.
- Know your limits.
- Wear protective gear.
- Know how to spot a hazard.

Accidents usually happen when people stop being safety conscious and become careless. Think back to the last accident you had. Try to remember the elements that led to the accident. These are the steps in the **accident chain**. The accident chain is a series of events that include a situation, an unsafe habit, and an unsafe action. For any accidents to occur, three elements must be present. These are the situation, the unsafe habit, and the unsafe act.

Breaking the Accident Chain

Tony's accident did not have to happen. Look at the links in Tony's accident chain. Breaking just one link would have kept Tony from being injured.

- **Change the Situation.** Tony could have gotten up earlier. He could have set his alarm for a reasonable time. He could have asked a family member to wake him if he overslept.
- **Change the Unsafe Habit.** Tony could have put his books on a bookshelf or in his book bag.
- **Change the Unsafe Action.** Tony could have paid attention to where he was going. He could have slowed down and watched his step. Being safety conscious might have kept Tony from tripping and falling.

By changing the situation, the unsafe habit, or the unsafe action, Tony could have prevented his accident. By becoming more safety conscious, Tony can take responsibility for preventing accidents in the future.

Reading Check

Explain What changes can break the accident chain?

1 The Situation
Tony has overslept. He wakes up in a panic. The bus is coming in 15 minutes.

3 The Unsafe Action
Without looking where he is going, Tony runs to the bathroom to wash up.

5 The Result
When Tony falls, he sprains his wrist. He misses his bus and is in a lot of pain.

Be safety conscious by thinking about how accidents happen and take steps to reduce them. **Explain how the steps in the accident chain can help you reduce accidents.**

2 The Unsafe Habit
Tony didn't put his books away from the night before. He just left them on the floor.

4 The Accident
Tony trips over his books and falls down.

Lesson 1 Review

What I Learned

1. **VOCABULARY** Define *accident*. Use the word in an original sentence.

2. **IDENTIFY** What are the five links in the accident chain?

3. **GIVE EXAMPLES** What are three ways to break the accident chain?

Thinking Critically

4. **DESCRIBE** What behavioral factors are associated with causes of death in the U.S.?

5. **APPLY** Benny has always had a bookshelf on the wall next to his bed. Now that he is taller, the bookshelf has become a problem. In fact, this year Benny has bumped his head on the shelf three times. What should he do to be safer?

6. **ANALYZE** Grant's friend dared him to walk across a narrow 12-foot-high fence. What should Grant do, and why?

Applying Health Skills

7. **DECISION MAKING** Tina wants to go bike riding with a friend, but she left her helmet in her dad's truck. What are Tina's options? Use the decision-making process to help Tina make a safe decision.

Safety at Home and School

BIG IDEA Following safety rules can keep you safe both at home and away from home.

Safety in the Home

MAIN IDEA Your home may be filled with many potential safety hazards, such as stairs or appliances.

Home is a place where everyone should feel safe and comfortable. Yet, homes can contain hazards. Stairways or spilled water can lead to falls, the most common type of home injury. Appliances can cause electrical shocks. Sharp tools in the kitchen or garage can lead to cuts. Other injuries can result from poisonings, choking, drowning, and guns. Fires are the third leading cause of unintentional injury and death in the home. Overall, about 45 percent of deaths from accidental injuries occur in and around the home. However, these hazards can be avoided. Following safety rules can reduce the risks of home hazards.

Preventing Falls

Many falls in the home can be prevented. One simple rule is to avoid running on slippery or waxed floors. Many times, falls in the home occur in the kitchen, the bathroom, or on the stairs. These safety rules can help you prevent falls in those areas of the home.

- **Kitchen safety.** Clean up spills right away. Use a stepstool, not a chair, to get items that are out of reach. Learn to handle knives correctly. If you are cooking, avoid leaving the stove unattended. When pots are on the stove, make sure the handles are turned inward. Keep young children away from the stove.
- **Bathroom safety.** Put a nonskid mat near the tub or shower. Use rugs that have a rubber backing to prevent the rug from slipping. Keep personal products in plastic bottles.
- **Staircase safety.** Keep staircases well-lit and clear of all objects. Apply nonslip treads to slippery stairs. Make sure handrails are secure and stable. If small children live in the house, put gates at the top and bottom of the stairs.

> **HOME IS A PLACE WHERE EVERYONE SHOULD FEEL SAFE AND COMFORTABLE.**

Preventing Poisonings

Many common household products are poisonous. Items such as cleaning products, insecticides, medications, and vitamins can be harmful if not used or taken as directed. Poisoning can happen by swallowing, absorbing through the skin, injection from a needle, or breathing poisonous fumes.

To help keep the people in your home safe from poisoning, take these steps:

- Label all household products so that they are clearly marked.
- Store cleaning products, insecticides, and other potential poisons out of reach of young children.
- Avoid referring to a child's medicine or vitamins as candy.
- Use childproof caps on medicines and keep the bottles out of reach of children.
- Always use a product or medication as directed on the label.
- Keep the Poison Control Center number handy.

Picking up toys left on the floor or stairs by younger siblings can prevent falls. **Describe other ways you can help prevent accidents at home.**

Gun Safety

In many states, it is illegal for most teens to own a gun. If you find a gun, do not touch it. Never handle a gun without a parent, guardian, or other trusted adult present. If you know that someone at school is carrying a gun or any other weapon, tell a school authority right away. The best way to prevent a gun accident in the home is to not have guns in the home. If a gun must be kept in the home, use the following safety precautions:

- Guns should have trigger locks and be stored unloaded in a locked cabinet.
- Ammunition should be stored in a separate locked cabinet.
- Anyone who handles a gun should be trained in gun safety.
- All guns should be handled as if they are loaded.
- *Never* play with a gun or point it at someone.
- Tell school authorities if someone brings a gun to school.

Electrical Safety

Improper use of electrical appliances or outlets can cause dangerous electrical shocks. To prevent electrical shocks,

- Avoid using electrical appliances around water or if you are wet.
- Unplug small appliances such as hair dryers when they are not in use.
- Gently pull the plug of cord to unplug them, not the cord.
- Repair appliances with frayed cords.
- Unplug any appliance that is not working properly.
- Avoid running cords under rugs.
- In homes with small children, cover unused outlets with plastic outlet protectors.

Reading Check

List What are two ways to prevent electrical shocks?

Fires in the home are preventable. **Name five ways that home fires can be prevented.**

Careless cooking Spattered grease and oil can cause kitchen fires. Unattended cooking pots can spill onto burners or in the oven.
Careless smoking Cigarettes can start fires if people leave them unattended or fall asleep while they are still burning. Cigarettes can also start fires if people toss them into the trash when they are still burning.
Incorrect storage of flammable materials Examples of flammable materials are paint, chemicals, oil, rags, and newspapers.
Damaged electrical systems or electrical overload Fires can start due to too much current flowing through overloaded circuits. Shredded wires or torn cords can also lead to fires. Broken appliances can cause fires as well.
Gas leaks Gas lines can leak and catch fire. Natural gas is odorless and colorless, so it has an additive that makes it smell. If you smell gas, first get out of the house, then call 911.

Fire Safety

MAIN IDEA It is a healthy practice to have a fire safety plan, a fire extinguisher, and working smoke alarms.

Fires happen in about 370,000 homes in the United States each year, killing approximately 2,800 people. The kitchen is the room in a home where most fires start. A fire needs three elements to start—fuel, heat, and air. Sources of heat include cigarettes, matches, or electrical wires. If heat sources come in contact with anything **flammable**, or substances that can catch fire easily, a fire can result. Household chemicals, rags, wood, or newspapers are all examples of flammable materials. Other fires start from **electrical overload**. It's a dangerous situation in which too much electric current flows along a single circuit. Here are some safety guidelines to help you prevent fires in the home:

- Keep stoves and ovens clean to prevent pieces of food or grease from catching fire.
- Keep flammable objects at least three feet away from stoves and portable heaters.
- Remind adults who smoke never to smoke in bed or on overstuffed furniture. Never let a smoker toss a cigarette into a trashcan before making sure it is completely extinguished.
- Regularly inspect electrical wires, outlets, and appliances to make sure they are in proper working order. Never pull on the cord to unplug an appliance. Never run cords under rugs or carpets. If a cord is damaged, replace it.
- Discard old newspapers, oily rags, and other materials that burn easily.
- Use and store matches and lighters properly. Keep them out of reach of young children. Don't leave candles burning unattended.
- Do this if your clothing catches fire: stop, drop, and roll. Rolling on the ground can put out the fire and protect you from serious injury.

If a fire does occur in the home, do the following:

- Leave quickly. Once safe, call 9-1-1 and let others know you are safe.
- Avoid opening closed doors. A closed door can prevent the fire from spreading.
- Crawl on the ground in a smoke-filled room. Smoke rises so there will be more oxygen near the floor.
- Stuff a blanket or towel along the cracks in a door to keep smoke out of a room.
- Stop, drop, and roll on the ground if your clothing catches fire.
- Never go back into a burning building.

> **FIRES HAPPEN IN ABOUT 370,000 HOMES IN THE UNITED STATES EACH YEAR.**

Being Prepared in Case of Fire

The earlier you receive warning of a fire, the better your chances of getting out of the building safely. Every level of a house should have smoke alarms. Smoke alarms are especially useful when you are sleeping and might not notice the early signs of a fire. As a result, you should install them as close to sleeping areas and bedrooms as possible. Test smoke alarms every month. If batteries power them, put in fresh batteries at least once a year.

Water will put out fires in which paper, wood, or cloth is burning. However, you should never use water to put out a fire that involves grease, oil, or electricity. That will actually make the fire worse. Instead, use a **fire extinguisher**. This is a device that sprays chemicals that put out fires. Every home should have a fire extinguisher. Read the fire extinguisher's directions, and make sure that you know how to use it properly.

Create a fire escape plan with your family. Most fires happen at night, so be sure to know escape routes from each bedroom. Choose a meeting point outside where everyone can gather in the event of a fire. Practice the escape plan with your family every six months.

October is National Fire Prevention Month. During this time, many news outlets have stories on their local fire departments. They also discuss fire safety tips and how to create a fire escape plan. June is National Fireworks Safety Month. Watching a fireworks display on the 4th of July can be fun. However, fireworks are explosives that burn at high temperatures. If mishandled, fireworks can burn clothes and skin. They can also cause eye injuries. Prior to the 4th of July, many news outlets have stories promoting the safe use of fireworks, including sparklers.

Reading Check

Give Examples What three ways to prevent fires in the home?

1. If possible, leave quickly. Get out of the building before calling 911 or the fire department.

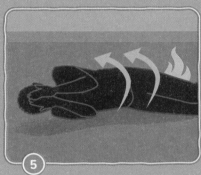

2. Before opening a closed door, feel it to see if it is hot. If it is hot, do not open it. There may be flames just outside the door.

3. If you must exit through smoke, crawl along the floor. Smoke and hot air rise, so it is important to stay as low as possible. The air you breathe will be cleaner. The smoke will not be as likely to overcome you.

4. If you can't get out, stay in the room with the door closed. Roll up a blanket or towel and put it across the bottom of the door to keep out smoke. If there is a telephone in the room, call 911 or the fire department. If possible, open the window and yell for help.

5. If your clothing catches fire, stop, drop, and roll. Rolling on the ground will smother the flames. Never run; the rush of air will fan the flames.

6. Once outside, go to the prearranged meeting point. Let everyone know that you are safe. Then someone should call 911 or the fire department. Never go back into a burning building.

> If you are in a fire, you need to know what to do to escape safely. **Explain why you should get out first and then call 911.**

> PRACTICE AN ESCAPE PLAN WITH YOUR FAMILY EVERY SIX MONTHS.

Reading Check

Recall What strategies can you and your peers use to stay safe at school?

Myth vs Fact

Myth Small fires are not a major concern and can be controlled.

Fact A fire of any size needs immediate attention. Small fires can become uncontrollable fires very quickly. A fire that starts with material burning in a wastebasket in the room of a house can spread to two rooms within four minutes. Smoke alarms and sprinkler systems play an important role in fire safety. In homes with smoke alarms, the risk of dying in a fire is reduced by 50 percent compared to homes without working smoke alarms. In homes with both smoke alarms and a sprinkler system, the risk of death is reduced by 82 percent.

Safety at School

MAIN IDEA Following rules at school helps you, other students, and teachers stay safe.

Your school probably has rules in place to keep students and teachers safe. Many accidental injuries at school can be avoided. Follow these strategies to protect the health and safety of students and teachers.

- **Play by the rules.** Rules are made to protect you and others. The cafeteria, classrooms, halls, gym, and auditorium may all have important safety rules to follow.
- **Report weapons or unsafe activities.** It's essential to follow rules prohibiting weapon possession at school. If you think that someone has brought a gun or other weapon to school, report it immediately to a teacher or principal.
- **Wear necessary safety gear.** Working in a science lab or playing sports are two places where appropriate gear will help keep you safe.

Safety in the Community

Safety at school also includes staying safe while you travel to and from school. Most students who walk or bike to school take the same route every day. To prevent crime, stay aware of your surroundings. Remember the following safety tips:

- **Make sure your cell phone is easy to reach.** Remember that 911 will connect you to emergency services.
- **Avoid walking alone.** Stay away from isolated places. Choose to walk on well-lit streets.
- **Walk with purpose.** Predators will target someone who appears to be lost or unfamiliar with their surroundings. If you are in an unfamiliar neighborhood, ask for directions.

Wear safety equipment while working in the chemistry lab or playing sports. **Describe why it is important to wear safety goggles when working with chemicals.**

- **Put your wallet or any money you're carrying in a place where people can't see it.** If you carry valuables, such as a phone, tablet, or computer, carry them in a pocket or case.
- **When traveling in a car, ask the driver to find a well-lit parking spot.** Make sure the car is locked before you leave it. Check the back seat before entering the car.
- **Avoid hitchhiking or giving rides to hitchhikers.** Asking for a ride or picking up someone who you don't know is extremely dangerous. Crimes such as assault, robbery, and rape can occur when teens hitch hike, or pick up a stranger who is seeking a ride.
- **When using public transportation, exit the bus only in busy, well-lit areas.** A person intent on hurting you can hide in dark doorways and other areas that are not well-lit.
- **Tell your parents or guardians where you are going and when you will be home.** Someone should know where you are, who you are with, and when you are expected to return home. If you are delayed, your parent or guardian can alert others to your absence.

Many communities struggle with the problems of crime and violence. Some have passed laws for stricter gun control. They have also made the punishments for violent crimes stricter. In some areas, people have formed **neighborhood watch programs**. These are programs in which residents are trained to identify and report suspicious activity. Communities may also try to protect teens by setting curfews. You can help protect yourself from dangerous situations by following these strategies:

- Walk with purpose to and from your home.
- Travel with another person or in a group whenever possible.
- Avoid taking shortcuts through unfamiliar or unsafe areas.

Some schools have security guards patrolling campus to help keep students safe. **Discuss what you can do to improve safety at school.**

Lesson 2 Review

What I Learned

1. **VOCABULARY** Define the term *fire extinguisher.*

2. **IDENTIFY** List three strategies for preventing poisoning.

3. **DESCRIBE** How can you be prepared for a fire that might happen in your home?

4. **EXPLAIN** Why is it important that the media have stories about fire safety and fireworks safety?

Thinking Critically

5. **APPLY** In what ways is a cluttered room a hazard?

6. **ANALYZE** Why is it a bad idea to call medicine "candy" to get children to take it?

7. **EXPLAIN** Why is it important to immediately report finding a weapon at school?

Applying Health Skills

8. **REFUSAL SKILLS** Will wants to see the hunting rifle that Troy's dad just bought. The rifle is in a locked case, but Troy knows where the key is. How could Troy refuse Will's request?

Safety on the Road and Outdoors

Before You Read

Quick Write Write down three actions you take to stay safe when participating in outdoor activities.

Vocabulary

pedestrian
defensive driving
heatstroke
heat cramps
hypothermia

BIG IDEA Following safety rules can help prevent injury on the road and outdoors.

Staying Safe on the Road

MAIN IDEA Safety in vehicles includes wearing a safety belt and not distracting the driver.

In the United States, motor vehicle crashes are one of the leading causes of accidental deaths in people 1 to 24 years old. To be a safety-conscious passenger, wear a safety belt whenever you ride in a vehicle. Safety belts help keep you in your seat if your vehicle gets into a crash. Do not distract the driver of the vehicle. In 49 U.S. states, the use of a safety belt by front seat passengers is required by law. Some states also require passengers in the back seat to wear safety belts. To learn more about the laws in your state, go to www.ghsa.org and search for *seat belt laws*.

In 2017, more than 3,000 people were killed in accidents involving a distracted driver. Another way to stay safe is to never get in a car with a driver who has been drinking alcohol or using drugs. Call a trusted adult to pick you up instead.

Many cars have air bags, too. Air bags can help keep people in the front seats from colliding with the steering wheel and dashboard. However, the force of air bags can hurt small children. The safest place for children to ride is in the back seat. Infants and small children should ride in an appropriate car seat or booster seat until they are large enough to use a safety belt.

If you take the bus to school, don't distract the bus driver while he or she is driving. Don't get up while the bus is moving or put your arms out the window. When you get off the bus, make sure the bus driver and all drivers of the vehicles around the bus can see you clearly. Don't cross behind the bus. If you are in a bus during an emergency, cooperate with the driver so that you and everyone else on the bus will remain safe.

> **MOTOR VEHICLE CRASHES ARE ONE OF THE LEADING CAUSES OF ACCIDENTAL DEATHS.**

Safety on Foot

Ever since you learned to walk, you have been a **pedestrian**, a person who travels on foot. Safety is important for pedestrians. Start by paying attention to what is happening around you. Follow these rules to become a safer pedestrian.

- Walk on the sidewalk if there is one. If there is no sidewalk, walk facing oncoming traffic, staying to the left side of the road.
- Cross streets only at crosswalks. Do not jaywalk, or cross the street in the middle of the block.
- Look both ways several times before crossing, and keep looking and listening for oncoming cars.
- If you cross in front of a stopped vehicle, be sure the driver can see you. Make eye contact with him or her before stepping in front of the vehicle.
- Obey all traffic signals.
- During the day, wear bright clothing. If you walk at night, take a well-lit route. Wear light-colored or reflective clothing and carry a flashlight.
- Do not talk on a cell phone or wear headphones as you walk. Be aware of your surroundings.

Reading Check

Name What are two safety guidelines for pedestrians?

Using safety belts increases your safety while riding in a car. Younger children should be in car seats. **Explain why infants and young children are safer riding in the back seat of a car.**

Safety on Wheels

Riding bicycles and using skates, in-line skates, skateboards, and scooters are activities many teens enjoy. One way to prevent injury while participating in these activities is to learn about the risks and then follow rules to avoid them. Always wear an appropriate helmet and other safety gear. Also, make sure your clothing fits well and does not interfere with your activities.

Head injuries cause 70 to 80 percent of the deaths from bicycle accidents. Wearing a helmet every time you get on your bike can reduce your risk of head injury by 85 percent. It is also important for bicycle riders to ride with the flow of traffic and obey traffic signs and signals. Bicyclists should never weave in and out of traffic. When riding with a friend, ride in single file, not side-by-side. Learn hand signals, and use them before you turn.

Both bicyclists and drivers must practice **defensive driving**. It means watching out for other people on the road and anticipating unsafe acts. To stay safe, bicyclists should be visible to others by wearing bright, reflective clothes. Bicycles should have lights and reflectors. To reduce risk of injury while bicycling, do not ride at night or in bad weather.

Skates, in-line skates, skateboards, and scooters can be a lot of fun, but only when they are used safely. Here are some guidelines for having fun while staying injury free.

- Wear protective gear, including a hard-shell helmet, wrist guards, gloves, elbow pads, and knee pads.
- Do not let your speed get out of control.
- Always follow your community's rules on where you can ride your skateboard or scooter. This may include not skating or riding in parking lots, streets, and other areas with traffic.
- Before you head downhill, practice a safe way to fall on a soft surface.
- Don't skate or ride a scooter after dark.
- Avoid riding or skating on wet, dirty, or uneven surfaces.
- In cold weather, avoid riding on slick, icy surfaces.

Reading Check

Give Examples Name one do and one don't for staying safe on wheels.

Bicycle riders need to obey the same traffic rules as drivers. **Describe what else bicycle riders should do to stay safe.**

Staying Safe Outdoors

MAIN IDEA Safety precautions make outdoor activities more fun.

Do you enjoy swimming or boating? How about hiking or camping? These and other outdoor activities are more fun when you "play it safe." To stay safe while enjoying outdoor activities, follow these tips:

Use good judgment. Before scheduling any outing, plan ahead. Check the weather forecast. Make sure you have the proper safety gear for each activity and that what you are doing is safe. If you're unsure, ask a trusted adult.

Take a buddy or two. When you spend time outdoors, be sure you are with at least one other person. If something happens to you and you are with a group, one friend can stay with you and another friend can go for help. Always tell your parent or guardian where you are going and when you expect to be home.

Warm up and cool down. This will help prevent injuries. Stretch after your warm-up and cool down.

Stay aware. Learn the signs of weather emergencies. When necessary, move quickly to shelter.

Know your limits. Be aware of your skills and abilities. Don't take on more than you can handle. For example, if you are a beginning swimmer, don't try to swim a long distance.

Protect your skin. Remember to wear bug protection and sunscreen. It is important to wear sunscreen to protect your skin from the sun's damaging rays, which can cause skin cancer later in life.

Planning for Weather

When planning your outdoor activity, always check the weather. One major risk present during outdoor activities is an electrical storm. If you are caught outdoors during an electrical storm, try to find shelter in a building or car. Also, there are certain safety tips to follow during both hot weather and cold weather.

During hot weather, your body can overheat when you are active outdoors. If you feel dizzy, out of breath, or have a headache, take a break. Keep cool by drinking plenty of water. Rest in the shade when you can. Overworking your body in the heat can lead to two dangerous conditions: heat exhaustion and **heatstroke**. It's a serious form of heat illness in which the body's normal processes for dealing with heat close down. Signs of heat exhaustion can include cold, clammy skin, dizziness, or nausea. Signs of heatstroke can include an increase in body temperature, difficulty breathing, and a loss of consciousness. Heatstroke can be deadly. If someone shows signs of heatstroke, get medical help right away. Strenuous physical activity during hot weather can also cause heat cramps. Resting, cooling down, and drinking water should help relieve **heat cramps**. Heat cramps are painful, involuntary muscle spasms that may occur during heavy exercise in hot weather. Gentle stretching and massage may also help.

Reading Check

Identify What safety items should you bring with you on a hike or camping trip?

WEARING A HELMET EVERY TIME YOU GET ON YOUR BIKE CAN REDUCE YOUR RISK OF HEAD INJURY BY 85%.

Cold weather can be dangerous if your body or parts of your body get too cold. This can lead to a condition called **hypothermia**. It's a sudden and dangerous drop in body temperature. When you are active outside in cold weather, dress in layers. Wear a hat, warm footwear, and gloves or mittens. Anyone who starts to feel very cold or shiver should go inside and get warm. Just like in warm weather, it is important to drink plenty of water and take a break when you feel tired. Also, wearing sunscreen is important in both warm and cold weather. While performing some winter activities, such as skiing, you are exposed to the sun. Wearing sunscreen and sunglasses help protect your skin from harmful UV rays.

WHEN PLANNING YOUR ACTIVITY ALWAYS CHECK THE WEATHER.

Planning your trip can make it safer and more fun. **Explain what else you can do to stay safe while participating in outdoor activities.**

Swimming in an area where a lifeguard is present is one way to stay safe while in the water. **Identify what else you can do to stay safe while in the water.**

Water Safety

Water activities can be a lot of fun. To avoid injury, you should learn and follow water safety rules. Know how to swim well. Good swimmers are less likely to panic in an emergency. If an emergency does occur in the water, stay in one place and try to conserve energy until help arrives. Follow these other tips to stay safe in the water.

- Swim only at beaches and pools if a lifeguard or other trusted adult is present.
- Swim with a buddy.
- Monitor yourself. Don't swim if you are tired or cold, or if you have been out in the sun for too long.
- Look around your environment often. Watch for signs of storms. If you are swimming when a storm begins, get out of the water right away.
- Never swim in water with strong currents.
- Dive only in areas that are marked as safe for diving. The American Red Cross suggests that water be at least nine feet deep for diving or jumping. Never dive into unfamiliar water or into above-ground pools.
- If you are responsible for children, take extra care. Don't let them near the water unless there is a trained lifeguard on duty. Accidents can happen even in small wading pools.
- When boating or waterskiing, wear a life jacket at all times. If the water is cold, wear a wetsuit.

Reading Check

Identify What safety items should you bring with you on a hike or camping trip?

Safety on the Trail

Preparation is the first step in a safe and enjoyable hike or camping trip. Following the tips listed below to help you stay safe while hiking and camping.

- **Never camp or hike alone.** Make sure family members know your route and your expected date and time of return. Carry a cell phone or long-range walkie-talkie if you can.
- **Dress properly.** Be aware of the weather and dress accordingly. Dress in layers and wear long pants to protect yourself against ticks. If you are hiking up a mountain, know that the weather may change as you change altitude. Wear sturdy footwear. Before you hike in any shoes or boots, break them in to avoid getting blisters.

Safe hiking and camping takes planning. **Explain why it is important to let people know where you are hiking and when you plan to return.**

- **Bring equipment and supplies.** You should have a map of the area in which you will be hiking or camping. Learn how to read a compass and carry one. Take along a first aid kit, flashlight and extra batteries. Be sure to bring an adequate supply of drinking water and food for your trip. Bring food that will not spoil.
- **Know the plants and animals.** Learn to recognize the dangerous plants and animals in your area so that you can avoid them. For example, learn what poison ivy and poison oak look like. To avoid insect bites and stings, tuck your pant legs into your socks and apply insect repellent. Learn first aid to treat reactions to poisonous plants, insects, and snakebites.
- **Use fire responsibly.** Learn the proper way to build a campfire. Light campfires only where allowed. Put out all campfires completely before you go to sleep or leave the campsite. To do so, soak the campfires with water or cover them completely with sand or dirt that is free of debris.

> **PREPARATION IS THE FIRST STEP IN A SAFE AND ENJOYABLE HIKE OR CAMPING TRIP.**

Lesson 3 Review

What I Learned

1. **DEFINE** What is a *pedestrian*?
2. **EXPLAIN** Why is it important to learn how to fall when skating or riding a bike?
3. **NAME** What are three ways to stay safe while in the water?

Thinking Critically

4. **ANALYZE** What are some safety factors that can reduce your risk of traffic injuries?
5. **EVALUATE** How does practicing the buddy system help keep you safe outdoors?
6. **APPLY** Research the safety belt laws for children and adults in your state. Explain why it is important to have safety belt laws.

Applying Health Skills

7. **ADVOCACY** Create a poster that displays the dangerous plants in your area. Show plants such as poison ivy or poison oak as well as plants that are poisonous if eaten. Post a phone number to call if someone has eaten a poisonous plant.

Personal Safety and Online Safety

Before You Read

Quick Write What steps do you take to keep yourself safe? Write a short paragraph about these strategies.

Vocabulary

precautions

BIG IDEA You can protect yourself from violence by avoiding dangerous situations.

Personal Safety

MAIN IDEA You can reduce your risk of becoming a victim of violence by avoiding unsafe situations.

Did you know that teens are the victims of violence more than any other age group? Violence is physical force used to harm people or damage property. Teens are more likely than children to go out at night and less likely to protect their personal safety than adults. You can reduce your risk of becoming a victim of violence by avoiding unsafe situations. Be alert to what is going on around you and trust your instincts. If a situation feels unsafe, it probably is.

Staying Safe at Home

Staying safe at home involves observing safety rules such as the ones listed below.

- When you're home, keep your doors and windows locked. Only open the door for someone you know, or not at all if your parents tell you not to answer the door.
- When you answer the phone or use the Internet, don't give out any personal information. Never tell a stranger that you are home alone. Say your parents are busy and can't come to the phone.
- When you come home, have your key ready before you reach the door.
- If someone comes to the door or window and you feel you are in danger, call 911.

Reading Check

Explain Why is it important to tell your family where you are going and when to expect you home?

IF A SITUATION FEELS UNSAFE, IT PROBABLY IS.

Staying Safe on the Street

If you are going out, tell your family where you are going and how you will get there. Make sure they also know when you expect to return. Don't walk by yourself, if possible. After dark, walk in well-lit areas. Stay in familiar neighborhoods; avoid deserted streets and dangerous shortcuts. If you think someone is following you, go into a public place, such as a store or a well-lit area where there are other people.

Don't look like an easy target. Stand tall and walk confidently. Never carry your wallet, purse, or backpack in a way that is easy for others to grab. If someone wants your money or possessions, give them up.

Avoid strangers. Never get into or go near a stranger's car or hitchhike. Do not enter a building with a stranger. Don't agree to run errands or do other tasks for strangers. If someone tries to grab you, scream and run away. Go to the nearest place with people. Ask them to call 911 or your parents.

Online Safety

MAIN IDEA Staying safe online is an important part of your overall safety.

You probably use online tools to complete your homework, shop, and communicate with friends. Most teens today have a smartphone or tablet of their own, and use the device to keep in touch with friends. Using text messaging and social media sites is now a common way that teens stay in touch with friends.

The use of online tools requires some care. Text messages with personal information can be forwarded without your knowledge or agreement. The people you meet on social media sites may not be who they say they are. Everyone who uses online tools must take **precautions** to avoid becoming a victim of online predators. Precautions are planned actions taken before an event to increase the chances of a safe outcome.

Using social media is a great way to stay in touch with family and friends. **Describe why you should take precautions when using social media.**

Taking precautions when using the Internet helps protect your safety. **What are ways you can protect your personal data?**

THE INTERNET IS LIKE ANY TOOL. YOU'VE GOT TO USE IT RIGHT TO STAY SAFE.

Protect Your Information

Protecting your personal information can prevent people from stealing your usernames and passwords. Protecting your information also means being careful about sending photos and other private information online. Use the following tips to help protect your data while using the Internet.

- **Avoid sending personal information online.** When you send a message with photos or personal information, you lose control of who else sees it. The person who receives your message can forward it to anyone without your permission. Cyberbullies might use the tactic to bully you.

- **Avoid opening attachments or email links.** Viruses that can ruin your computer are frequently sent as email links or attachments. As a rule, if you don't know the sender, don't open the email or the links.

- **Be careful about downloads.** Unless a site has a solid reputation, don't use it to download applications. The applications could come with viruses and spyware attached.

- **Keep passwords private.** Your passwords are private. They keep your information safe and prevent others from seeing it without your permission. Never reveal them to people online.

- **Use filtering software.** Search engines and browsers have a filter function that can block information that's not meant for you. Check out the "preferences" tab to set it up.

- **Get permission.** Make sure to get the okay from a trusted adult before heading online—and especially before filling out any forms on the Internet.

- **Establish rules with dating partners for online sharing.** It is important to set personal boundaries. Talk to dating partners about the types of photos, videos, and other posts that you are comfortable to be shared online. Make sure the other person agrees to your limits.

Avoiding Internet Predators

Internet predators might be cyberbullies from your school, or they might be older adults who target teens. To avoid falling victim to Internet predators, follow the general guidelines for online safety.

- **Avoid sending photos to strangers online.** Remember that you lose control of who sees the photo when you send it to someone else.

- **Avoid responding to inappropriate messages.** If someone online sends you a message that makes you feel uncomfortable, avoid responding. He or she may be an internet predator. Responding gives the person an opening to communicate more with you. Tell a parent or other trusted adult and report the sender to your internet service provider (ISP).

- **Be cautious about meeting an online friend.** Meeting new friends can be fun. However, if you first met the person online, you can't be sure that the person is truthful about his or her age and other information. If you decide to meet an online friend in person, plan to meet in a public place. Tell a parent or guardian where you will go, who you are meeting, and when you will be home.

- **Talk to a trusted adult.** If you feel uncomfortable or scared about something that happened online, tell a trusted adult about it immediately.

Reading Check

List What are three things you can do to protect yourself from online predators?

Lesson 4 Review

What I Learned

1. VOCABULARY Define *precautions*.

2. DESCRIBE What strategies should you use to keep yourself safe if you are walking home and you think you are being followed?

3. IDENTIFY If you were approached by a stranger in person, on the phone, or on the Internet, how would you protect yourself?

Thinking Critically

4. APPLY Imagine that you are home alone and two people you don't know come to the door. They tell you that their car broke down and ask if they can come in and use your phone. What should you do to avoid possible risks to your safety?

5. EVALUATE Greg has been studying at a friend's house all day. Now, it is dark outside and he is uncomfortable walking home alone. What could Greg do?

6. APPLY Irena has started dating someone. The dating partner wants to post some photos of them at the park online. Why is it important for Irena to set personal boundaries and discuss these boundaries with her dating partner?

APPLYING HEALTH SKILLS

7. COMMUNICATION SKILLS Suppose your friend has told you she has been chatting with a stranger online. At first the communication was harmless, but now she is uncomfortable with the messages she has been receiving. She is afraid to tell her parents about what is happening because she thinks she will get into trouble. Write a short dialogue that your friend could use to tell her parents about this issue.

Weather Safety and Natural Disasters

BIG IDEA Weather emergencies include thunderstorms, tornadoes, hurricanes, and blizzards. Natural disasters include flood and earthquakes.

What Are Weather Emergencies?

MAIN IDEA Weather emergencies are dangerous situations brought on by changes in the atmosphere.

Weather events make the news on a fairly regular basis. These events often happen with little warning. People cannot prevent them. A **weather emergency** is a dangerous situation brought on by changes in the atmosphere. Examples include thunderstorms, tornadoes, hurricanes, and blizzards.

Weather emergencies can impact a person's safety and health. As a result, the National Weather Service (NWS) works to track the progress of storms. The NWS sends out bulletins to the public. The bulletins keep people informed about possible weather emergencies. This helps keep people and communities safe. Storm bulletins may involve watches or warnings. A storm watch indicates that a storm is likely to develop. A storm warning indicates that a severe storm has already developed and a weather emergency is happening. As a result, people in the area are in danger. If your area is under a storm warning, turn on the television or radio. Follow the instructions of the NWS and local officials.

Technology has helped scientists who watch the weather. Satellites gather data very quickly and feed it into powerful computers. Computers can also help predict the paths of storms. Television and the Internet can warn the public of danger very quickly. These early warnings give people more time to plan and stay safe.

> **WEATHER EMERGENCIES CAN IMPACT A PERSON'S SAFETY AND HEALTH.**

Thunderstorms and Lightning

Thunderstorms can be frightening sometimes. They can occur during any season, though they are more common during warm weather. Lightning is the most dangerous part of a thunderstorm. It is caused by clouds releasing stored-up electrical energy.

How can you protect yourself during thunderstorms? Whenever you see lightning or hear thunder, seek shelter. If you are indoors, stay there. Do not use the telephone, unless it is a cordless or cell phone. Be prepared for a power loss. If you are outdoors, look for the nearest building. An alternative is an enclosed metal vehicle with the windows completely shut. If you are in an open field with no shelter nearby, lie down. Wait for the storm to pass. Avoid all metal objects including electric wires, fences, machinery, motors, and power tools. Unsafe places include underneath canopies, small picnic or rain shelters, or near trees.

> THUNDERSTORMS CAN OCCUR ANY SEASON, THOUGH THEY ARE MORE COMMON DURING WARM WEATHER.

> Thunderstorms occur all over the United States. **List two ways you can protect yourself during a thunderstorm.**

Reading Check

Identify What are some precautions to take during a tornado?

Weather experts track the path of a hurricane by recording the coordinates of its location at regular intervals. **Describe how tracking a hurricane might help people who live in a coastal area.**

Tornadoes

A **tornado** is a whirling, funnel-shaped windstorm that drops from storm clouds to the ground. It's a weather emergency. Tornadoes can happen all over the United States. However, states in the Midwest and those nearest the Gulf of Mexico experience more tornadoes than other states do. In fact, this region is often called "Tornado Alley."

Tornadoes typically happen in the spring. Most tornadoes move at about 25 to 40 miles per hour, although some speed along as fast as 60 miles per hour. Tornadoes are extremely dangerous storms, often destroying everything in their path. If a tornado watch is issued for your area, listen to the radio for updates. Prepare to take shelter if you need to protect yourself. If a tornado warning is issued for your area, get to this shelter right away.

You are safest underground in a cellar or basement. If you cannot go underground, take shelter in an inner hallway or any central, windowless room such as a bathroom or closet. If you are outdoors, lie in a ditch or flat on the ground. Stay away from trees, cars, and anything that could fall on you.

Cover yourself with whatever protection you can find. If you are in the basement, try to get under a workbench. If you are in a room with furniture, stay under a heavy table. Lying in a bathtub under a cushion, mattress, or blanket may also offer good protection. Stay where you are. The storm will pass quickly.

Hurricanes

A **hurricane** is a strong windstorm with driving rain that forms over the sea. Wind speeds during a hurricane can reach or exceed 100 miles per hour. Hurricane-force winds can turn over cars and knock down buildings. As they blow over water, the strong winds can create large waves, which in turn can produce flooding.

The storm clouds of a hurricane can extend over hundreds of miles and a swirling air mass revolves around a calm center called an eye. The faster the air mass swirls, the faster the winds and the more powerful the hurricane. Most hurricanes happen in the late summer or early fall. Compared to tornadoes, hurricanes form and move slowly. The National Weather Service tracks hurricanes and can estimate when and where a hurricane will hit land. This gives people time to plan ahead. Take the following steps to stay safe in a hurricane.

- Secure your home by boarding up windows. Close storm shutters before the winds start blowing. Bring inside items such as furniture and bikes that wind could smash into houses.
- Staying alert to TV or radio reports is important. Sometimes residents will be instructed to leave their homes, or evacuate, and head inland. It is necessary to follow these safety instructions.
- If no evacuation is called for, shelter-in-place by staying indoors. Stay away from windows and doors to avoid injury if debris breaks through the barriers.

Blizzards

Do you live in an area hit by snow in the winter? If you do, you may have experienced a **blizzard**. A blizzard is a very heavy snowstorm with winds up to 45 miles per hour. Blizzards can last from an hour or two to several days. Blizzards make travel difficult, often shutting down roads. Blizzards also make it hard for food and other daily needs to reach consumers. You can protect yourself during blizzards and winter storms by following these precautions:

- **Stay indoors.** The safest place during a blizzard is inside. Visibility in a blizzard is usually is reduced to less than 500 feet, making it easy to get lost or disoriented. A person may get lost even if he or she is only a few yards away from home.
- **Bundle up.** One danger in a blizzard is hypothermia. Hypothermia occurs when the skin freezes. Another health risk is frostbite. Frostbite can cause severe injury to the skin and sometimes to deeper tissues. If you must go out, wear layers of loose-fitting lightweight clothing under layers of outerwear that is both wind- and waterproof. Add a scarf, hat, gloves and boots.
- **Don't wander.** Use landmarks to avoid getting lost, or stay put until help arrives. Be careful of downed power lines, which can be dangerous to people on foot.
- **Watch for ice.** Ice is slippery. People can fall while trying to walk on it. On roadways, ice may be difficult to see. Wearing shoes with a non-slip tread on the bottom can help keep a person from slipping on ice.

> BE CAREFUL OF DOWNED POWER LINES, WHICH CAN BE DANGEROUS.

Reading Check

Recall What are three ways to stay safe during a blizzard?

What Are Natural Disasters?

MAIN IDEA Natural Disasters are dramatic events caused by Earth's processes.

Like weather emergencies, a **natural disaster** are events caused by nature that result in widespread damage, destruction, and loss. Natural disasters include floods and earthquakes. One way to stay safe during a natural disaster is to plan ahead. Keep some basic supplies on hand such as fresh water, a radio, a flashlight, batteries, blankets, canned food, a can opener, and a first-aid kit.

Floods

The most common natural disasters are floods, the rising of a body of water and it is overflowing onto normally dry land. Floods can occur almost anywhere. As noted previously, hurricanes can cause floods. Another cause of flooding is heavy rainfall.

Flash floods, floods that occur with little or no warning, are the most dangerous of all. Flashflood waters rise very quickly and are surprisingly powerful. Two feet of moving water has enough force to sweep away cars. More water than that can carry away trucks and houses.

If the NWS issues a flood watch for your area, take your emergency kit, and go to the highest place in your home. Listen to a battery-powered radio for a flood warning. If a flood warning is issued and you are told to evacuate, do so immediately. The following tips can help you survive a flood:

- Head for higher ground. The home of a relative or neighbor who lives outside the warning area on higher ground is a good choice.
- Never walk, swim, ride a bike, or drive a car through flooding water. You could be swept away, be electrocuted by downed power lines, or drown.
- Drink only bottled water. Floodwater is easily polluted by garbage and other waste.
- If you have evacuated the area, return home only after you are told it is safe for you to do so.
- Once the floodwaters go down, make sure that everything that came in contact with the floodwater is cleaned and disinfected. Wear rubber or latex gloves during the cleanup. Throw out all contaminated food. Make sure the water supply is safe before drinking any.

TWO FEET OF MOVING WATER HAS ENOUGH FORCE TO SWEEP AWAY CARS.

The most common natural disasters are floods. **What are some ways you can prepare in case of a flood?**

Reading Check

Describe What are the actions you can take in a flood?

Earthquakes

An **earthquake** is a shifting of the Earth's plates resulting in a shaking of the Earth's surface. A large earthquake typically is followed by a series of **aftershocks**. These are smaller earthquakes as the Earth readjusts after the main earthquake. Earthquakes can occur anywhere in the United States, but happen most often west of the Rocky Mountains. Collapsing walls and falling debris are responsible for causing most injuries in an earthquake.

Injuries from earthquakes can be serious and can cause death. Most injuries and deaths occur because of falling debris and walls. The glass windows and panels of high-rise buildings may also break and fall to the ground, hurting or killing people below. Crushing injuries may also occur as a result of earthquakes.

Protecting yourself during an earthquake involves indoor and outdoor safety. Both indoors and outdoors, if gas is smelled after an earthquake, leave the area. A gas line might have ruptured. Avoid striking matches or a lighter to avoid explosions and fire. Leave the area immediately.

If an earthquake occurs when indoors, items such as lamps can be thrown violently across a room. Other items such as bookcases and heavy furniture can fall and cause crushing injuries. Outdoor injuries might be caused by falling debris from buildings, bridges or overpasses, and utility poles and electrical wires.

To protect while indoors during an earthquake, do the following:

1. Drop down to your hands and knees.

2. Cover your head and neck. Try crawling under a sturdy table.

3. Hold onto your shelter until the shaking stops.

Do not stand in doorways. Many people believe that doorways are stronger than other parts of the house. They are not. Also, standing in a doorway does not protect you from flying debris.

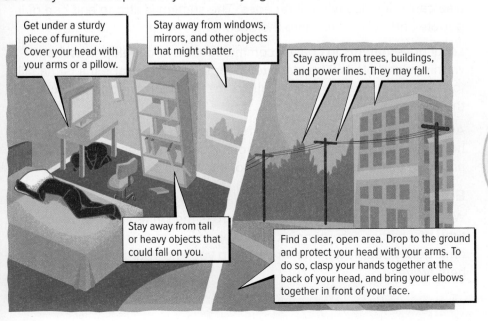

Get under a sturdy piece of furniture. Cover your head with your arms or a pillow.

Stay away from windows, mirrors, and other objects that might shatter.

Stay away from trees, buildings, and power lines. They may fall.

Stay away from tall or heavy objects that could fall on you.

Find a clear, open area. Drop to the ground and protect your head with your arms. To do so, clasp your hands together at the back of your head, and bring your elbows together in front of your face.

During an earthquake, stay clear of falling objects. **Name two ways to stay safe during an earthquake.**

An earthquake occurs with very little warning. **Describe how you would react differently if you were in a car versus being outside during an earthquake.**

To protect yourself while outdoors during an earthquake, do the following:

1. Move away from buildings, utility wires, sinkholes, or fuel and gas lines.

2. Once in an open area, crouch down low to avoid being knocked off your feet by the shaking.

3. Look for hazards such as downed power lines and stay away from them.

In an automobile, do the following:

1. If you're in a car, pull off to the side of the road.

2. Avoid parking under bridges and overpasses.

3. Stay in the car and set the brake.

4. If a powerline falls in the car, stay in the car until emergency services arrive and remove it.

Tsunami

Tsunamis are giant waves that are created underwater. They affect coastal areas. People who live close to a coastline may be affected by a tsunami. A tsunami wave is powerful. It can cause broken bones and other severe injuries. Deaths from drowning can also occur.

A tsunami is caused by an earthquake, landslide, volcanic eruption, or a meteorite that hits the Earth. Giant waves are formed underwater. When they come ashore, they destroy everything in their path. The wave can also cause injuries when it recedes. The suction is strong and can pull people and debris back into the ocean.

Local governments in coastal communities have developed tsunami warnings and plans. The plans tell people how to escape if a tsunami is forecast. They identify safe locations that are high enough to escape the waves.

After a wave recedes, the area affected by a tsunami is dangerous to health. When the wave comes ashore, it contaminates the local land and water. After a tsunami the risk of infectious disease increases.

Emergency Supply Kits

Weather emergencies can happen across the United States. Learn what type of weather emergencies can happen where you live. Prepare an emergency supply kit. The kit can help you and your family survive until help arrives.

Weather emergency kits should contain enough supplies for each member of the family. A kit should also be prepared for each of the family pets. Some items to include in emergency supply kits are:

- Food and water
- Extra clothing
- Medicines and prescription drugs
- First Aid items

Weather emergency kits should be kept in a safe place that everyone in the family can reach if an emergency occurs. Families should also prepare an emergency contact plan. This plan will help members of a family contact each other if an emergency occurs when someone is away from the home.

The CDC offers information on preparing emergency supply kits for each type of weather emergency. Go to www.cdc.gov and search for natural disasters and severe weather. Learn more about the types of weather emergencies that affect your area. The information will also help you identify how you can stay safe.

> **LEARNING ABOUT THE TYPES OF WEATHER EMERGENCIES THAT AFFECT YOUR ARE CAN HELP YOU IDENTIFY HOW TO STAY SAFE.**

Lesson 5 Review

What I Learned

1. **VOCABULARY** Define *hurricane*, and use it in a sentence.

2. **DESCRIBE** How can you protect yourself during a flood?

3. **COMPARE** What is the difference between a weather emergency and a natural disaster?

Thinking Critically

4. **APPLY** You are playing soccer in a field, and you see a flash of lightning from an approaching thunderstorm. What should you do?

5. **SYNTHESIZE** After a major earthquake, your friend wants you go with him to inspect a damaged building. How would you respond? Explain your answer.

Applying Health Skills

6. **COMMUNICATION SKILLS** With a partner, choose a weather emergency from this lesson. Write and perform a skit to demonstrate strategies for staying safe during that event.

First Aid and Emergencies

BIG IDEA Knowing how to administer basic first aid can save a person's life in an emergency.

Giving First Aid

MAIN IDEA First aid is the immediate care given to someone who becomes injured or ill until regular medical care can be provided.

Some emergencies are minor. You cut your fingertip and it bleeds. A friend falls while skateboarding and injures his or her knee. These types of injuries should be cleaned with soap and warm water. They can also be wrapped in a breathable bandage.

Other emergencies can be life-threatening. Taking immediate action can mean the difference between life and death. Knowing basic **first aid** may help you deal with some emergencies while you wait for help to arrive. First aid is the immediate care given to someone who becomes injured or ill until regular medical care can be provided. You can prevent further injury and may even speed recovery if you know what to do in an emergency. Knowing what not to do is equally important. Anyone who has received first aid should be taken to a medical provider as soon as possible.

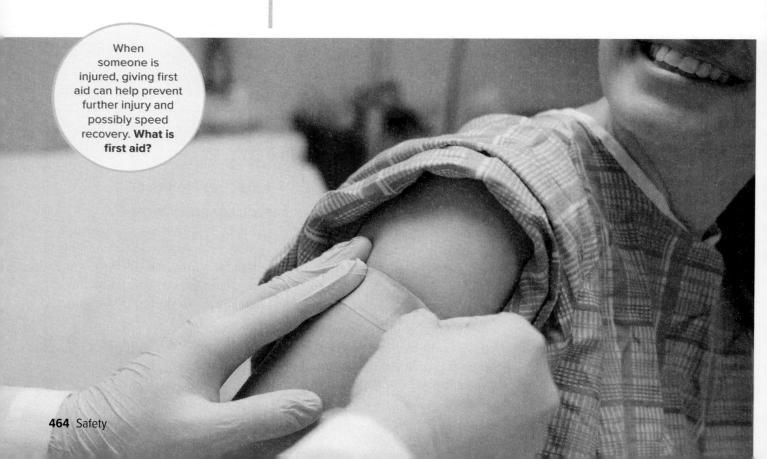

When someone is injured, giving first aid can help prevent further injury and possibly speed recovery. **What is first aid?**

Steps to Take in an Emergency

How can you tell if an emergency is life-threatening? A victim's life is considered in danger if the person: (1) has stopped breathing, (2) has no heartbeat, (3) is bleeding severely, (4) is choking, (5) has swallowed poison, (6) has been severely burned, or (7) cannot respond to you. People in these situations need help immediately. Call for help and then begin to treat the victim. Proper training is needed to give first aid. In an emergency, the American Red Cross suggests the following strategy: Check-Call-Care.

- **Check the scene and the victim.** Often something you see, hear, or smell will alert you to an emergency. Is someone calling out in trouble? Have you heard glass shattering? Do you smell smoke or anything unusual that makes your eyes sting or causes you to cough or have difficulty breathing? These sensations can signal a chemical spill or toxic gas release. Make sure the area is safe for you and the victim. Move the victim only if he or she is in danger.

- **Call for help.** Call 911 or the local EMS number. *EMS* stands for "emergency medical service." When making a call for help, stay calm. Describe the emergency to the operator and give a street address or describe the location by using landmarks. The operator will notify the police, fire, or emergency medical service departments. Stay on the phone until the operator tells you to hang up.

- **Care for the person until help arrives.** After you have called for help, stay with the victim until help arrives. Carefully loosen any tight clothing on the victim. Use a coat or blanket to keep the person warm or provide shade if the weather is warm. This will help the person maintain a normal body temperature. Avoid moving the victim to prevent further pain or injury. Only move the victim if he or she is in danger, such as in the path of traffic. Hands-Only™ Cardiopulmonary Resuscitation (CPR) may be necessary if the victim is unconscious and unresponsive. This lifesaving technique is described later on in this lesson.

> **HOW CAN YOU TELL IF AN EMERGENCY IS LIFE-THREATENING?**

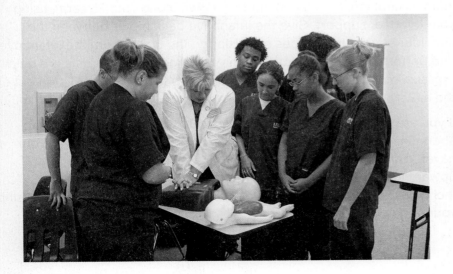

> CPR lessons are available through local branches of the American Red Cross or American Heart Association. **Describe the process to use if you encounter a person who is injured.**

Reading Check

Explain What information should you give when calling 911 or another emergency number?

Universal Precautions

Viruses such as HIV, hepatitis B, and hepatitis C can be spread through contact with an infected person's blood. As a result, steps should be taken to minimize contact with another person's blood. To protect yourself when giving first aid, follow **universal precautions**. They are actions taken to prevent the spread of disease by treating all blood as if it were contaminated. Wear protective gloves while treating a victim. If possible, use a facemask or shield, when giving first aid for breathing emergencies. Cover any open wounds on your body with sterile dressings. Avoid touching any object that was in contact with the victim's blood. Always wash hands thoroughly after giving first aid.

First Aid for Choking

MAIN IDEA Abdominal thrusts can help save someone who is choking.

Choking kills more than 3,000 people every year in the United States. When a piece of food or some other object blocks a person's airway, oxygen cannot reach the lungs. If a person is clutching his or her throat, that is the universal sign for choking. Symptoms of choking include gasping or wheezing, a reddish-purple coloration, bulging eyes, and an inability to speak. If a person can speak or cough, it is not a choking emergency. However, if the choking person makes no sound and cannot speak or cough, give first aid immediately. A person can die from choking within minutes.

If an adult or child is choking, give the person five quick blows to the back. To perform back blows, stand slightly behind the person who is choking. Place one of your arms diagonally across the person's chest and lean him or her forward. Strike the person between the shoulder blades five times. If this does not dislodge the object, use **abdominal thrusts**. They are quick inward and upward pulls into the diaphragm to force an obstruction out of the airway. The proper technique for performing abdominal thrusts is shown in the illustration in the lesson.

Infants who are choking require a different first-aid procedure. If an infant is choking, hold the infant face down along your forearm, using your thigh for support. Give the infant five back blows between the shoulder blades. If this does not dislodge the object, turn the infant over and perform five **chest thrusts** with your fingers. Chest trusts are quick presses into the middle of the breastbone to force an obstruction out of the airway.

If you are alone and choking, there are two ways to give yourself an abdominal thrust. First, make a fist and position it slightly above your navel. With your other hand, grasp your fist and thrust inward and upward into your abdomen until the object dislodges. Another technique is to lean over the back of a chair, or any firm object, pressing your abdomen into it.

> CHOKING KILLS MORE THAN 3,000 PEOPLE EVERY YEAR IN THE UNITED STATES.

Reading Check

Identify When does a person need first aid for choking?

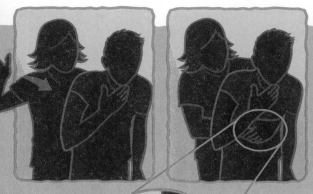

For adults and children

1. Give 5 back blows with the heel of your hand.

2. Place the thumb of your fist against the person's abdomen, just above the navel. Grasp your fist with your other hand. Give quick, inward and upward thrusts until the person coughs up the object. If the person becomes unconscious, call 911 or the local emergency number. Begin CPR.

For infants

1. Hold the infant facedown on your forearm. Support the child's head and neck with your hand. Point the head downward so that it is lower than the chest. With the heel of your free hand, give the child five blows between the shoulder blades. If the child doesn't cough up the object, move on to chest thrusts (step 2).

2. Turn the infant over onto his or her back. Support the head with one hand. With two or three fingers, press into the middle of the child's breastbone—directly between and just below the nipples—five times. Repeat chest thrusts until the object comes out or the infant begins to breathe, cry, or cough. Make sure a health care professional checks the infant. If the infant becomes unconscious, call 911.

Follow these steps to help a person who is choking. **Why do you think chest thrusts are used to help a choking infant?**

Rescue Breathing and CPR

MAIN IDEA Hands-Only™ CPR is for older children and adults who have stopped breathing.

All organs need oxygen-rich blood to work properly. If the heart stops beating, the flow of blood to the brain stops, too. When the brain stops functioning, breathing also stops. If you are confident that the victim is not breathing, it is necessary to begin **cardiopulmonary resuscitation (CPR)**. It's a first-aid procedure to restore breathing and circulation.

The American Heart Association (AHA) recommends two forms of CPR. A trained person will perform CPR that combines chest compressions with **rescue breathing**, and focuses on chest compressions. A first-aid procedure where someone forces air into the lungs of a person who cannot breathe on his or own. A person who is untrained in giving CPR can perform *Hands-Only*™ *CPR*. This form of CPR focuses only on chest compressions. In an emergency, if no trained person is present, an untrained person should begin Hands-Only™ CPR before medical professionals arrive.

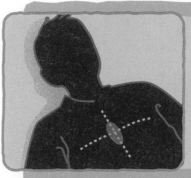

1. Use your fingers to find the end of the victim's sternum (breastbone), where the ribs come together.

2. Place two fingers over the end of the sternum.

3. Place the heel of your other hand against the sternum, directly above your fingers (on the side closest to the victim's face).

4. Place your other hand on top of the one you just put in position. Interlock the fingers of your hands and raise your fingers so they do not touch the person's chest.

1. Straighten your arms, lock your elbows, and line up your shoulders so that they are directly above your hands.

2. Press downward firmly on the person's chest, forcing the breastbone down by 1.5 to 2 inches (3.8 to 5 cm).

3. Begin compressions at a steady pace. You can maintain a rhythm by counting. "One and two and three and..." Press down each time you say a number. Emergency medical experts recommend pressing down 100 times a minute.

Follow these steps to help a person whose heart has stopped beating. **In what type of situation should you use CPR?**

Hands-Only™ CPR for Adults

The first step is to call 911. Before performing CPR, tap the victim and shout, "Are you OK?" Check the victim for signs of movement and normal breathing. Put your ear and cheek close to the victim's nose and mouth. Listen and feel for exhaled air. Look to see if the chest is rising and falling. Performing chest compressions on a victim who is still breathing normally can cause injury. If the victim is unconscious and there is no response, begin Hands-Only™ CPR.

The Steps of Hands-Only™ CPR for Adults, shows where to position your hands over the victim's chest and begin compressions. Try to give 100 chest compressions each minute, until the victim responds or until paramedics arrive. A trained person will then begin CPR that includes chest compressions and rescue breathing.

The AHA recommends taking a class in conventional CPR, which includes training in rescue breathing and giving chest compressions. Conventional CPR may be better than Hands-Only™ CPR for some people including:

- Infants and young children
- An adult who has collapsed
- A person who is drowning or suffering life-threating breathing problems

Shock

MAIN IDEA Shock is a life-threatening emergency that requires immediate medical attention.

Shock is a life-threatening condition in which the circulatory system fails to deliver enough blood to vital tissues and organs. Injury, burns, and severe infection can cause a person to go into shock, as can heat, poisoning, blood loss, and heart attack. Always look for the signs of shock when providing first aid because it can result from a medical emergency.

Signs to watch for include cool, clammy, pale, or gray skin; weak and rapid pulse; and slow, shallow breathing. The eyes may have a dull look with the pupils dilated. The victim, if conscious, may feel faint, weak, confused, and anxious. If you think someone is in shock or about to go into shock, call for medical help and take these precautions:

- Help the person to lie down on his or her back with feet raised slightly higher than the head. Try to keep the person as still as possible.
- Loosen tight clothing.
- Use a blanket, coat, or any available cover to help keep the person warm.
- Do not give the person anything to drink.
- Roll the person onto his or her side to help prevent choking in the event of vomiting or bleeding from the mouth.

First Aid for Severe Bleeding

MAIN IDEA To control bleeding, apply a cloth and direct pressure to the wound and elevate the wound, if possible.

Severe bleeding can be a life-threatening emergency. Blood loss prevents oxygen from getting to the body's organs. When providing first aid to a person who is bleeding severely, follow universal precautions. Avoid touching the person's blood or wear gloves, if possible. Always wash your hands when you are finished. If the person has a wound that is bleeding severely or needs other medical help, call 911 before taking action. Wash the wound with mild soap and water to remove dirt and debris. Then follow these steps to control the bleeding:

- If possible, raise the wounded body part above the level of the heart.
- Cover the wound with sterile gauze or a clean cloth.
- Press the palm of your hand firmly against the gauze. Apply steady pressure to the wound for five minutes, or until help arrives. Do not stop to check the wound; you may interrupt the clotting of the blood.
- If blood soaks through the gauze, do not remove it. Instead, add another gauze pad on top of the first and continue to apply pressure.
- Once the bleeding slows or stops, secure the pad firmly in place with a bandage or strips of gauze or other material. The pad should be snug, but not so tight that you cannot feel the victim's pulse.
- Stay with the victim until help arrives.

BLOOD LOSS PREVENTS OXYGEN FROM GETTING TO THE BODY'S ORGANS.

Reading Check

Explain What should you do if the cloth you have used to cover a wound is soaked with blood?

First Aid for Burns

MAIN IDEA Major burns require medical attention as soon as possible.

Burns can affect all parts of the body, including the eyes and the airways. Burns to the skin are rated by degree. Minor burns to the skin may not require professional medical attention. Burns affecting the eyes and airway always require medical attention.

Being safety conscious can help you avoid burns. For example, never play with matches or fire. Handle hot foods carefully. Avoid making the water too hot in the shower. Sunburns can be serious, too. Protect yourself by wearing sunscreen, staying covered, and limiting time in the sun. If you do get burned, make sure the burn gets treated.

A first-degree burn is a burn in which only the outer layer of skin has burned and turned red. There may be pain and swelling. To treat this type of burn, flush the burned area with cold water for at least 20 minutes. Do not use ice. Then loosely wrap the burn in a clean, dry dressing. Most sunburns are first-degree burns.

A second-degree burn is a moderately serious burn in which the burned area blisters. It is also called a partial-thickness burn. With a second-degree burn, there is usually severe pain and swelling. A burn no longer than 2 to 3 inches in diameter can be treated as a first-degree burn. Flush the burned area with cold water (not ice) for at least 20 minutes. Elevate the burned area. Loosely wrap the cooled burn in a clean, dry dressing. Do not pop blisters or peel loose skin. If the burn is larger, or is on the hands, feet, face, groin, buttocks, or a major joint, get medical help immediately.

A third-degree burn, or full-thickness burn, is a very serious burn in which all the layers of skin are damaged. There may be little or no pain felt at this stage. Third-degree burns usually result from fire, electricity, or chemicals. Third-degree burns require immediate medical attention. Call 911 or another emergency number immediately. Do not try to remove burned clothing. Reduce the heat on the affected area and then cover with a cool, clean, moist cloth. Only a medical professional should treat full-thickness burns.

Reading Check

Describe How should you treat all first-degree burns and second-degree burns that are less than 3 inches in diameter?

The severity of burns is determined by degree. **Explain the difference between a first-degree burn and a second-degree burn?**

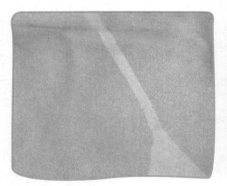
First-degree burn

Second-degree burn

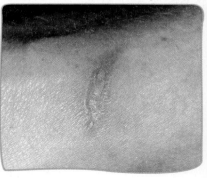
Third-degree burn

First Aid for Other Emergencies

MAIN IDEA Animal bites, bruises and sprains, broken or dislocated bones and poisonings require different types of treatment.

Other common emergencies include insect and animal bites, bruises and sprains, broken or dislocated bones, and poisonings.

Insect and Animal Bites

Insect bites and stings can be painful but are not usually dangerous unless the person is allergic to the venom of the insect. If an allergic person has been stung, get medical help immediately. For all other bites and stings follow these steps:

- Remove the stinger by scraping it off with a firm, straight-edged object. Do not use tweezers.
- Wash the site thoroughly with mild soap and water.
- Apply ice (wrapped in a cloth) to the site for ten minutes to reduce pain and swelling. Alternate ten minutes on and off.
- To treat animal bites, wash the bite with soap and water. Apply pressure to stop any bleeding. Apply antibiotic ointment and a sterile dressing. For any bite that has broken the skin, contact your doctor.

Bruises and Sprains

A bruise forms when an impact breaks blood vessels below the surface of the skin. This allows blood to leak from the vessels into the tissues under the skin, leaving a black or bluish mark. A sprain is a condition in which the ligaments that hold the joints in position are stretched or torn. The most commonly sprained joints are ankles and knees. Symptoms of sprains include swelling and bruising. While a doctor should evaluate serious sprains, minor sprains can be treated using the P.R.I.C.E. method:

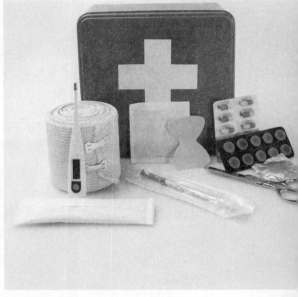

- *Protect* the injured part by keeping it still. Moving it could cause further injury.
- *Rest* the affected joint for 24 to 48 hours.
- *Ice* the injured part to reduce swelling and pain. A cloth between the skin and ice bag will reduce discomfort. Be sure to remove the ice every 15–20 minutes so that it does not become too cold.
- *Compress* the injured area by wrapping it in an elastic bandage.
- *Elevate* the injured part above the level of the heart to reduce swelling.

> **A BRUISE FORMS WHEN AN IMPACT BREAKS BLOOD VESSELS BELOW THE SURFACE OF THE SKIN.**

Developing Good Character

Citizenship A good neighbor and citizen is prepared to report accidents, fires, serious illnesses, injuries, and crimes. Familiarize yourself with emergency phone numbers to call in your community. Make a list to keep handy by the telephone.

Broken or Dislocated Bones

A fracture usually happens along the length of a bone. An open fracture is a complete break with one or both sides of the bone piercing the skin. A closed fracture does not break the skin and may be difficult to identify. Pain, swelling and a misshapen appearance are typical symptoms of a closed fracture. However, not all broken bones cause immediate pain. An X ray is the only way to be sure if a bone is broken.

Problems can also develop where bones meet at a joint. For example, a dislocation happens if your upper arm bone is pulled out of your shoulder socket. Moving a broken bone or dislocated joint could cause further injury. For both fractures and dislocations, call for help at once. While you wait for help, keep the victim still. Once a trained medical professional arrives, he or she can then immobilize the fracture or dislocation.

Ice helps slow swelling after a sprain. **How can you reduce the risk of sprains during physical activity?**

> **PAIN, SWELLING AND MISSHAPEN APPEARANCE ARE TYPICAL SYMPTOMS OF A CLOSED FRACTURE.**

Poisoning

A poison is a substance that causes harm when swallowed, inhaled, absorbed by the skin, or injected into the body. Medicines and household products play a role in about half of all poisonings. All poisonings require immediate treatment. In the event of a poisoning, call 911, EMS or the nearest **poison control center**. It's a community agency that helps people deal with poisoning emergencies. Be ready to provide information about the victim and the suspected poison. The poison control center will advise you about how to proceed. The victim might need to drink water or milk to dilute the poison, or a dose of syrup of ipecac, a medication that causes vomiting. While waiting for help to arrive, keep the person warm and breathing. Look for extra traces of poison around the victim's mouth. Remove these with a damp, clean cloth wrapped around your finger. Make sure to save the container of poison. Show it to the ambulance team. Tell them all you know about what happened.

If a poisonous chemical such as a pesticide or household cleaning agent has made contact with someone's skin, first remove all clothing that has touched the chemical and rinse the skin with water for 15 minutes. Then wash gently with soap and water. Call the poison control center while the skin is being washed.

Some cases of poisoning are caused by contact with a poisonous plant. Poison ivy, poison oak, and poison sumac are three such plants. Contact with these plants can cause redness, itching, and swelling. Most of these injuries can be easily treated at home using soap and water, rubbing alcohol, and over-the counter creams. For severe cases, see a doctor for treatment.

Reading Check

List Give two ways poisons can enter the body.

Some plants can cause an allergic reaction. **List the three symptoms of coming into contact with a poisonous plant.**

Lesson 6 Review

What I Learned

1. **VOCABULARY** Define *first aid*. Use the term in a sentence.

2. **NAME** What are the four universal precautions to taken when administering first aid?

3. **LIST** Briefly give the steps in controlling severe bleeding.

Thinking Critically

4. **INFER** Why can you infer that a person who cannot speak or cough?

5. **APPLY** If you come upon an injured person on a hiking trail, should you try to move the person off the trail? Why or why not?

APPLYING HEALTH SKILLS

6. **STRESS MANAGEMENT** Emergency situations are often very stressful. With classmates, discuss strategies for reducing stress while dealing with a medical emergency.

Green Schools and Environmental Health

LESSONS

1 Pollution and Health

2 Preventing and Reducing Pollution

3 Green Schools

20

Pollution and Health

Before You Read

Quick Write Give three examples of types of pollution that affect your local community.

Vocabulary

environment
pollute
pollution
fossil fuels
acid rain
ozone
smog
ozone layer
greenhouse effect
climate change
sewage
landfills
biodegradable
hazardous waste

BIG IDEA Pollution harms the environment and your health, and is often ugly.

Your Environment

MAIN IDEA Pollution is made up of dirty or harmful substances in the environment.

The **environment** is all the living and nonliving things around you. It includes forests, mountains, rivers, and oceans. It also includes your home, school, and community. The air you breathe, the water you drink, the plants and animals that live nearby, and the climate you live in are all part of the environment. All living things are affected by the health of the environment.

Sometimes people **pollute** the environment. They make it unfit or harmful for living things. The result is **pollution**, which are dirty or harmful substances in the environment. Pollution can also harm your health. It affects everything in your environment. On days when air pollution is heavy, you may have to limit the time you spend outside. Everyone's health depends upon the environment, so it's important for each person to do his or her part to keep the environment clean.

Your environment includes all the living and nonliving things around you. **Identify how the environment in this photo is being damaged.**

Air Pollution

MAIN IDEA Burning fossil fuels such as coal, oil, and natural gas pollutes the air.

Some air pollution is caused by natural events. For example, gases and ash from erupting volcanoes pollute the air. The main cause of air pollution, however, is the burning of **fossil fuels**, which are oil, coal, and natural gas that are used to provide energy. The energy from fossil fuels provides heat for homes and electricity to power factories, towns, and cities. Fossil fuels also power most motor vehicles. Burning fossil fuels releases toxic gases into the atmosphere.

Acid Rain

Certain chemicals in gases released from burning fossil fuels include sulfur dioxides and nitrogen oxides. They mix with moisture in the air to form **acid rain**, which is rain that is more acidic than normal rain. Over time, acid rain can harm plants and even whole forests. It can contaminate water supplies, too. Acid rain can even eat away at rock and stone.

> IT'S IMPORTANT FOR EACH PERSON TO DO HIS OR HER PART TO KEEP THE ENVIRONMENT CLEAN.

Smog

Fossil fuels create other gases when they burn. Some of these gases are changed by heat and sunlight into ozone. High up in the atmosphere, **ozone** forms. It's a gas made of three oxygen atoms that occurs naturally and helps protect you from the sun's harmful rays. Closer to ground level, ozone mixes with other gases to form **smog**. It's a yellow-brown haze that forms when sunlight reacts with air pollution.

Ozone and smog can cause health problems or make existing health problems worse. For example, people who have bronchitis, asthma, or emphysema have a very hard time breathing when smog is in the air. Many cities issue warnings on days when there is too much smog or ozone in the air. On such days, people sensitive to smog or ozone should limit the time they spend outside.

Reading Check
Identify What is pollution?

Damage to the Ozone Layer

The naturally occurring **ozone layer** is in the upper atmosphere. It is a shield above the Earth's surface that protects living things from ultraviolet (UV) radiation. In the 1970s, scientists discovered that the ozone layer was breaking down. Chemicals such as the propellants used in aerosol cans, emissions from automobiles, and the chemical that keeps refrigerators and air conditioners cool were damaging the ozone layer.

The ozone layer provides protection from the sun. Without it, people are more likely to develop skin cancer and eye damage. It is important to protect yourself from the sun's rays by using sunscreen and sunglasses. Today, many countries are working to help restore the ozone layer by banning the use of the chemicals that damage it. You can do your part by using products that won't cause more damage to the ozone layer.

The heating of Earth by gases in the atmosphere is similar to how a greenhouse warms. **Identify how the greenhouse effect can be stopped.**

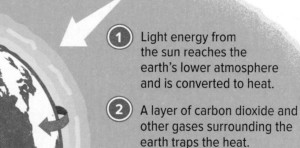

1. Light energy from the sun reaches the earth's lower atmosphere and is converted to heat.

2. A layer of carbon dioxide and other gases surrounding the earth traps the heat.

3. The surface of the earth and the lower atmosphere become warmer because of the trapped heat.

Climate Change

The trapping of heat by carbon dioxide and other gases in the air is known as the **greenhouse effect**. The greenhouse effect warms the Earth to support life. However, the release of carbon dioxide and other gases from burning fossil fuels increases the greenhouse effect. Many scientists agree that the increase in greenhouse gases in the atmosphere is the cause of **climate change**. Climate change is a rise in the Earth's temperatures. It can affect weather patterns worldwide as well as ocean water levels.

Fossil fuels are burned every day in communities. Burning fossil fuels provides electricity and heat for homes and buildings. The gasoline that makes your family's car run is a fossil fuel. Burning fossil fuels causes the level of carbon dioxide in the atmosphere to rise. The rise in carbon dioxide affects the entire planet.

Water Pollution

MAIN IDEA Chemicals used on land are the primary source of water pollution.

Water is vital to all forms of life. However, wastes, chemicals, and other harmful substances pollute Earth's water. Water pollution is a widespread problem. Forty percent of all the nation's rivers, lakes, streams, and oceans are too polluted to use for swimming, fishing, or drinking. One type of pollution is **sewage**. It consists of human waste, garbage, detergents, and other household wastes washed down drains and toilets. Sewage in the United States is treated. However, many countries do not properly treat water.

Chemicals used in industry also contribute to water pollution. Some enter the water from factories. Agriculture also contributes to water pollution. Oil spills from large tanker ships kill plants and animals and harm delicate habitats. Oil spills also run off into nearby lakes, rivers, and wetlands.

Water polluted with sewage can spread diseases. Eating shellfish from polluted water can cause hepatitis, a disease of the liver. Drinking water contaminated by metals such as lead or mercury can damage the liver, the kidneys, and the brain. It can also cause birth defects.

Reading Check

Identify Name two environmental problems caused by air pollution?

Reading Check

Describe How could water pollution affect your health?

WATER IS VITAL TO ALL FORMS OF LIFE.

Solid Waste

MAIN IDEA Solid wastes include items that are disposed of in landfills.

Land may also become polluted. Many of the items we use in daily life are made of plastic and metal. When they are thrown away, these materials take a long time to break down. Much of this waste goes into **landfills**, which are huge specially designed pits where waste materials are dumped and buried.

Landfills may have walls or linings of clay or plastic so that water flowing through the landfill does not carry chemicals or other material into water supplies. In time, all landfills get filled up. When this happens, they are capped and sealed. A new landfill is made somewhere else.

Biodegradable Waste

Not all solid waste ends up in landfills. Many discarded items are **biodegradable**, or easily broken down in the environment. Food waste, paper, and wood all break down naturally. Some people set up a compost pile, a place where biodegradable wastes can break down naturally and turn into fertilizer. Leaves, grass, shredded newspaper, and some food wastes are items that can be composted.

Water pollution is a widespread problem. **Name the percentage of U.S. rivers, lakes, streams, and oceans that are too polluted to use.**

Hazardous Wastes

Some wastes are hazardous to the health of all living things. These wastes should never go into a landfill. **Hazardous wastes** are human-made liquid, solid, sludge, or radioactive wastes. They may endanger human health or the environment. Some examples of hazardous waste items are industrial chemicals, asbestos, radioactive materials, and some medical wastes. Others are hazardous substances from our homes such as motor oil, paint, insecticides, nail polish remover, antifreeze, bleach, and drain cleaner. Batteries, computers, and air conditions also contain hazardous wastes.

Hazardous wastes should be disposed of safely. Many are stored in facilities where they will not be released into the environment. Some communities schedule hazardous waste drop-off days. The community collects these materials and disposes of them. This prevents them from reaching the environment.

> SOME WASTES ARE HAZARDOUS TO THE HEALTH OF ALL LIVING THINGS. THESE WASTES SHOULD NEVER GO INTO A LANDFILL.

Lesson 1 Review

What I Learned

1. **VOCABULARY** Define *fossil fuels.*

2. **RECALL** Name two sources of air pollution and two sources of water pollution.

3. **IDENTIFY** How is smog formed?

Thinking Critically

4. **EXPLAIN** How do fossil fuels contribute to climate change?

5. **ANALYZE** What is the difference between ozone in the upper atmosphere and ozone nearer to ground level?

Applying Health Skills

6. **ACCESSING INFORMATION** Use reliable sources to research the dangers of exposure to lead in water and explain how to avoid this potentially harmful substance. Report your findings to the class.

Preventing and Reducing Pollution

Before You Read

Quick Write Make a list of actions you already take to reduce pollution.

Vocabulary

Environmental Protection Agency (EPA)

Occupational Safety and Health Administration (OSHA)

conservation

nonrenewable resources

groundwater

precycling

recycle

Reading Check

Describe What can you do to promote cleaner air?

BIG IDEA You have the power to prevent and reduce pollution.

Keeping the Environment Clean

MAIN IDEA The Environmental Protection Agency is a government agency committed to protecting the environment.

We can all do our part to help reduce pollution. When we work together as a community, we can do even more. Governments around the world are committed to reducing and preventing pollution. In the United States, the **Environmental Protection Agency (EPA)** is an agency of the U.S. government that is responsible for protecting the environment. The **Occupational Safety and Health Administration (OSHA)** is a branch of the U.S. Department of Labor that protects American workers. People who work at OSHA make sure that work environments are safe and free of hazardous materials. Many local governments maintain air and water quality. Some methods they use are waste management strategies and controlling auto emissions. Waste management is the disposal of wastes in a way that protects the health of the environment and the people.

Every person can take actions to help protect the environment. **List two ways you can reduce air pollution.**

Reducing Air Pollution

MAIN IDEA Using alternative transportation, staying tobacco free, and planting trees can all help reduce air pollution.

Any time you use an electrical appliance, ride in a car, or run a power lawn mower, you are burning fossil fuels to produce energy. You are also contributing to air pollution. Reducing air pollution can help keep you and your community safe. Here are some strategies to help reduce air pollution in your community.

- **Walk or ride your bike.** When you ride a bike or walk rather than ride in a vehicle, you save fuel and reduce pollution. You can also get the benefit from some physical activity.
- **Use public transportation or carpool.** Carpooling, or taking a bus, train, or subway, cuts down on the number of cars producing exhaust fumes. That helps cut down on pollution.
- **Stay tobacco free.** Tobacco smoke is not only unhealthy for people who smoke, it pollutes the air.
- **Plant trees and other plants.** Plants convert and remove carbon dioxide from the air during photosynthesis, helping reduce the amount of carbon dioxide in the atmosphere.
- **Don't burn trash, leaves, and brush.** Let the local waste management facility dispose of your trash.
- **Advocate for the use of renewable energy resources in your community.** Advocate for the development of energy resources that rely on renewable energy, such as wind power and hydroelectric power.

When you use less of a resource, such as fossil fuels, you are practicing **conservation**, or the saving of resources. You can conserve energy resources in your own home for example, by turning off lights when you leave a room. When you conserve energy, you are also using fewer fossil fuels.

Riding a bike rather than using a motor vehicle that burns fossil fuels is better for the environment. **Describe the health benefits of riding on a bicycle over in a car.**

Reducing Water Pollution

MAIN IDEA Picking up after pets, picking up litter, using environment-friendly products, and disposing of chemicals properly can all reduce water pollution.

. . . .

Reading Check

List What are two ways to reduce water pollution?

. . . .

We all need clean drinking water to maintain health. Clean water is important for all plants and animals. The industries and farms that use water to produce the foods and beverages we eat and drink need clean water. We also need clean water for water recreation activities. To help keep water clean, follow these tips.

- Pick up pet waste from public areas to reduce toxic runoff.
- Use soaps, detergents, and cleaners that are biodegradable.
- Pick up any litter that is not hazardous.
- Dispose of chemicals properly and legally. Never pour them into a drain.
- Take hazardous wastes to the appropriate collection sites.

CLEAN WATER IS IMPORTANT FOR ALL PLANTS AND ANIMALS.

Clean water is an important part of a healthy environment. **List three ways you can help reduce water pollution.**

Protecting Natural Resources

MAIN IDEA Some resources, such as fossil fuels, are nonrenewable, meaning that they can be used only once.

Fossil fuels are natural materials known as **nonrenewable resources**. When fossil fuels, such as oil, natural gas, or coal are used, they cannot be replaced. Other resources are always being renewed. For example, the supply of freshwater is constantly being renewed through the water cycle. The water cycle is the movement of water through, around, and over the earth. Even renewable resources, however, need to be protected. There is a limited amount of freshwater. Pollution makes freshwater more expensive because polluted water has to be cleaned before it is used. Trees are cut down to make paper and lumber. Removing too many trees upsets the balance of nature. By upsetting this balance, the lives of all living things are endangered. Conservation is a good way to protect resources such as water and trees.

Conserving Water Resources

As a teen, you may wonder how you can conserve water resources. How can one person conserve water? You can do several things to conserve water both at home and elsewhere.

Some household products contain hazardous materials. Dumping them down the drain would pollute the water. **Explain how you can safely dispose of hazardous wastes in your community.**

Inside the House

- Never let water run unnecessarily. Turn off the faucet while brushing your teeth.
- Wash clothes in warm or cold water, which uses less energy than hot water.
- Run the washing machine or dishwasher only when you have a full load, and use the short cycle when appropriate.
- If you have an older toilet, place a 1-liter bottle filled with water inside your toilet tank. This will reduce the amount of water used for flushing. Another option is to replace an older toilet with a newer model that requires less water per flush.
- Fix leaky faucets.
- Install water-saving showerheads or take shorter showers.

Outside the House

- Turn the hose off when you are washing the car. Use the hose only for rinsing the car.
- Water lawns only when needed. Use soaker hoses for watering gardens.
- Garden with plants that conserve water.

Doing a full load of laundry conserves water. **Explain why we need to conserve water even though it's a renewable resource.**

Old computers are considered a type of hazardous waste. **Tell how you can learn how to dispose of hazardous wastes in your community.**

Dealing with Wastes

MAIN IDEA Recycling and conservation have a positive impact on the environment.

Land pollution results from littering. It also occurs as a result of the careless disposal of household and industrial garbage. Many of the items we use in daily life are made of plastic and metal. When they are thrown away, these materials take a long time to break down, if they ever do. This affects not only the soil but also the air and **groundwater.** It's the water that collects under the Earth's surface.

Types of Wastes

The average U.S. citizen produces about 4 pounds of trash, or solid waste, daily. The solid waste produced by households and businesses usually ends up in a landfill. Landfills may have walls or linings of clay or plastic so that water flowing through the landfill does not carry chemicals or other material into water supplies. In time, all landfills get filled up. When this happens, they are capped and sealed. A new landfill is made somewhere else.

Some wastes are hazardous to the health of all living things. All hazardous wastes require careful handling and special disposal. If you need to dispose of household hazardous waste, contact your local health department or environmental agency. They will explain how to get rid of it safely. Many communities have drop-off centers to collect household hazardous waste. Never put household hazardous wastes in the regular trash or pour chemicals down the drain.

Reading Check

Recall What are two types of wastes?

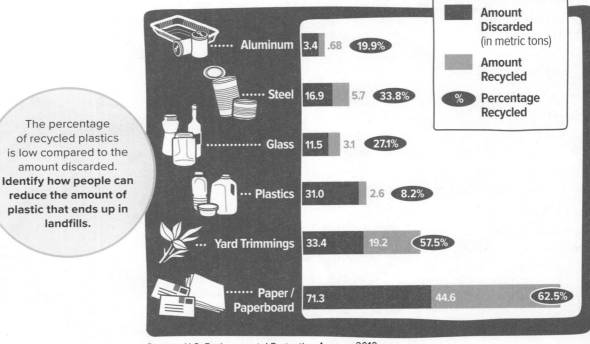

The percentage of recycled plastics is low compared to the amount discarded. **Identify how people can reduce the amount of plastic that ends up in landfills.**

Source: U.S. Environmental Protection Agency, 2010.

Reducing, Reusing, and Recycling

MAIN IDEA Precycling is reducing waste before it is used, and recycling conserves energy and natural resources.

What can one person do to reduce solid waste going into landfills? The answer is *plenty*—especially if you are willing to take positive action. The key to reducing solid waste is by following the three Rs: reduce, reuse, and recycle.

Reducing wastes by **precycling** is reducing waste before it occurs. Below are some basic guidelines for precycling:

- Buy products in packages made of glass, metal, or paper. You can reuse or recycle these materials.
- Look for products in refillable containers.
- Carry store-bought purchases home in your own reusable cloth sack or plastic bags.
- Avoid using paper plates and plastic cups, knives, forks, or spoons.
- Buy products in bulk to reduce the amount of packaging you throw away.

• • • • • • • • • •

Reading Check

Recall What are the three Rs?

• • • • • • • • • •

Reusing objects is another way to cut down on waste. Think of other ways to use items you would otherwise throw away. You can buy reusable food containers. Reuse plastic grocery bags as trash bags or to clean up after pets. Donate unwanted clothes to charity rather than throwing them out.

To **recycle** is to change items in some way so that they can be used again. This practice conserves natural resources while it helps reduce solid waste. Paper, aluminum, glass, plastics, and yard waste are the most commonly collected recycling materials. When aluminum cans are recycled, they are changed back into sheets of aluminum that can be used to make new cans or other products.

A symbol with three curved arrows appears on many kinds of products. It indicates that an item can be recycled, or that it is made of recycled materials. More and more people are becoming involved in recycling through drop-off centers and curbside programs.

THE KEY TO REDUCING SOLID WASTE IS BY FOLLOWING THE THREE RS: REDUCE, REUSE, AND RECYCLE.

Lesson 2 Review

What I Learned

1. **DEFINE** What is the *Environmental Protection Agency?*

2. **GIVE EXAMPLES** What can you do to reduce air pollution?

3. **IDENTIFY** Name the three Rs and tell how they are related to your health.

Thinking Critically

4. **INFER** Why is it a good idea to turn off lights when you leave a room?

5. **ANALYZE** If conservation is a good idea, why do you think people might still need to be reminded to conserve resources?

6. **ANALYZE** How does properly disposing of hazardous waste affect your environment as well as your personal health?

Applying Health Skills

7. **ADVOCACY** Write and illustrate a comic book that encourages teens to conserve electricity and water. In your comic book, be sure to explain why conservation of these resources is important.

Green Schools

Before You Read

Quick Write What actions do you take at school to help protect the environment?

Vocabulary
green school
pesticide

BIG IDEA You can help your school become green.

What is a Green School?

MAIN IDEA Green schools are environmentally friendly and provide a healthy environment for students.

You may have heard the term *green* used to refer to programs or actions that are environmentally friendly. A **green school** is a school that is environmentally-friendly in several different ways. The school has made changes to conserve energy, water, and other resources.

Green schools provide a healthy environment for students. They eliminate toxins such as mold, certain cleaning chemicals, and **pesticides**. A pesticide is a product used on crops to kill insects and other pests.

Green schools also make sustainable food choices, such as having a school garden, and reducing solid wastes. Solid wastes include food and paper. Green schools also work to educate students, parents, and the community about conserving resources and protecting the environment.

Volunteering in the community to clean-up your school grounds and parks helps the environment. **Name another way you can help your community's environment**

Being Green at School

MAIN IDEA You can protect the environment while at school.

You have already read about actions you can take to protect the environment while you are at home. Are you wondering what you can do to help protect the environment at school? The tips listed below may give you some ideas.

- **Don't be a litterbug.** Throw trash into trashcans and recycling bins. Pick up litter when you see it. Organize litter clean-up events and talk to school officials about having enough trash cans available around school grounds.
- **Conserve energy.** Walk or ride your bicycle to school, take the bus, or carpool. While at school, use natural light when possible. Turn off lights in unoccupied areas. Turn of computers and other electronic equipment when they are not in use.
- **Conserve water.** Turn off the faucet while you are soaping your hands. Do not let a drinking fountain run unless you are drinking from it. Report leaking or dripping faucets or toilets that run continuously to your teacher or another school staff member.
- **Reduce solid waste.** Reuse paper by writing on both sides. Make double-sided copies or printouts. When paper can no longer be used, recycle waste paper and other items. Pack a lunch from home in reusable containers and carry it in a reusable lunch bag. This helps cut down on disposable packaging.

GREEN SCHOOLS PROVIDE A HEALTHY ENVIRONMENT FOR STUDENTS.

Some schools provide bins for collecting different kinds of recyclable materials. **Name the ways that your school encourages recycling.**

Reading Check

Describe Identify two actions from the list on this page that you already take to help protect the environment while you are at school?

Helping Your School Go Green

You can do many things to conserve resources and help protect the environment. Getting your whole school involved can really have an impact. The following list contains steps you and your school can take to help conserve resources and become greener.

- **Reduce use and exposure to hazardous chemicals.** Ask your school principal to consider switching to environmentally friendly cleaning products. Follow all rules when working with chemicals in the science laboratory. Set up a program to properly dispose of hazardous wastes including batteries, fluorescent light bulbs, and electronic waste.

- **Start a recycling and/or composting program.** Set up re-use stations for paper, file folders, envelopes, and other paper products. Ask school officials to place recycling containers in every classroom, as well as in the kitchen and cafeteria. Try to recycle paper, plastic bottles, glass bottles, and aluminum cans. Build a compost pile so that leaves, grass clippings, and other yard wastes, as well as food wastes from the kitchen, can be properly composted. At the end of the year, request that any leftover edible food from the kitchen, such as canned goods or other dry goods, be donated to a local food bank.

- **Start a school garden.** Grow vegetables and fruit for your school's cafeteria on school grounds. This cuts down on the use of fossil fuels because the food does not have to be transported. Consider using the compost from your compost pile as fertilizer for your garden.
- **Make the cafeteria green.** Replace paper or plastic disposable dishes and utensils with dishes and utensils made from material that can be washed and reused all year. Set up a compost collection bin for leftover food wastes that are appropriate for composting. Have the kitchen incorporate food that you grow in the school garden into lunches
- **Make the bathrooms green.** Ask your school officials to switch to foam soap dispensers and air dryers to replace paper towels. Check into having your school purchase toilet paper made from recycled paper.
- **Promote environmental education.** Form a school club that discusses conservation and environmental issues and projects. Create posters, pamphlets, web sites or campaigns that explain how environmentally friendly actions can be practiced at school.

> **YOU CAN DO MANY THINGS TO CONSERVE ENERGY AND HELP PROTECT THE ENVIRONMENT.**

Lesson 3 Review

What I Learned

1. DEFINE Define *green school*.

2. DESCRIBE What actions can you take to protect the environment at school?

3. IDENTIFY List three actions you can take to help your school be greener?

Thinking Critically

4. APPLY Briefly describe how each of the following strategies to help your school become more green helps protect the environment: walking to school, eliminating the use of toxic cleaning materials, replacing disposable plastic dishes and utensils with ones that can be reused, recycling paper and plastic products, planting a school garden.

5. EVALUATE Why is it important for schools to become actively involved in protecting the environment?

Applying Health Skills

6. COMMUNICATION SKILLS Choose one action your school could take to become greener. Write a presentation that could be given at a school board meeting to advocate for this change. Include facts and statistics to support your request.

Glossary/Glosario

English

Español

Abdominal thrust Quick inward and upward pull into the diaphragm to force an obstruction out of the airway.

presión abdominal Presión rápida, hacia adentro y arriba sobre el diafragma, para desalojar un objeto que bloquea la vía respiratoria de una persona.

Abstinence (AB stuh nuhns) The conscious, active choice not to participate in high-risk behaviors.

abstinencia Opción activa y conciente de no participar en comportamientos de alto riesgo.

Abuse (uh BYOOS) The physical, emotional, or mental mistreatment of another person.

abuso Maltrato físico, emocional o mental de otra persona.

Accident Any event that was not intended to happen.

accidente Suceso que ocurre de manera no intencional.

Accident chain A series of events that include a situation, an unsafe habit, and an unsafe action.

cadena del accidente Serie de sucesos que incluye una situación, un hábito peligroso y un acto peligroso.

Accidental injuries Injuries resulting from an accident.

lesiones accidentales Lesiones que resultan de un accidente.

Accountability A willingness to answer for your actions and decisions.

responsabilidad Voluntad de responder de tus acciones y decisiones.

Acid rain Rain that is more acidic than normal rain.

lluvia ácida Lluvia que es más ácida de lo normal.

Acne (AK nee) Skin condition caused by active oil glands that clog hair follicles.

acné Condición de la piel causada por glándulas de aceite activas que obstruyen folículos de cabello.

Acquaintance Someone you see occasionally or know casually.

conocido Alguien a quien ves ocasionalmente o conoces casualmente.

Active listening Hearing, thinking about, and responding to another person's message.

escucha activa Oír el mensaje de otra persona, pensar en el mensaje y responder.

Adapt Adjust to new situations.

adaptarse Acostumbrarse a nuevas situaciones.

Addictive Capable of causing a user to develop intense cravings.

adictivo Capaz de ocasionar que el consumidor desarrolle una necesidad repentina intensa.

Adolescence (a duhl EH suhns) The stage of life between childhood and adulthood, usually beginning somewhere between the ages of 11 and 15.

adolescencia Periodo de vida entre la niñez y la adultez que empieza generalmente entre los 11 y los 15 años.

Adrenaline (uh DRE nuhl in) A hormone that increases the level of sugar in the blood, giving your body extra energy.

adrenalina Hormona que aumenta el nivel de azúcar en la sangre, y por lo tanto, proporciona energía adicional al cuerpo.

Advertisement Messages designed to influence people to buy a product or service.

anuncio publicitario Mensaje diseñado para influenciar a los consumidores para que compren un producto o servicio.

Advocacy Taking action in support of a cause.

promoción de hábitos saludables Actuar en apoyo de una causa.

Aerobic (ah ROH bik) exercise Rhythmic, moderate-to vigorous activity that uses large amounts of oxygen and works the heart and lungs.

ejercicio aeróbico Actividad rítmica, moderada o fuerte que usa grandes cantidades de oxígeno y trabaja el corazón y los pulmones.

English

Affection Feelings of love for another person.

Affordable Care Act (ACA) A law that was enacted in 2010 to ensure that all U.S. citizens have access to affordable health care.

Aftershocks Smaller earthquakes as the earth readjusts after the main earthquake.

Age of consent A law that defines the age at which a person is considered mature enough to become sexually active.

Aggressive Overly forceful, pushy, hostile, or otherwise attacking in approach.

AIDS (acquired immunodeficiency syndrome) A deadly disease that interferes with the body's natural ability to fight infection.

Air Quality Index (AQI) A measure of ozone, sulfur dioxide, carbon monoxide, and fine particles close to the ground.

Alcohol (AL kuh hawl) A drug created by a chemical reaction in some foods, especially fruits and grains.

Alcohol abuse Using alcohol in ways that are unhealthy, illegal, or both.

Alcohol poisoning A dangerous condition that results when a person drinks excessive amounts of alcohol over a short time period.

Alcoholism A disease in which a person has a physical and psychological need for alcohol.

Allergens (AL er juhnz) Substances that cause allergic responses.

Allergy Extreme sensitivity to a substance.

Allied health professionals Medical professionals who perform duties which would otherwise have to be performed by doctors or nurses.

Alveoli (al VEE oh lye) The tiny air sacs in the lungs.

Amino acids Small units that make up a protein.

Amnesia Partial or total loss of memory.

Amphetamines Strong stimulant drugs that speed up the nervous system.

Español

afecto Sentimiento de amor hacia otra persona.

Ley de Asistencia Asequible La Ley de Cuidado de Salud Asequible (ACA) es una ley que se promulgó en 2010 para garantizar que todos los ciudadanos de los EE. UU. Tengan acceso a atención médica asequible.

réplicas Temblores mas pequeños que ocurren mientras la tierra se reajusta después de un terremoto principal.

edad de consentimiento Una ley que define la edad a la que una persona se considera suficientemente madura para volverse sexualmente activa.

agresivo(a) Excesivamente forzoso, hostil o de otra manera, que ataca durante el acercamiento.

SIDA (síndrome de inmunodeficiencia adquirida) Enfermedad mortal que interfiere con la habilidad natural del cuerpo de combatir infecciones.

Índice de calidad del aire (ICA) Medida de ozono, dióxido de sulfuro, monoxido de carbono y finas partículas que hay cerca de la tierra.

alcohol Droga producida por una reacción química en algunos alimentos, especialmente frutas y granos.

abuso del alcohol Uso de alcohol en formas que son no saludables, ilegales, o ambas.

intoxicación por alcohol Condición peligrosa que ocurre cuando una persona consume cantidades de alcohol excesivas en un corto periodo de tiempo.

alcoholismo Enfermedad que se caracteriza por la necesidad física y psicológica de consumir alcohol.

alérgenos Sustancias que causan reacciones alérgicas.

alergia Sensibilidad extrema a una sustancia.

profesionales médicos aliados Profesionales médicos que desarrollan deberes que de otra manera serian desarrollados por doctores o enfermeras.

alvéolos Pequeños sacos de aire en los pulmones.

aminoácidos Unidades pequeñas que forman las proteínas.

amnesia Pérdida total o parcial de la memoria.

anfetaminas Drogas fuertemente estimulante que acelera el sistema nervioso.

Glossary/Glosario

English

Anabolic steroids (a nuh BAH lik STAIR oydz) Substances that cause muscle tissue to develop at an abnormally high rate.

Anaerobic exercise Intense physical activity that builds muscle but does not use large amounts of oxygen.

Angioplasty A surgical procedure in which an instrument with a tiny balloon, drill bit, or laser attached is inserted into a blocked artery to clear a blockage.

Anorexia nervosa (a nuh REK see ah ner VOH sah) An eating disorder in which a person strongly fears gaining weight and starves herself or himself.

Antibiotics (an ti by AH tiks) Medicines that reduce or kill harmful bacteria in the body.

Antibodies Proteins that attach to antigens, keeping them from harming the body.

Antigens (AN ti genz) Substances that send the immune system into action.

Antihistamines Medicines that reduce the production of histamines.

Anxiety A state of uneasiness, usually associated with a future uncertainty.

Anxiety disorder Extreme fears of real or imaginary situations that get in the way of normal activities.

Appetite The psychological desire for food.

Appropriate weight The weight that is best for your body.

Arrhythmia Irregular heartbeat.

Arteries Blood vessels that carry blood away from the heart to various parts of the body.

Arteriosclerosis (ar TIR ee oh skluh ROH sis) A group of disorders that cause a thickening and hardening of the arteries.

Arthritis (ar THRY tus) A disease of the joints marked by painful swelling and stiffness.

Assault An attack on another person in order to hurt him or her.

Assertive response A response that declares your position strongly and confidently.

Assertive Willing to stand up for yourself in a firm but positive way.

Español

esteroides anabólicos Sustancias que causan que los tejidos musculares se desarrollen rápida y anormalmente.

ejercicio anaeróbico Actividad física intensa que desarrolla músculos pero no usa mucho oxígeno.

angioplastia Proceso quirúrgico en el cual un instrumento con un globo pequeño, un pedacito de taladro, o láser es insertado en una arteria bloqueada para desbloquearla.

anorexia nerviosa Trastorno alimenticio en el cual una persona teme mucho subir de peso y se mata de hambre.

antibióticos Medicinas que disminuyen o matan bacterias dañinas en el cuerpo.

anticuerpos Proteínas que se pegan a los antígenos, impidiendo que éstos le hagan daño al cuerpo.

antígenos Sustancias que provocan el funcionamiento del sistema inmunológico.

antihistamínicos Medicinas que reducen la producción de histaminas.

ansiedad Estado de intranquilidad, usualmente asociado con una incertidumbre futura.

trastorno de ansiedad Temores extremos a situaciones reales o imaginarias que se interponen en el desarrollo de actividades normales.

apetito Deseo psicológico de alimentarse.

peso apropriado Peso más adecuado para tu cuerpo.

arritmia Latido del corazón irregular.

arterias Vasos sanguíneos que transportan sangre desde al corazón hacia otras partes del cuerpo.

arteriosclerosis Conjunto de trastornos que provoca el engrosamiento y endurecimiento de las arterias.

artritis Enfermedad de las articulaciones caracterizada por inflamación dolorosa y anquilosamiento.

asalto Ataque hacia otra persona con la intención de herirla.

respuesta asertiva Reacción que establece tu posición con fuerza y confianza.

asertivo(a) Dispuesto a defenderse de manera resuelta y positiva.

English

Asthma Chronic inflammatory disorder of the airways that causes air passages to become narrow or blocked, making breathing difficult.

Astigmatism (uh STIG muh tiz uhm) An eye condition in which images appear wavy or blurry.

Atherosclerosis (a thuh roh skluh ROH sis) A condition of arteriosclerosis that occurs when fatty substances build up on the inner lining of the arteries.

Attitude (AT ih tood) A personal feeling or belief.

Autonomic system Part of the nervous system that deals with actions you do not usually control.

B

Bacteria (bak TIR ee uh) Simple one-celled organisms.

Bacterial vaginosis A condition that happens when there is too much of a certain bacteria in the vagina, changing the normal balance of bacteria.

Balance The feeling of stability and control.

Battery The beating, hitting, or kicking of another person.

Behavior The way you act in the many different situations and events in your life.

Benign (bi NYN) Not cancerous.

Binge drinking Having several drinks in a short period of time.

Binge eating An eating disorder in which a person repeatedly eats too much food at one time.

Biodegradable (by oh di GRAY duh buhl) Easily broken down in the environment.

Biological age Age determined by how well various body parts are working.

Biopsy The removal of a sample of tissue from a person for examination.

Birth defects An abnormality present at birth that causes physical or mental disability or death.

Blizzard A very heavy snowstorm with winds up to 45 miles per hour.

Blood alcohol concentration (BAC) The amount of alcohol in the blood.

Español

asma Trastorno crónico inflamatorio que causa que los pasajes de aire se hagan más pequeños o que se bloqueen, haciendo que la respiración se dificulte.

astigmatismo Afección del ojo que causa que las imágenes se vean distorsionadas y los objetos aparezcan ondulados o borrosos.

aterosclerosis Condición de arteriosclerosis que ocurre cuando sustancias grasosas se forman en el interior de las arterias.

actitud Sentimiento o creencia.

sistema autónomo Sistema que envuelve las acciones que no controlas.

bacterias Organismos simples de una sola célula.

vaginosis bacteriana Una condición que ocurre cuando hay demasiada cantidad de ciertas bacterias en la vagina, cambiando el equilibrio normal de las bacterias.

equilibrio Sentimiento de estabilidad y control.

agresión Dar palizas, golpear o dar puntapiés a otra persona.

comportamiento Forma en la cual actúas en diferentes situaciones y eventos en tu vida.

benigno No canceroso.

consumo excesivo de alcohol Ingerir varias bebidas en un periodo de tiempo corto.

alimentación compulsiva Trastorno en la alimentación por el cual una persona repetidamente come grandes cantidades de alimentos de una vez.

biodegradable Que se descompone fácilmente en el medio ambiente.

edad biológica Medida de la edad, determinada según el funcionamiento de varias partes del cuerpo.

biopsia Quitar una muestra de tejido de una persona para examinarlo.

defecto de nacimiento Anormalidad presente al momento del nacimiento que causa incapacidad física o mental o la muerte.

ventisca Tormenta de nieve fuerte, con vientos que llegan a 45 millas por hora.

concentración de alcohol en la sangre Cantidad de alcohol en la sangre.

Glossary/Glosario

English

Body composition The proportions of fat, bone, muscle, and fluid that make up body weight.

Body image The way you see your body.

Body language Postures, gestures, and facial expressions.

Body Mass Index (BMI) A method for assessing your body size by taking your height and weight into account.

Body odor The smell from the body.

Body system A group of organs that work together to carry out related tasks.

Brain The command center, or coordinator, of the nervous system.

Bronchi (BRAHNG ky) Two passageways that branch from the trachea, one to each lung.

Bronchodilator (brahng koh DY lay tur) A medication that relaxes the muscles around the air passages.

Bulimia (boo LEE mee ah) nervosa An eating disorder in which a person repeatedly eats large amounts of food and then purges.

Bullying A type of violence in which one person uses threats, taunts, or violence to intimidate another again and again.

Bullying behavior Actions or words that are designed to hurt another person.

C

Calorie (KA luh ree) A unit of heat that measures the energy available in foods.

Cancer A disease that occurs when abnormal cells multiply out of control.

Capillaries Tiny blood vessels that carry blood to and from almost all body cells and connect arteries and veins.

Carbohydrates The starches and sugars found in foods.

Carbon monoxide (KAR buhn muh NAHK syd) A colorless, odorless, poisonous gas produced when tobacco burns.

Carcinogen (kar SIN un juhn) A substance that can cause cancer.

Español

composición corporal Proporción de grasa, hueso, músculo y líquidos que componen el peso del cuerpo.

imagen corporal Forma en la que ves tu cuerpo.

lenguaje corporal Posturas, gestos, y expresiones faciales.

Índice de masa corporal (IMC) Método que evalúa el tamaño indicado de tu cuerpo utilizando tu peso y tu estatura.

olor corporal El olor del cuerpo.

sistema del cuerpo Grupo de órganos que trabajan juntos para ejecutar funciones relacionadas.

cerebro Centro de mando, o el coordinador, del Sistema nervioso.

bronquios Dos pasajes que se ramifican desde la tráquea hacia los dos pulmones.

broncodilatador Medicina que relaja los músculos alrededor de los bronquios.

bulimia nerviosa Trastorno en la alimentación por el cual una persona come grandes cantidades y después se induce el vómito.

acoso escolar Tipo de violencia en la cual una persona usa amenazas, burlas, o actos violentos para intimidar a otra persona una y otra vez.

comportamiento intimidatorio Cualquier comportamiento que se dirige a otra persona que use los comentarios hirientes, amenazas o violencia.

caloría Unidad de calor que mide la energía que contienen los alimentos.

cáncer Enfermedad causada por células anormales cuyo crecimiento está fuera de control.

vasos capilares Péqueños vasos sanguíneos que transportan sangre desde y hacia casi todas las células del cuerpo y conectan arterias y venas.

carbohidratos Almidones y azúcares que proporcionan energía.

monóxido de carbono Gas incoloro, inodoro y tóxico que produce el tabaco al quemarse.

carcinógeno Sustancia en el medio ambiente que produce cáncer.

English

Cardiac muscle Muscle found in the walls of your heart.

Cardiopulmonary resuscitation (CPR) A first-aid procedure to restore breathing and circulation.

Cardiovascular disease A disease of the heart and blood vessels.

Cardiovascular (KAR dee oh VAS kyoo ler) system Organs and tissues that transport essential materials to body cells and remove their waste products.

Carrier A person who is infected with a virus and who can pass it on to others.

Cartilage (KAHR tuhl ij) A strong, flexible tissue that allows joints to move easily, cushions bones, and supports soft tissues.

Cataracts An eye condition in which the lens becomes cloudy as a person ages.

Cell The basic unit of life.

Cell respiration The process in which the body's cells are nourished and energized.

Centers for Disease Control and Prevention (CDC) An agency of the Federal government that protects the health, safety, and security of Americans.

Central nervous system (CNS) The brain and the spinal cord.

Cervix The entrance to the uterus.

Character The way a person thinks, feels, and acts.

Character trait Qualities that demonstrate how a person thinks, feels, and acts.

Chemotherapy The use of powerful medicines to destroy cancer cells.

Chest thrusts Quick presses into the middle of the breastbone to force an obstruction out of the airway.

Chlamydia (kluh MI dee uh) A bacterial STD that may affect the reproductive organs, urethra, and anus.

Cholesterol A waxy, fat-like substance that the body uses to build cells and make other substances.

Chromosomes Threadlike structures that carry genes.

Español

músculo cardiaco Músculo de las paredes del corazón.

resucitación cardiopulmonar Procedimiento de primeros auxilios para restaurar la respiración y la circulación de la sangre.

enfermedad cardiovascular Enfermedad del corazón y de los vasos sanguíneos.

sistema cardiovascular Órganos y tejidos que transportan materia esencial a las células del cuerpo y eliminan los desechos.

portador Persona que parece saludable pero esta infectada con el VIH y puede transmitirlo a otros.

cartílago Tejido fuerte y flexible que permite que las articulaciones se muevan fácilmente, amortigua huesos y sirve de soporte para tejidos suaves.

cataratas Condición del ojo en la cual la lente se pone borrosa a medida que la persona envejece.

célula Unidad basica de la vida.

respiración cellular Proceso en el cual las células del cuerpo se nutren y energizan.

Centros para el Control y la Prevención de Enfermedades (CDC) Una agencia del gobierno federal que protege la salud, la seguridad y la seguridad de los estadounidenses.

Sistema nervioso central Cerebro y médula espinal.

cuello uterino La entrada al útero.

carácter Manera en que piensas, sientes y actúas.

rasgo de carácter Cualidad que demuestra la manera en que una persona piensa, siente y actúa.

quimioterapia Uso de medicina poderosa para destruir células cancerosas.

compresiones torácicas Presión rápida en el centro del esternón para desalojar un objeto que bloquea la vía respiratoria de una persona.

clamidia Infección de transmisión sexual bacterial que puede afectar los órganos de reproducción, la uretra y el ano.

colesterol Sustancia grasosa y cerosa que el cuerpo utiliza para crear células y otras sustancias.

cromosomas Estructuras fibrosas que transportan genes.

Glossary/Glosario

English

Chronic diseases Diseases that are present either continuously or off and on over a long time.

Chronic obstructive pulmonary disease (COPD) The condition in which passages in the lungs become swollen and irritated, eventually losing their elasticity.

Chronological (krah nuh LAH ji kuhl) age Age measured in years.

Circulatory (SER kyuh luh tohr ee) system The group of organs and tissues that carry needed materials to cells and remove their waste products.

Cirrhosis (suh ROH suhs) The scarring and destruction of liver tissue.

Citizenship The way you conduct yourself as a member of a community

Climate change A rise in the Earth's temperatures.

Clinical social worker (CSW) A licensed, certified mental health professional with a master's degree in social work.

Clique A group of friends who hang out together and act in similar ways.

Club drugs Illegal drugs that are found mostly in nightclubs or at all-night dance parties called raves.

CNS depressants Substances that slow down the body's functions, including heart and breathing rates.

Cold turkey Stopping all use of tobacco products immediately.

Collaborate Work together.

Colon The large intestine.

Commitment A pledge or a promise.

Communicable (kuh MYOO nih kuh buhl) disease A disease that can be spread to a person from another person, an animal, or an object.

Communication The exchange of information through the use of words or actions.

Community service Volunteer programs whose goal is to improve the community and the life of its residents.

Español

enfermedades crónicas Enfermedades que están siempre presentes o reaparecen repetidamente durante un largo period de tiempo.

Enfermedad pulmonar obstructiva crónica (EPOC) La condición en la cual los conductos en los pulmones se hinchan e irritan, perdiendo finalmente su elasticidad.

edad cronológica Edad medida en años.

sistema circulatorio Grupo de órganos y tejidos que transportan materiales necesitados hacia células las y eliminam los desperdicios.

cirrosis Cicatrización y destrucción del tejido del hígado.

civismo Forma en que te comportas como miembro de una comunidad.

cambio climático Aumento en las temperaturas de la Tierra.

trabajador social clínico Profesional en salud mental licenciado y certificado en trabajo social.

camarilla Grupo de amigos que salen juntos y que se comportan de manera similar.

drogas de clubs Drogas ilegales que normalmente son utilizadas en discotecas y otras fiestas que duran toda la noche llamadas raves.

depresores del SNC Sustancias que hacen que las funciones corporales sean más lentas, incluyendo al corázon y la respiración.

parar en seco Acto de parar inmediatamente el uso de productos que contienen tabaco inmediatamente.

colaborar Trabajar juntos.

colon Intestino grueso.

compromiso Promesa o voto.

enfermedad contagiosa Enfermedad que se puede propagar de una persona a otra persona, un animal o un objeto.

comunicación Intercambio de información a través del uso de palabras y acciones.

servicio comunitario Programas voluntarios desarrollados con la meta de mejorar la comunidad y la vida de los residentes.

English

Comparison shopping Collecting information, comparing products, evaluating their benefits, and choosing products with the best value.

Compromise When both sides in a conflict agree to give up something to reach a solution that will satisfy everyone.

Concussion A jarring injury to the brain that can cause unconsciousness.

Conditioning Training to get into shape for physical activity or a sport.

Confidence Belief in your ability to do what you set out to do.

Conflict A disagreement between people with opposing viewpoints, interests, or needs.

Conflict resolution A life skill that involves solving a disagreement in a way that satisfies both sides.

Conflict-resolution skills The ability to end a disagreement or keep it from becoming a larger conflict.

Congenital disorders All disorders that are present when the baby is born.

Consequences The results of actions.

Conservation The saving of resources.

Constructive criticism Using a positive message to make a suggestion.

Consumer A person who buys products and services.

Consumer skills Techniques that enable you to make wise, informed purchases.

Contagious (kuhn TA juhs) Able to spread to others by direct or indirect contact

Contagious period Length of time that a particular disease can be spread from person to person.

Cool-down Gentle exercises that let the body adjust to ending a workout.

Cooperation Working together for the common good.

Coordination The smooth and effective working together of your muscles and bones.

Español

comparación de productos Recolectar información, comparer productos, evaluar sus beneficios, y escoger el producto que tiene mejor valor.

acuerdo Cuando los dos lados de un conflicto concuerdan con dejar algo de lado para alcanzar una solución que satisfaga a todos.

conmoción cerebral Una lesión discordante en el cerebro que puede causar la pérdida del conocimiento.

acondicionamiento Entrenamiento para ponerse en forma para alguna actividad física o deporte.

confianza Creer en tu habilidad de hacer lo que te propones a hacer.

conflicto Desacuerdo entre dos personas con puntos de vista, intereses o necesidades opuestas.

resolución de conflictos Habilidad que implica el hecho de resolver un desacuerdo satisfaciendo a los dos lados.

destrezas para resolver conflictos Habilidad de poder solucionar un desacuerdo o hacer que el desacuerdo no se convierta en algo más grande.

trastornos congénitos Todos los desórdenes que se presentan cuando el bebé nace.

consecuencias Resultados de los actos.

conservación Protección de los recursos naturales.

crítica constructiva Spanish Definition: Con un mensaje positive a hacer una sugerencia.

consumidor Persona que compra productos y servicios.

destrezas del consumidor Técnicas que te permiten hacer compras inteligentes e informadas.

contagiosa Capaz de propagarse a otros por contacto directo o indirecto.

periodo de contagio Periodo de tiempo en que se puede transmitir una enfermedad determinada de una persona a otra.

enfriamiento Ejercicios moderados que permiten que el cuerpo se ajuste al ir finalizando el plan de ejercicios.

cooperación Trabajar juntos por el bienestar común.

coordinación El funcionamiento conjunto de los músculos y los huesos de manera eficiente y sin complicaciones.

Glossary/Glosario

English

Coping strategies Ways of dealing with the sense of loss people feel when someone close to them dies.

Cornea A clear, protective structure of the eye that lets in light.

Coupons Digital codes or slips of paper that reduce the price of a product.

Crisis hot line A toll-free telephone service where abuse victims can get help and information.

Criteria Standard on which to base decisions.

Cross-training Switching between different forms of physical exercise.

Cultural background The beliefs, customs, and traditions of a specific group of people.

Culture The collected beliefs, customs, and behaviors of a group

Cumulative (KYOO myuh luh tiv) risk When one risk factor adds to another to increase danger.

Cuticle A fold of epidermis around the fingernails and toenails.

Cyberbullying The electronic posting of mean-spirited messages about a person often done anonymously.

Cycle of abuse Pattern of repeating abuse from one generation to the next.

Español

estrategias de superación Formas de tratar con el sentido de pérdida que las personas sienten cuando alguien cercano fallece.

córnea Una estructura protectora clara del ojo que deja pasar la luz.

cupones Códigos digitales o tiras de papel que reducen el precio de un product.

línea para casos de crisis Servicio telefónico sin pago en cual víctimas de abusos pueden recibir ayuda e información.

criterio Estándar sobre el cual basar las decisiones.

entrenamiento combinado Cambiar de un ejercicio físico a otro.

bagaje cultural Creencias, costumbres y tradiciones de un grupo específico de personas.

cultura Colección de creencias, costumbres y comportamientos de un grupo.

riesgo acumulativo Cuando un factor riesgoso se suma a otro e incrementa el peligro.

cutícula Doblez de epidermis alrededor de las uñas de los pies y las manos.

ciberacoso La publicación electronica de la media mensajes animados sobre una persona hace a menudo anónimamente.

ciclo de abuso Patrón de repetición del abuso de una generación a la siguiente.

D

Dandruff When too many dead skin cells flake off the outer layer of the scalp.

Date rape When one person in a dating relationship forces the other person to take part in sexual activity.

Dating violence When a person uses violence in a dating relationship to control his or her partner.

Deafness A condition in which someone has difficulty hearing sounds or has complete hearing loss.

Decibel The unit for measuring the loudness of sound.

Decision making The process of making a choice or solving a problem.

Decisions Choices.

caspa Cuando demasiadas células de piel muertas se descaman de la capa exterior del cuero cabelludo.

violación de pareja Una persona en una relación de citas obliga a la otra persona a participar en la actividad sexual.

relacion violenta Cuando una persona usa violencia en una relación amorosa para poder controlar a su pareja.

sordera Condición en la cual una persona tiene dificultad para escuchar sonidos o ha perdido completamente la capacidad de escuchar.

decibel Unidad que se usa para medir el volumen del sonido.

tomar decisiones Proceso de hacer una selección o de resolver un problema.

decisiónes Opciones que eliges.

English

Defensive driving Watching out for other people on the road and anticipating unsafe acts.

Degenerative diseases Diseases that cause further breakdown in body cells, tissues, and organs as they progress.

Dehydration The excessive loss of water from the body.

Depressant (di PRE suhnt) A drug that slows down the body's functions and reactions, including heart and breathing rates.

Depression An emotional problem marked by long periods of hopelessness and despair.

Dermatologist (DER muh TAHL uh jist) A physician who treats skin disorders.

Dermis (DER mis) The skin's inner layer.

Detoxification (dee tahk si fi KAY shuhn) The physical process of freeing the body of an addictive substance.

Developmental tasks Events that need to happen in order for you to continue growing toward becoming a healthy, mature adult.

Diabetes (dy uh BEE teez) A disease that prevents the body from converting food into energy.

Diaphragm (DY uh fram) A large, dome-shaped muscle below the lungs that expands and compresses the lungs, enabling breathing.

Digestion (di JES chuhn) The process by which the body breaks down food into smaller pieces that can be absorbed by the blood and sent to each cell in your body.

Digestive (dy JES tiv) system The group of organs that work together to break down foods into substances that your cells can use.

Disease (dih ZEEZ) Any condition that interferes with the proper functioning of the body or mind.

Dislocation A major injury that happens when a bone is forced from its normal position within a joint.

Disorder A disturbance in the normal function of a part of the body.

Distress Negative stress.

Domestic violence Physical abuse that occurs within a family.

Español

conducción defensiva Estar atento a las otras personas en la carretera y anticipar acciones peligrosas.

enfermedades degenerativas Enfermedades que causan la destrucción progresiva de las células, tejidos y órganos del cuerpo a medida que avanzan.

deshidratación Pérdida excesiva de agua del cuerpo.

depresor Droga que disminuye las funciones y reacciones del cuerpo, incluso el ritmo cardiaco y la respiración.

depresión Un problema emocional marcado por largos periodos de desesperación.

dermatólogo Médico que trata trastornos de la piel.

dermis Capa interior de la piel.

desintoxicación Proceso físico de liberar al cuerpo de una sustancia adictiva.

tareas de desarrollo Sucesos que deben ocurrir para que continúes desarrollándote hasta llegar a convertirte en un adulto saludable y maduro.

diabetes Enfermedad que impide que el cuerpo convierta los alimentos en energía.

diafragma Un músculo grande en forma de cúpula debajo de los pulmones que expande y comprime los pulmones, posibilitando la respiración.

digestión Proceso por el cual el cuerpo desintegra las comidas en pedazos mãs pequeños que pueden ser absorbidos por la sangre y enviados a cada célula en el cuerpo.

sistema digestivo Conjunto de órganos que trabajan juntos para descomponer los alimentos en sustancias que tus células puedan usar.

enfermedad Toda afección que interfiere con el buen funcionamiento del cuerpo o de la mente.

dislocación Daño mayor que ocurre cuando un hueso es forzadofuera de lugar en la articulación.

trastorno Desequilibrio en el funcionamiento normal de una parte del cuerpo.

aflicción Estrés negative.

violencia doméstica Abuso físico que ocurre dentro deuna familia.

Glossary/Glosario

English

Driving while intoxicated (DWI) A person driving a car with a blood alcohol concentration (BAQ) of 0.08% or greater.

Drug A substance other than food that changes the structure or function of the body or mind.

Drug abuse Intentionally using drugs in a way that is unhealthful or illegal.

Drug free A characteristic of a person not taking illegal drugs or of a place where no illegal drugs are used.

Drug misuse Taking or using medicine in a way that is not intended.

Drug possession A person has or keeps illegal drugs.

Drug rehabilitation A process where the person relearns how to live without the abused drug.

Drug trafficking Buying or selling of drugs.

Drug-free zone A 1,000-yard distance around a school where anyone caught with drugs can be arrested.

E

Earthquake A shifting of the earth's plates resulting in a shaking of the earth's surface.

Eating disorder Extreme eating behavior that can lead to serious illness or even death.

Egg cell A reproductive cell from the female that joins with a sperm cell to make a new life.

Electrical overload A dangerous situation in which too much electric current flows along a single circuit.

Embryo The developing organism from two weeks until the end of the eighth week of development.

Emotional needs Needs that affect a person's feelings and sense of well-being.

Emotions Feelings such as love, joy, or fear.

Empathy Identifying with and sharing another person's feelings.

Emphysema (em fuh SEE muh) A disease that results in the destruction of the alveoli in the lungs.

Español

conducir en estado de embriaguez Persona que maneja un auto con una concentración de alcohol en la sangre de 008% o mayor.

droga Sustancia no alimenticia que causa cambios en la estructura o el funcionamiento del cuerpo o la mente.

abuso de drogas Uso de drogas intencionalmente en una forma no saludable o ilegal.

libros de drogas Característícas de una persona que no consume drogas ilícitas, o establecimiento donde no se consumen.

uso indebido de drogas Tomar o usar medicina de una forma que no es la indicada.

posesión de drogas Cuando una persona tiene drogas ilegales.

rehabilitacion de drogas Proceso por el cual una persona vuelve a aprender como vivir sin el abuso de una droga.

tráfico de drogas La compra o venta de drogas.

zona libre de drogas Distancia de 1,000 yardas alrededor de una escuela en la cual cualquier persona que posee drogas puede ser arrestada.

terremoto Cambio de las placas terrestres que hace que la superficie terrestre tiemble.

trastorno alimenticio Conducta de alimentación extrema que puede causar enfermedades graves o la muerte.

óvulo Célula reproductora feminina que se une con el espermatooide para crear una nueva vida.

sobrecarga eléctrica Situación peligrosa en que demasiada corriente eléctrica fluye a través de un solo circuito.

embrión Organismo en desarrollo desde las dos semanas hasta las ocho semanas de desarrollo.

necesidades emocionales Necesidades que afectan los sentimientos y el bienestar de una persona.

emociones Sentimientos como el amor, la alegría o el miedo.

empatía Identificar y compartir sentimientos de otra persona.

enfisema Enfermedad que resulta de la destrucción de los alvéolos en los pulmones.

English

Empty-calorie foods Foods that offer few, if any, nutrients but do supply calories.

Enablers Persons who create an atmosphere in which the alcoholic can comfortably continue his or her unacceptable behavior.

Endocrine (EN duh krin) system The system of glands throughout the body that regulate body functions.

Endorsement A statement of approval.

Endurance (en DUR uhnce) The ability to perform difficult physical activity without getting overly tired.

Energy equation The balance between the calories you take in from food and the calories your body uses through activity.

Environment (en VY ruhn muhnt) All the living and nonliving things around you.

Environmental Protection Agency (EPA) An agency of the U.S. government that is dedicated to protecting the environment.

Enzymes Proteins that affect the many body processes.

Epidermis The outermost layer of skin.

Epiglottis A flap of tissues in the back of your mouth.

Escalate To become more serious.

Ethical Choosing to take the right action.

Euphoria A feeling of well-being or elation.

Excretion The process the body uses to get rid of waste.

Excretory (EKS kru to ree) system The group of organs that work together to remove wastes.

Exercise Planned physical activity done regularly to build or maintain one's fitness.

Extended family A person's immediate family plus other relatives such as grandparents, aunts, uncles, and cousins.

F

Family The basic unit of society and includes two or more people joined by blood, marriage, adoption, or a desire to support each other.

Español

comidas sin valor calórico Comidas que ofrecen pocos nutrientes pero sí contienen calorías.

habilitadores Personas que crean una atmósfera en la cual el alcohólico puede continuar su comportamiento inaceptable de una forma cómoda.

sistema endocrino Sistema de glándulas a través del cuerpo que regulan las funciones corporales.

aprobación Declaración de aceptación.

resistencia Habilidad de desempeñar actividades físicas sin cansarse demasiado.

ecuación de energía El equilibrio entre las calorías que ingieres de los alimentos y las calorías que usa tu cuerpo a través de la actividad.

medioambiente Todas las cosas vivas y no vivas que te rodean.

Agencia de Protección Ambiental (EPA) Agencia del gobierno de Estados Unidos a cargo de la protección del medio ambiente.

enzimas Proteínas que afectan varios processos corporals.

epidermis Capa más externa de la piel.

epiglotis Pedazo de tejido en forma de solapa que se encuentra en la parte trasera de la boca que no dejar que la comida pase a la tráquea.

intensificar Llegar a ser más grave.

ético Escoger la acción correcta.

euforia Sentimiento de bienestar o regocijo.

excreción Proceso mediante el cual el cuerpo se deshace de los desechos.

sistema excretor El grupo de órganos que trabaja en conjunto para eliminar desperdicios del cuerpo.

ejercicio Actividad física planeada realizada regularmente para crear o mantener un buen estado físico.

familia extensa Familia nuclear y otros parientes tales como abuelos, tías, tíos y primos.

familia Unidad básica de la sociedad que incluye dos o mas personas unidas por sangre, matrimonio, adopción o el deseo de ser soporte el uno del otro.

Glossary/Glosario

English

Family therapy Counseling that seeks to improve troubled family relationships.

Famine A widespread shortage of food.

Farsightedness A condition in which faraway objects appear clear while near objects look blurry.

Fatigue Tiredness.

Fats Nutrients that promote normal growth, give you energy, and keep your skin healthy.

Fatty liver A condition in which fats build up in the liver and cannot be broken down.

Fertilization The joining of a male sperm cell and a female egg cell to form a fertilized egg.

Fetal (FEE tuhl) alcohol syndrome (FAS) A group of alcohol-related birth defects that include both physical and mental problems.

Fetus The developing organism from the end of the eighth week until birth.

Fiber A complex carbohydrate that the body cannot break down or use for energy.

Fight-or-flight response The body's way of responding to threats.

Fire extinguisher A device that sprays chemicals that put out fires.

First aid The immediate care given to someone who becomes injured or ill until regular medical care can be provided.

FITT principle A method for safely increasing aspects of your workout without injuring yourself.

First-degree burn A burn in which only the outer layer of skin has burned and turned red.

Fitness Being able to handle physical work and play each day without getting overly tired.

Flammable Substances that catch fire easily.

Flexibility The ability to move joints fully and easily through a full range of motion.

Fluoride A chemical that helps prevent tooth decay.

Español

terapia familiar Asesoramiento cuyo propósito es mejorar relaciones problemáticas entre familiares.

hambruna Falta general de alimentos.

hipermetropía Capacidad de ver claramente los objetos a la distancia, mientras los objetos cercanos se ven borrosos.

fatiga Cansancio.

grasas Nutrientes que promueven crecimiento normal, dan energía, y mantienen la piel saludable.

hígado graso Condición en la cual la grasa se forma en el hígado y no puede ser deshecha.

fecundación Unión de un espermatozoide masculino con una óvulo femenino para forma un óvulo fertilizado.

síndrome de alcoholismo fetal Conjunto de defectos de nacimiento causados por el alcohol que incluyen problemas físicos y mentales.

feto Organismo en desarrollo desde el final de la octava semana hasta el nacimiento.

fibra Carbohidratos complejos que el cuerpo no puede deshacer para usarlos como energía.

respuesta de lucha o huida Forma del cuerpo de responder a las amenazas.

extintor de fuego Aparato que rocia productos químicos que apagan fuegos.

primeros auxilios Cuidado inmediato que se da a una persona herida o enferma hasta que sea posible proporcionarle ayuda médica normal.

principio FITT Método mediante el cual es posible incrementar con seguridad aspectos del ejercicio sin hacerte daño a ti mismo.

quemadura de primer grado Quemadura en que sólo la capa exterior de la piel se quema y se enrojece.

buena condición física Poder resistir trabajo físico y jugar cada día sin cansarse demasiado.

inflamable Sustancias que se encienden fácilmente.

flexibilidad La habilidad de poder mover articulaciones completa y fácilmente a través de una extensión completa de movimiento.

fluoruro Un producto químico que ayuda a prevenir caries en los dientes.

English

Food and Drug Administration (FDA) A department of the Federal government that regulates foods, drugs, and other substances.

Foodborne illness A sickness that results from eating food that is not safe to eat.

Fossil (FAH suhl) fuels The oil, coal, and natural gas that are used to provide energy.

Fracture A break in a bone.

Fraud A calculated effort to trick or fool others.

Frequency The number of days you work out each week.

Friendship A relationship with someone you know, trust, and regard with affection.

Frostbite Freezing of the skin.

Fungi (FUHN jy) Organisms that are more complex than bacteria, but cannot make their own food.

G

Gallbladder A small, saclike organ that stores bile.

Gang A group of young people that comes together to take part in illegal activities.

Gender discrimination The singling out or excluding a person based on gender.

Generic products Products sold in plain packages at lower prices than brand name products.

Genes The basic units of heredity.

Genetic disorders Disorders that are caused partially or completely by a defect in genes.

Genital herpes (HER peez) A viral STD that produces painful blisters on the genital area.

Genital warts Growths or bumps in the genital area caused by certain types of the human papillomavirus (HPV).

Gingivitis (jin juh VY tis) A common disorder in which the gums are red and sore and bleed easily.

Gland A group of cells, or an organ, that secretes a chemical substance.

Español

Administración de Alimentos y Medicamentos (FDA) Un departamento del Gobierno Federal que regula alimentos, fármacos y otras sustancias.

intoxicación alimentaria Enfermedad que proviene de haber comido un alimento no sano.

combustible fósil Petróleo, carbón y gas natural que se usan para proporcionar energía.

fractura Rotura de un hueso.

fraude Esfuerzo calculado para engañar a otros.

frecuencia Número de horas en las que haces ejercicios cada semana.

amistad Relación con una persona que conoces, en la que confías, y aprecias con afecto.

congelación Congelamiento de la piel.

hongos Organismos que son más complejos que bacterias, pero no pueden producir su propio alemento.

vesícula biliar Pequeño órgano en forma de bolsa que almacena bilis.

pandilla Grupo de jóvenes que se juntan para participar en actividades ilegales.

discriminación sexual Distinguir o excluir a personas de acuerdo a su sexo.

productos genéricos Productos que se venden en envases comunes y a menor precio que los de marca.

genes Unidades básicas de la herencia.

trastorno genético Trastorno causado parcial o totalmente por defectos en los genes.

herpes genital Infección de transmisión sexual, causada por un virus, que produce ampollas dolorosas en el área genital.

verrugas genitales Erupciones o protuberancias en el área genital causadas por ciertos tipos del virus papiloma de los seres humanos.

gingivitis Trastorno común que se caracteriza por el enrojecimiento y dolor de las encías que sangran con facilidad.

glándula Grupo de células, o un órgano, que secreta una sustancia química.

Glossary/Glosario

English

Glaucoma An eye condition in which fluid pressure builds up inside the eye.

Goal Something you hope to accomplish.

Goal setting The process of working toward something you want to accomplish.

Gonorrhea (gah nuh REE uh) A bacterial STD that affects the mucous membranes of the body, particularly in the genital area.

Greenhouse effect The trapping of heat by carbon dioxide and other gases in the air.

Green School A school that is environmentally-friendly in several different ways.

Grief The sorrow caused by loss of a loved one.

Grief reaction The process of dealing with strong feelings following any loss.

Groundwater Water that collects under the earth's surface.

Guarantee A promise to refund you money if the product doesn't work as claimed.

Gynecologist A doctor who specializes in the female reproductive system.

Español

glaucoma Condición del ojo en la cual la presión de fluido crece dentro del ojo.

meta Algo que esperas lograr.

establecer metas Proceso de esforzarte para lograr algo que quieres.

gonorrea Infección de transmisión sexual causada por bacterias que afecta las membranas mucosas del cuerpo, en particular en el área genital.

efecto invernadero Retención del calor por la presencia de dióxido de carbono y otros gases en el aire.

escuela verde Una escuela que es respetuoso con el medio ambiente de varias maneras diferentes.

duelo Pesar provocado por la muerte de un ser querido.

reacción al duelo El proceso de tratar con algún sentimientos fuertes que se ocasionan por alguna perdida.

agua subterránea Agua acumulada debajo de la superficie de la tierra.

garantía legal Promesa de que en caso de que el producto no trabaje como dicho tu dinero será devuelto.

ginecólogo Médico que se especializa en el aparato reproductor femenino.

H

Habits Patterns of behavior that you follow almost without thinking.

Hair follicles Small sacs on the dermis from which hair grows.

Hallucinogens (huh LOO suhn uh jenz) Drugs that distort moods, thoughts, and senses.

Hangnail A split in the cuticle along the edge of a fingernail.

Harassment (huh RAS muhnt) Ongoing conduct that offends another person by criticizing his or her race, color, religion, physical disability, or gender.

Hate crime A crime committed against another person or group based on racial, religious, or sexual background.

Hazard Potential source of danger.

Hazardous wastes Human-made liquid, solid, sludge or radioactive wastes that may endanger human health or the environment.

hábito Patrónes de conducta que sigues casi sin pensarlo.

foliculos pilosos Saquito menbranoso en las dermis donde nace el pelo.

alucinógenos Droga que altera el estado de ánimo, los pensamientos y los sentidos.

padrastro Grieta en la cutícula al lado del borde de la uña.

hostigamiento Conducta frecuente que ofende a otra persona con críticas sobre su raza, color, religión, incapacidad física, o sexo.

crimen de odio Un delito cometido contra otra persona porque él o ella es miembro de un determinado grupo social, étnico o cultural.

peligro Fuente posible de peligro.

desperdicios peligrosos Líquidos, sólidos, sedimentos, o desperdicios radiactivos producidos por humanos que ponen en peligro la salud humana o el medio ambiente.

English

Health The combination of physical, mental/emotional, and social well-being.

Health care Any services provided to individuals or communities that promote, maintain, or restore health.

Health care system All the medical care available to a nation's people, the way they receive the care, and the way the care is paid for.

Health fraud The selling of products or services to prevent diseases or cure health problems which have not been scientifically proven safe or effective for such purposes.

Health insurance A plan in which a person pays a set fee to an insurance company in return for the company's agreement to pay some or all medical expenses when needed.

Health maintenance organization (HMO) A health insurance plan that contracts with selected physicians and specialists to provide medical services.

Health skills Skills that help you become and stay healthy.

Heart The muscle that acts as the pump for the circulatory system.

Heart and lung endurance A measure of how efficiently your heart and lungs work when you exercise and how quickly they return to normal when you stop.

Heart attack A serious condition that occurs when the blood supply to the heart slows or stops and the heart muscle is damaged.

Heat cramps Painful, involuntary muscle spasms that may occur during heavy exercise in hot weather.

Heat exhaustion An overheating of the body that can result from dehydration.

Heatstroke A serious form of heat illness in which the body's normal processes for dealing with heat close down.

Hepatitis (hep uh TY tis) A viral disease characterized by an inflammation of the liver and yellowing of the skin and the whites of the eyes.

Hepatitis B A disease caused by the hepatitis B virus that affects the liver.

Heredity (huh RED I tee) The passing of traits from parents to their biological children.

Español

salud Combinación de bienestar físico, mental/emocional y social.

cuidado medico Cualquier servicio proporcionado a individuos o comunidades que promueve, mantiene y les hace recobrar la salud.

sistema de salud Servicios médicos disponibles para a la gente de una nación y las formas en las cuales estos son pagados.

fraude médico Venta de productos o servicios para prevenir enfermedades o curar problemas desalud que no han sido aprobados científicamente o hechos efectivos para ese uso.

seguro médico Plan en el que una persona paga una cantidad fija a una compañía de seguros que acuerda cubrir parte o la totalidad de los gastos médicos.

Organización para el mantenimiento de la salud Plan de seguro de salud que contrata a ciertos médicos y especialistas para dar servicios médicos.

destrezas para conservar la salud Habilidades que ayudan a ser y mantenerte saludable.

corazón Músculo que funciona como una bomba para el aparato circulatorio.

resistencia del corazón y los pulmones Medida de qué tan eficientemente tu corazón y tus pulmones cuando haces ejercicios y qué tan rápido regresan a lo normal cuando paras.

ataque cardiaco Afección seria que se presenta cuando el flujo de sangre al corazón disminuye o cesa, dañando el músculo cardiaco.

calambres por calor Dolorosos espasmos involuntarios de los músculos que pueden ocurrir al realizar ejercicio vigoroso cuando hace mucho calor.

agotamiento por calor Recalentamiento del cuerpo que puede dar como resultado una deshidratación.

insolación Tipo de enfermedad debido al calor grave en que los procesos normales del cuerpo que controlan los efectos del calor dejan de funcionar.

hepatitis Enfermedad viral caracterizada por inflamación del hígado, el torno amarillo de la piel y la parte blanca del ojo.

hepatitis B Enfermedad causada por el virus de la hepatitis B que afecta el hígado.

herencia Transferencia de características de los padres biológicos a sus hijos.

Glossary/Glosario

English

Hernia An internal organ pushing against or through a surrounding cavity wall.

Histamines (HIS tuh meenz) The chemicals that the immune cells release to draw more blood and lymph to the area affected by the allergen.

HIV (human immunodeficiency virus) The virus that causes AIDS.

Hives Raised bumps on the skin that are very itchy.

Homicide A violent crime that results in the death of another person.

Hormones (HOR mohnz) Chemical substances produced in certain glands that help to regulate the way your body functions.

Hospice care Care provided to the terminally ill that focuses on comfort, not cure.

Host The organism that is either carrying the disease or is sick.

Human trafficking The recruitment, transportation, transfer, harboring, or receipt of persons by improper means for the purpose of using those people as forced labor for sexual exploitation.

Hunger The physical need for food.

Hurricane A strong windstorm with driving rain that forms over the sea.

Hygiene (HY jeen) Cleanliness.

Hypertension A condition in which the pressure of the blood on the walls of the blood vessels stays at a level that is higher than normal.

Hypothermia (hy poh THER mee uh) A sudden and dangerous drop in body temperature.

"I" message A statement that presents a situation from the speaker's personal viewpoint.

Illegal drugs Drugs that are made and used purely for their effects.

Immune (i MYOON) system A combination of body defenses made up of the cells tissues and organs that fight pathogens in the body.

Immunity The ability to resist the pathogens that cause a particular disease.

Español

hernia Órgano interno que está empujando contra o a través de una pared de cavidad.

histaminas Substancias químicas que las células inmunes suelta sueltan para atraer mas sangre y linfa hacia el area afectada por los alergenos.

VIH (virus de inmunodeficiencia humana) Virus que causa el SIDA.

urticaria Granos en la piel que pican mucho.

homicidio Crimen violento que resulta en la muerte de otra persona.

hormonas Sustancias químicas, producidas por ciertas glándulas que ayudan a regular las funciones del cuerpo.

atención paliativa Asistencia para personas con enfermedades incurables que apunta a brindar comodidad, no a la cura.

huésped El organismo que lleva la enfermedad o está enfermo.

trata de personas El reclutamiento, transporte, transferencia, alojamiento o recepción de personas por medios inapropiados con el propósito de utilizar a esas personas como fuerza laboral o para la explotación sexual.

hambre Necesidad física de alimentos.

huracán Tormenta de vientos y lluvia torrencial que se origina en alta mar.

higiene Limpieza.

hipertensión Afección en que la presión arterial de una persona se mantiene a niveles más altos de lo normal.

hipotermia Descenso rápido y peligroso de la temperatura del cuerpo.

I

mensaje en primera persona Declaración que presenta una situación desde el punto de vista personal del orador.

drogas ilegales Drogas que son hechas y usadas sólo por sus efectos.

sistema inmunitario Combinación de las defensas del cuerpo, compuesta de células, tejidos y órganos que combaten agentes patógenos.

inmunidad Habilidad del cuerpo de resistir los agentes patógenos que causan una enfermedad en particular.

English

Infancy The first year of life after birth.

Infection A condition that happens when pathogens enter the body, multiply, and cause harm.

Inflammation The body's response to injury or disease, resulting in a condition of swelling, pain, heat and redness.

Influenza A highly communicable viral disease characterized by fever, chills, fatigue, headache, muscle aches and respiratory symptoms.

Infomercial An ad that looks like a news story or television show.

Ingrown toenail A condition in which the nail pushes into the skin on the side of the toe.

Inhalants (in HAY luhntz) The vapors of chemicals that are sniffed or inhaled to get a "high."

Inhibition A conscious or unconscious restraint on his or her behaviors or actions.

Insulin (in suh lin) A protein made in the pancreas that regulates the level of glucose in the blood.

Integrity Being true to your ethical values.

Intensity How much energy you use when you work out.

Interpersonal communication Sharing thoughts and feelings with other people.

Intervention A gathering in which family and friends get the problem drinker to agree to seek help.

Intimidation Purposely frightening another person through threatening words, looks, or body language.

Intoxicated (in TAHK suh kay tuhd) Being drunk.

J

Joints The places where two or more bones meet.

K

Kidneys Organs that filter water and dissolved wastes from the blood and help maintain proper levels of water and salts in the body.

L

Labeling Name-calling.

Español

infancia Primer año de vida después del nacimiento.

infección Afección que se produce cuando agentes patógenos invaden el cuerpo, se multiplican y dañan las células.

inflamación Reacción del cuerpo a lesiones o enfermedades que resulta en hinchazón, dolor, calor y enrojecimiento.

influenza Enfermedad viral muy contagiosa que se caracteriza por fiebres, escalofríos, dolores de cabeza, dolor muscular, y síntomas respiratorios.

infomercial Anuncio de televisión largo cuyo propósito principal parece ser proveer información acerca de un producto en vez de vender el producto.

uña encarnada Afección en la cual la uña se introduce en la piel al lado del dedo del pie.

inhalantes Los vapores de substancias químicas que son olidos o inhalados para drogarse.

inhibición Reprimir comportamientos o acciones consiente o inconscientemente.

insulina Proteína hecha en el páncreas que regula el nivel de glucosa en la sangre.

integridad Ser fiel a tus valores éticos.

intensidad En cuanto al estado físico, la cantidad de energía que usas cuando haces ejercicio.

comunicación interpersonal Compartir pensamientos y sentimientos con otra persona.

intervención Reunión en la cual familia y amigos hacen que la persona con problemas alcohólicos busque ayuda.

intimidación Asustar a otra persona a propósito con palabras amenazantes, miradas o lenguaje corporal.

intoxicado(a) Estar borracho.

articulación Lugar en donde se unen dos o más huesos.

riñones Órganos que filtran el agua y los desechos disueltos de la sangre y contribuyen a mantener los niveles adecuados de agua y sales en el cuerpo.

etiquetas Dar nombre.

Glossary/Glosario

English

Landfill Huge specially designed pit where waste materials are dumped and buried.

Larynx (LA ringks) The upper part of the respiratory system, which contains the vocal cords.

Lifestyle activities Physical activities that are part of your day-to-day routine or recreation.

Lifestyle factors Behaviors and habits that help determine a person's level of health.

Ligament A type of connecting tissue that holds bones to other bones at the joint.

Limits Invisible boundaries that protect you.

Liver A digestive gland that secretes a substance called bile, which helps to digest fats.

Long-term goal A goal that you plan to reach over an extended period of time.

Loyal Faithful.

Lungs Two large organs that exchange oxygen and carbon dioxide.

Lymphatic system A secondary circulatory system that helps the body fight pathogens and maintains its fluid balance.

Lymphocytes (LIM fuh sytes) Special white blood cells in the lymphatic system.

Español

vertedero Pozo enorme con diseño específico donde se arrojan y se entierran desechos.

laringe Parte superior del aparato respiratorio que contiene las cuerdas vocales.

actividades de la vida diaria Actividades físicas que son parte de la rutina diaria o recreación.

factores de estilo de vida Conductas y hábitos que ayudan a determinar el nivel de salud de una persona.

ligamento Tipo de tejido conjuntivo que mantiene en su lugar los huesos en las articulaciones.

límites Barreras invisibles que te protegen.

hígado Glándula digestiva que secreta una sustancia llamada bilis, que ayuda a digerir las grasas.

meta a largo plazo Objetivo que planeas alcanzar en un largo periodo de tiempo.

leal Fiel.

pulmones Dos órganos grandes que intercambian oxígeno y monóxido de carbono.

sistema linfático Aparato circulatorio secundario que le ayuda al cuerpo a defenderse de agentes patógenos y a mantener el equilibrio de los líquidos.

linfocitos Glóbulos blancos especiales en la linfa.

Mainstream smoke The smoke that is inhaled and then exhaled by a smoker.

Major depression A very serious mood disorder in which people lose interest in life and can no longer find enjoyment in anything.

Malignant (muh LIG nuht) Cancerous.

Malnutrition A condition in which the body doesn't get the nutrients it needs to grow and function properly.

Managed care A health insurance plan that saves money by encouraging patients and providers to select less costly forms of care.

Marijuana Dried leaves and flowers of the hemp plant, called cannabis sativa.

Marrow A tissue in the center of some bones.

humo directo Humo que el fumador aspira y exhala.

depresión mayor Desorden del estado deanimo en el cual las personas pierden interés en su vida y no encuentran placer ennada.

maligno Canceroso.

desnutrición Cual es una afección en la que el cuerpo no recibe los nutrientes que necesita para crecer y funcionar de forma adecuada.

cuidado controlado Plan de seguro médico que ahorra dinero al limitar la selección de doctores de las personas.

marihuana Hojas y flores secas de la planta de cáñamo, llamada cannabis sativa.

médula Tejido del centro de algunos huesos.

English

Media Various methods for communicating information.

Media literacy The ability to understand the goals of advertising and the media.

Mediation Resolving conflicts by using another person or persons to help reach a solution that is acceptable to both sides.

Medicine A drug that prevents or cures an illness or eases its symptoms.

Medicine abuse Intentionally using medicines in ways that are unhealthful and illegal.

Medicine misuse Taking medicine in a way that is not intended.

Melanin Substance that gives skin its color.

Menstrual (MEN stroo uhl) cycle The beginning of one menstruation to the next.

Menstruation (men stroo AY shuhn) The flow from the body of blood, tissues, and fluids that result from the breakdown of the lining of the uterus.

Mental and emotional disorders Illnesses that affect a person's thoughts, feelings, and behavior.

Mental and emotional health The ability to handle the stresses and changes of everyday life in a reasonable way.

Metabolism The process by which the body gets energy from food.

Methamphetamine A stimulant similar to amphetamines.

Mind-body connection How your emotions affect your physical and overall health and how your overall health affects your emotions.

Minerals (MIN uh ruhls) Substances the body uses to form healthy bones and teeth, keep blood healthy, and keep the heart and other organs working properly.

Minor A person under the age of adult rights and responsibilities.

Mixed message A situation in which your words say one thing, but your body language says another.

Mob mentality Acting or behaving in a certain and often negative manner because others are doing it.

Español

medios de comunicación Diversos métodos de comunicar información.

alfabetización mediática Habilidad de entender las metas de la publicidad en los medios publicitarios.

mediación Resolución de conflictos por medio de otra persona que ayuda a llegar a una solución aceptable para ambas partes.

medicamento Droga que previene o cura enfermedades o que Alivia sus síntomas.

abuso de medicamentos Intencionalmente el uso de medicamentos en formas que son insalubres e ilegales.

uso indebido de medicamentos Tomar un medicamento de una manera que no es la intención.

melanina Sustancia que le da el color a la piel.

ciclo menstrual Cambio hormonal que ocurre en las mujeres desde el comienzo de una menstruación hasta el siguiente.

menstruación Fluido de sangre, tejidos y fluidos del cuerpo que resulta del desprendimiento del forro del útero.

trastornos mentales y emocionales Enfermedades que afectan los pensamientos, sentimientos y comportamiento de una persona.

salud mental y emocional Habilidad de hacerle frente de manera razonable al estrés y a los cambios de la vida diaria.

metabolismo Proceso mediante el cual el cuerpo obtiene energía de los alimentos.

metanfetamina Estimulantes parecidos a las anfetaminas.

conexión de la mente con el cuerpo Forma en la cual tus emociones afectan tu salud física y general, y como tu salud general afecta tus emociones.

minerales Sustancias que el cuerpo utiliza para formar huesos y dientes saludables, mantener la sangre saludable y mantener el corazón y otros órganos funcionando como deben.

menor Persona que es menor de en que se tienen la edad de derechos y responsabilidades de adultos.

mensaje contradictorio Situación en que tus palabras expresan algo pero tu lenguaje corporal lo contradice.

mentalidad de rebaño Actuar o comportarse de cierta manera, normalmente maneras negativas, sólo porque otros lo están haciendo.

Glossary/Glosario

English

Mononucleosis (MAH noh nook klee OH sis) A viral disease characterized by a severe sore throat and swelling of the lymph glands in the neck and around the throat area.

Mood disorder A mental and emotional problem in which a person undergoes mood swings that seem extreme, inappropriate, or last a long time.

Mood swings Frequent changes in emotional state.

Muscle endurance The ability of a muscle to repeatedly use force over a period of time.

Muscle strength The most weight you can lift or the most force you can exert at one time.

Muscular system Tissues that move parts of the body and control the organs.

MyPlate A visual reminder to help consumers make healthier food choices.

Narcotics (nar KAH tics) Drugs that get rid of pain and dull the senses.

National Institutes of Health (NIH) An agency of the Federal government that funds medical research to enhance health, lengthen life, and reduce illness and disability.

Natural disaster An event caused by nature that result in widespread damage, destruction, and loss.

Nearsightedness Objects that are close appear clear while those faraway look blurry.

Negative peer pressure Pressure you feel to go along with harmful behaviors or beliefs of others your age.

Neglect Failure to provide for the basic physical and emotional needs of a dependent.

Negotiation (neh GOH shee AY shuhn) The process of talking directly to the other person to resolve a conflict.

Neighborhood watch programs Programs in which residents are trained to identify and report suspicious activity.

Nervous system The body's message and control center.

Español

mononucleosis Una enfermedad viral que se caracteriza pordolores de garganta severos y hinchazón de las glándulas linfasen el cuello y alrededor del area de la garganta.

trastorno del estado de ánimo Problema emocional y mental en el cualla persona tiene cambios de humor que parecen extremosinapropiados, o que duran mucho tiempo.

cambios de humor Cambios frecuentes en elestado emocional.

resistencia muscular Habilidad de un músculo de usar fuerzarepetitivamente en un periodo de tiempo.

fuerza muscular Medida del peso máximo que puedes cargar ola fuerza máxima que puedes emplear a un mismo tiempo.

sistema muscular Tejidos que mueven partes del cuerpo ycontrolan los órganos.

MiPlato Un recuerdo visual para ayudar a los consumidoreselegir alimentos más saludables.

narcóticos Droga que alivia el dolor y entorpece los sentidos.

Institutos Nacionales de Salud Una agencia del gobierno federal que financia la investigación médica para mejorar la salud, prolongar la vida y reducir la morbilidad y la discapacidad.

desastre natural Evento causado por la naturaleza que resulta en daños extensos, destrucción y pérdida.

miopía Capacidad de ver claramente los objetos cercanos, mientras los objetos lejanos se ven borrosos.

presión negativa de compañeros Presión que sientes de seguir comportamientos malos o creencias de otras personas de tu misma edad.

abandono Fallas en el proceso de proveer las necesidades físicas y emocionales de una persona considerada como dependiente.

negociación Proceso de hablar directamente con otra persona para solucionar un conflicto.

programas de vigilancia comunitaria Programas en el cual los residentes están entrenados para identificar y reportar actividades sospechosas.

sistema nervioso Centro de mensajes y control del cuerpo.

English

Neurons (NOO rahnz) Cells that make up the nervous system.

Neutrality A promise not to take sides.

Nicotine (NIH kuh teen) An addictive, or habit-forming, drug found in tobacco.

Nicotine replacement therapies (NRT) Products that assist a person in breaking a tobacco habit.

Noncommunicable disease A disease that cannot be spread from person to person.

Nonrenewable resources Substances that cannot be replaced once they are used.

Nonverbal Communication Getting messages across without using words.

Nurture Fulfill physical, mental, emotional, and social needs.

Nutrient deficiency A shortage of a nutrient.

Nutrient dense Foods having a high amount of nutrients relative to the number of calories.

Nutrients (NOO tree ents) Substances in foods that your body needs to grow, have energy, and stay healthy.

Nutrition (noo TRIH shuhn) The process of taking in food and using it for energy, growth, and good health.

Español

neuronas Células que forman el sistema nervioso.

neutralidad Promesa de no tomar partido durante un conflict entre otros.

nicotina Droga adictiva, que forma hábito, encontrada en el tabaco.

terapias de reemplazo de nicotina Productos que ayudan a las personas a romper el hábito del tabaco.

enfermedad no contagiosa Enfermedad que no se puede transmitir de una persona a otra.

recursos no renovables Sustancias que no se pueden reemplazar después de usarse.

comunicación no verbal Trasmitir mensajes sin el uso de palabras.

criar Satisfacer necesidades físicas, mentales, emocionales y sociales.

deficiencia nutricional Escasez de un nutriente.

rico en nutrientes Tener una alta cantidad de sustancias nutritivas en comparación con la cantidad de calorías.

nutrientes Sustancias en los alimentos que el cuerpo necesita para desarrollarse, tener energía y mantenerse saludable.

nutrición Proceso de ingerir alimentos y usarlos para la energía, el desarrollo y el mantenimiento de la buena salud.

O

Obese A person is more than 20 percent higher than what is appropriate for their height, age, and body frame.

Occupational Safety and Health Administration (OSHA) A branch of the U.S. Department of Labor that protects American workers.

Ophthalmologist (ahf thahl MAH luh jist) A physician who specializes in the structure, functions, and diseases of the eye.

Opium A liquid from the poppy plant containing substances that numb the body.

Opportunistic infection A disease that attacks a person with a weakened immune system and rarely occurs in a healthy person.

Optimistic Having a positive attitude about the future.

obeso(a) Más del 20 por ciento del peso que es adecuado para tu estatura edad, y tipo de cuerpo.

Administración de Salud y Seguridad Ocupacional Rama del Ministerio del Trabajo que protege la seguridad de los trabajadores estadounidenses.

oftalmólogo Médico que se especializa en la estructura, funciones y enfermedades del ojo.

opio Líquido extraído de la planta de amapola que contiene sustancias que adormecen el cuerpo.

infección oportunista Infección que ocurre raramente en personas saludables.

optimista Tener una actitud positiva con respecto al futuro.

Glossary/Glosario

English

Optometrist (ahp TAH muh trist) A health care professional who is trained to examine the eyes for vision problems and to prescribe corrective lenses.

Organ A body part made up of different tissues joined to perform a particular function.

Orthodontist (or thuh DAHN tist) A dentist who prevents or corrects problems with the alignment or spacing of teeth.

Osteoarthritis (ahs tee oh ahr THRY tuhs) A chronic disease that is common in older adults and results from a breakdown in cartilage in the joints.

Ovaries The female endocrine glands that release mature eggs and produce the hormones estrogen and progesterone.

Overdose Taking more of a drug than the body can tolerate.

Over-the-counter (OTC) medicine A medicine that you can buy without a doctor's permission.

Overweight More than the appropriate weight for gender, height, age, body frame, and growth pattern.

Overworking Conditioning too hard or too often without enough rest between sessions.

Ovulation The process by which the ovaries release mature eggs, usually one each menstrual cycle.

Ozone (OH zohn) A gas made of three oxygen atoms.

Ozone layer A shield above the earth's surface that protects living things from ultraviolet (UV) radiation.

Español

optómetra Profesional de la salud que está preparado para examinar la vista y recetar lentes correctivas.

órgano Parte del cuerpo que comprende distintos tejidos unidos para cumplir una función particular.

ortodoncista Dentista que previene o corrige problemas del alineamiento o del espacio entre los dientes.

osteoartritis Enfermedad crónica, común en los ancianos, que es el resultado de la degeneración del cartílago de las articulaciones.

ovarios Glándulas endocrinas femeninas que sueltan huevos maduros y producen hormonas como estrógeno y progesterona.

sobredosis Tomar una cantidad de droga que supera lo que el cuerpo puede tolerar.

medicamentos de venta libre Medicina que se puede comprar sin receta de un médico.

sobrepeso Más del peso apropiado de acuerdo al sexo, estatura, edad, estructura corporal y ritmo de crecimiento.

ejercicio excesivo Acondicionamiento muy fuerte o muy seguido sin descanso suficiente entre sesiones.

ovulación Proceso en el cual los ovarios sueltan huevos maduros, normalmente uno en cada ciclo menstrual.

ozono Gas hecho de tres átomos de oxígeno.

capa de ozono Capa protectora sobre la superficie de la Tierra que protege a los seres vivos de la radiación ultravioleta.

Pacemaker A small device that sends steady electrical impulses to the heart to make it beat regularly.

Pancreas (PAN kree uhs) A gland that helps the small intestine by producing pancreatic juice, a blend of enzymes that breaks down proteins, carbohydrates, and fats.

Panic A feeling of sudden, intense fear.

Passive A tendency to give up, give in, or back down without standing up for rights and needs.

Passive smoker A nonsmoker who breathes in secondhand smoke.

Pathogens Germs that cause diseases.

marcapasos Pequeño aparato que envía pulsaciones eléctricas constantes al corazón, para que los latidos sean regulares.

páncreas Glándula que le ayuda al intestino delgado, a través de la producción de jugo pancreático, que está formado por una mezcla de varias enzimas que descomponen las proteínas, hidratos de carbono y grasas.

pánico Sentimiento repentino de miedo intenso.

pasivo Tendencia a renuncia, dejar de lado, o hacerse atrás sin reclamar derechos o necesidades.

fumador pasivo Persona que no fuma pero que inhala humo secundario.

patógenos Gérmenes que causan enfermedades.

English

Pedestrian A person who travels on foot.

Peers People close to you in age who are a lot like you.

Peer mediation (mee dee AY shuhn) A process in which a specially trained student listens to both sides of an argument to help the people reach a solution.

Peer pressure The influence that your peer group has on you.

Pelvic inflammatory disease An infection of the female's reproductive organs.

Peripheral nervous system (PNS) The nerves that connect the central nervous system to all parts of the body.

Personality A combination of your feelings, likes, dislikes, attitudes, abilities, and habits.

Personality disorder A variety of psychological conditions that affect a person's ability to get along with others.

Pesticide Product used on crops to kill insects and other pests.

Pharmacist A person trained to prepare and distribute medicines.

Phobia Intense and exaggerated fear of a specific situation or object.

Physical abuse The use of physical force.

Physical activity Any movement that makes your body use extra energy.

Physical dependence An addiction in which the body develops a chemical need for a drug.

Physical fitness The ability to handle the physical demands of everyday life without becoming overly tired.

Pituitary gland A gland that signals other endocrine glands to produce hormones when needed.

Plaque (PLAK) A thin, sticky film that builds up on teeth and leads to tooth decay.

Plasma The yellowish, watery part of blood.

Pneumonia A serious inflammation of the lungs.

Point of sale promotion Advertising campaigns in which a product is promoted at a store's checkout counter.

Español

peatón Persona que se traslada a pie.

compañeros Personas de tu grupo de edad que se parecen a ti de muchas maneras.

mediación entre compañeros Proceso en el cual un estudiante especialmente capacitado escucha los dos lados de un argumento para ayudar a las personas a llegar a un acuerdo.

presión de compañeros Influencia que tu grupo de compañeros tiene sobre ti.

enfermedad inflamatoria pélvica Una infección de los órganos reproductivos de una mujer.

sistema nervioso periférico (SNP) Nervios que unen el Sistema nervioso central con todas partes del cuerpo.

personalidad Combinación de tus sentimientos, gustos, disgustos, actitudes, habilidades y hábitos.

trastorno de personalidad Condición sicológica que afecta la habilidad de una persona de actuar normalmente con otras.

pesticida Producto que se usa en las cosechas para matar insectos y otras plagas.

farmaceuta Persona capacitada para preparar y distribuir medicinas.

fobia Miedo exagerado hacia una situación o objeto específico.

abuso físico Uso de fuerza física.

actividad física Todo movimiento que cause que el cuerpo use energía.

dependencia física Adicción por la cual el cuerpo llega a tener una necesidad química de una droga.

buen estado físico Capacidad de cumplir con las exigencies físicas de la vida diaria sin cansarse demasiado.

hipófisis Glándula que indica a otras glándulas endocrinas la necesidad de producir hormonas.

placa bacteriana Película delgada y pegajosa que se acumula en los dientes y contribuye a las caries dentales.

plasma Líquido amarillento, la parte líquida de la sangre.

neumonía Inflamación seria de los pulmones.

promoción en el punto de venta Campañas de publicidad en las cuales un producto puede adquirirse en la caja.

Glossary/Glosario

English

Point-of-service (POS) plans A health insurance plan that combines the features of HMOs and PPOs.

Poison control center A community agency that helps people deal with poisoning emergencies.

Pollen A powdery substance released by the flowers of some plants.

Pollute (puh LOOT) To make unfit or harmful for living things.

Pollution Dirty or harmful substances in the environment.

Pores Tiny opening in the skin that allow perspiration to escape.

Positive stress Stress that can help you reach your goals.

Precautions Planned actions taken before an event to increase the chances of a safe outcome.

Precycling Reducing waste before it occurs.

Preferred provider organization (PPO) A health insurance plan that allows it members to select a physical who participates in the plan or visit a physician of their choice.

Prejudice (PREH juh dis) A negative and unjustly formed opinion.

Preschooler A child between the ages of three and five.

Prescription (prih SKRIP shuhn) medicine A medicine that can be obtained legally only with a doctor's written permission.

Prevention Taking steps to avoid something.

Preventive care Steps taken to keep disease or injury from happening or getting worse.

P.R.I.C.E. formula Protect, rest, ice, compress, and elevate.

Primary care provider Health care professional who provides checkups and general care.

Probation Regularly check in with a court officer.

Español

Planes de Punto de servicio (POS) Un plan de seguro médico que combina las características de las HMO y las PPO.

centro toxicológico Agencia de la comunidad que ayuda a personas con emergencias relacionadas con venenos.

polen Sustancia en forma de polvo, que despiden las flores de ciertas plantas.

contaminar Hacer lo impropio o dañoso para las cosas vivientes.

contaminación Sustancias sucias o dañinas en el medio ambiente.

poros Aperturas pequeñas en la piel que permiten el proceso de transpiración.

estrés positive Estrés que te puede ayudar a alcanzar tus metas.

precauciones Acciones planeadas que son tomadas antes de un evento para incrementar la seguridad.

preciclaje Proceso de reducir los desechos antes de que se produzcan.

Organización de proveedores preferidos Plan de seguro de salud que permite que sus miembros seleccionen a un médico que participa en el plan o a uno de su preferencia.

prejuicio Opinión formada negativa e injustamente.

niño de preescolar Niño entre las edades de tres y cinco años.

medicamentos bajo receta Medicina que puede ser obtenida legalmente solo con el permiso escrito de un doctor.

prevención Tomar pasos para evitar algo.

cuidado preventivo Medidas que se toman para evitar que ocurran enfermedades o daños o que empeoren.

fórmula P.R.I.C.E protección/pulso, reposo, hielo, compresión, y elevación.

médico de cabecera Profesional de la salud que proporciona exámenes médicos y cuidado general.

libertad condicional Periodo de tiempo en el cual una persona que la sido arrestada tiene que presentarse ante un oficial de la corte regularmente.

English

Product placement A paid arrangement a company has made to show its products in media such as television or film.

Proteins (PROH teenz) The nutrient group used to build and repair cells.

Protozoa (proh tuh ZOH uh) One-celled organisms that are more complex than bacteria.

Psychiatrist (sy KY uh trist) A medical doctor who treats mental health problems.

Psychological dependence A person's belief that he or she needs a drug to feel good or function normally.

Psychologist (sy KAH luh jist) A mental health professional who is trained and licensed by the state to counsel.

Puberty (PYOO bur tee) The time when you develop physical characteristics of adults of your own gender.

Public health System to monitor and promote the welfare of the population.

Pulmonary circulation Blood travels from the heart, through the lungs, and back to the heart.

Pupil The dark opening in the center of the iris.

R

Radiation therapy A treatment that uses X rays or other forms of radiation to kill cancer cells.

Rape Forced sexual intercourse.

Reaction time The ability of the body to respond quickly and appropriately to situations.

Recovery The process of learning to live an alcohol-free life.

Recovery heart rate How quickly your heart rate returns to normal right after exercise is stopped.

Recurrence The return of cancer after a remission.

Recycle To change items in some way so that they can be used again.

Referral A suggestion to seek help or information from another person or place.

Refusal skills Strategies that help you say no effectively.

Español

colocación de un producto Acuerdo pagado hecho por una compañía para mostrar su producto en los medios de publicidad.

proteínas Nutrientes que se usan para reparar las células y tejidos del cuerpo.

protozoarios Organismos unicelulares que tienen una estructura más compleja que las bacterias.

psiquiatra Médico que trata trastornos de la salud mental.

dependencia psicológica Cuando una persona cree que necesita una droga para sentirse bien o para trabajar normalmente.

psicólogo Profesional de la salud mental que está licenciado por el estado para hacer terapias.

pubertad Etapa de la vida en la cual una persona comienza a desarrollar ciertas características físicas propias de los adultos del mismo sexo.

salud pública Esfuerzos para comprobar y promover el bienestar de la población.

circulación pulmonar Circulación que lleva la sangre desde el corazón, a través de los pulmones y de regreso al corazón.

pupila Abertura oscura en el centro del iris.

terapia de radiación Tratamiento que usa rayos X u otra forma de radiación para matar células cancerosas.

violación Relaciones sexuales forzadas.

tiempo de reacción Habilidad del cuerpo de responder rápida y apropiadamente a diferentes situaciones.

recuperación Proceso de aprender a vivir una vida libre de alcohol.

ritmo cardiaco de recuperación Que tan rápido tu corazón regresa a lo normal después de haber parado el ejercicio.

reaparición Regreso del cáncer después de una remisión.

reciclar Cambiar un objeto de alguna manera para que se pueda volver a usar.

referencia Sugerencia de buscar ayuda o información de otra persona o lugar.

destrezas de rechazo Estrategias que ayudan a decir no efectivamente.

Glossary/Glosario

English

Relapse A return to the use of a drug after attempting to stop.

Relationships The connections you have with other people and groups in your life.

Reliable Trustworthy and dependable.

Remission A period during which cancer signs and symptoms disappear.

Reproduction The process by which living organisms produce others of their own kind.

Reproductive (ree pruh DUHK tiv) system The body organs and structures that make it possible to produce children.

Rescue breathing A first-aid procedure where someone forces air into the lungs of a person who cannot breathe on his or her own.

Resilience The ability to recover from problems or loss.

Respiration The exchange of gases between your body and the air.

Respiratory system The organs that supply your blood with oxygen.

Resting heart rate The number of times your heart beats per minute when you are relaxing.

Revenge Punishment, injury, or insult to the person seen as the cause of the strong emotion.

Rheumatoid (ROO muh toyd) arthritis A chronic disease characterized by pain, inflammation, swelling, and stiffness of the joints.

Risk The chance that something harmful may happen to your health and wellness.

Risk behavior An action or behavior that might cause injury or harm to you or others.

Role A part you play when you interact with another person.

Role model A person who inspires you to think or act a certain way.

Español

recaída Regresar al uso de la droga después de haber intentado parar.

relaciones Conexiones que tienes con otras personas o grupos en tu vida.

confiable Confiable y seguro.

remisión Periodo durante el cual desaparecen las señales y síntomas del cáncer.

reproducción Proceso mediante el cual los organismos vivos producen otros de su especie.

sistema reproductor Órganos y estructuras corporales que hacen posible la producción de niños.

respiración de rescate Procedimiento de primeros auxilios en el que una persona llena de aire los pulmones de una persona que no está respirando.

resiliencia Habilidad de recuperarse de un problema o perdida.

respiración Intercambio de gases entre el cuerpo y el aire.

sistema respiratorio Órganos que suministran el oxígeno en lasangre.

ritmo cardiaco en reposo Número de veces el corazón late por minuto cuando estás relajado.

venganza Castigo, daño, o insulto hacia una persona por causa de una emoción fuerte.

artritis reumatoide Enfermedad crónica caracterizada por dolor, inflamación, hinchazón y anquilosamiento de las articulaciones.

riesgo Posibilidad de que algo dañino pueda ocurrir en tu salud y bienestar.

conducta arriesgada Acto o conducta que puede causarte daño o perjudicarte a ti o a otros.

papel Parte que tú desempeñas cuando actúas con otra persona.

ejemplo a seguir Persona que inspira a otras a que se comporten o piensen de cierta manera.

S

Safety conscious Being aware that safety is important and careful to act in a safe manner.

consciente de la seguridad Que se da cuenta de la importancia de la seguridad y actúa con cuidado.

English

Saliva (suh LY vuh) A digestive juice produced by the salivary glands in your mouth.

Saturated fats Fats that are usually solid at room temperature.

Schizophrenia (skit zoh FREE nee uh) A severe mental disorder in which a person loses contact with reality.

Second-degree burn A moderately serious burn in which the burned area blisters.

Secondhand smoke Air that has been contaminated by tobacco smoke.

Sedentary lifestyle A way of life that involves little physical activity.

Self-concept The way you view yourself overall.

Self-esteem How you feel about yourself.

Semen (SEE muhn) A mixture of sperm and fluids that protect sperm and carry them through the tubes of the male reproductive system.

Sewage Human waste, garbage, detergents, and other household wastes washed down drains and toilets.

Sexting The practice of sending someone sexually explicit photographs or messages via a mobile phone.

Sexual abuse Sexual contact that is forced upon another person.

Sexual harassment Uninvited and unwelcome sexual conduct directed at another person.

Sexually transmitted diseases (STDs) Infections that are spread from person to person through sexual contact.

Shock A life-threatening condition in which the circulatory system fails to deliver enough blood to vital tissues and organs.

Short-term goal A goal that you can achieve in a short length of time.

Side effect A reaction to a medicine other than the one intended.

Sidestream smoke Smoke that comes from the burning end of a cigarette, pipe, or cigar.

Skeletal muscle Muscle attached to bones that enables you to move your body.

Español

saliva Líquido digestivo producido por las glándulas salivales de la boca.

grasas saturadas Grasas que son sólidas a la temperatura ambiente.

esquizofrenia Trastorno mental grave por el cual una persona pierde contacto con la realidad.

quemadura de segundo grado Quemadura moderadamente seria en la que se forman ampollas en el área quemada.

humo de segunda mano Aire que está contaminado por el humo del tabaco.

estilo de vida sedentario Forma de vida que incluye poca actividad física.

autoconcepto Manera en que te ves a ti mismo.

autoestima Como te sientes sobre ti mismo.

semen Mezcla de esperma y fluidos que protegen a los espermatozoides y los transportan a través de los tubos del sistema reproductivo masculino.

aguas cloacales Basura, detergentes y otros desechos caseros que se llevan las tuberías de desagüe.

sexteo La práctica de enviar a alguien fotografías o mensajes sexualmente explícitos.

abuso sexual Contacto sexual forzado por una persona.

acoso sexual Conducta sexual no solicitada y fuera de lugar dirigida a otra persona.

enfermedades de transmisión sexual (ETS) Enfermedades que se propagan de una persona a otra, a través del contacto sexual.

choque Afección que puede causar la muerte en la cual el aparato circulatorio no lleva la suficiente cantidad de sangre a los tejidos y órganos vitales.

meta a corto plazo Meta que uno puede alcanzar dentro de un breve periodo de tiempo.

efecto secundario Reacción inesperada de una medicina.

humo indirecto Humo que procede de un cigarrillo, pipa o cigarro encendido.

músculo del sistema osteoarticular Músculo ligado a huesos que te permiten mover el cuerpo.

Glossary/Glosario

English

Skeletal system The framework of bones and other tissues that supports the body.

Small intestine A coiled tube from 20 to 23 feet long, in which about 90 percent of digestion takes place.

Smog Yellow-brown haze that forms when sunlight reacts with air pollution.

Smokeless tobacco Ground tobacco that is chewed or inhaled through the nose.

Smooth muscle Type of muscle found in organs and in blood vessels and glands.

Snuff Finely ground tobacco that is inhaled or held in the mouth or cheeks.

Social age Age measured by your lifestyle and the connections you have with others.

Social health Your ability to get along with the people around you.

Sodium A nutrient that helps control the amount of fluid in your body.

Somatic system Part of the nervous system that deals with actions that you control.

Specialist (SPEH shuh list) Health care professional trained to treat a special category of patients or specific health problems.

Sperm Male reproductive cells.

Spinal cord A long bundle of neurons that sends messages to and from the brain and all parts of the body.

Sports gear Sports clothing and safety equipment.

Sprain An injury to the ligament connecting bones at a joint.

Stamina Your ability to stick with a task or activity for a long period of time.

Stereotype Belief about people who belong to a certain group.

Stimulant (STIM yuh luhnt) A drug that speeds up the body's functions.

Strength The ability of your muscles to use force.

Español

sistema osteoarticular Armazón de huesos y otros tejidos que sostiene el cuerpo.

intestino delgado Tubo enrollado que mide entre veinte y veintitrés pies, en el cual ocurre el noventa por ciento de la digestión.

smog Neblina de color amarillo-café que se forma cuando la luz solar reacciona con la contaminación del aire.

tabaco sin humo Tabaco molido que es masticado o inhalado a través de la nariz.

músculo liso Tipo de músculo que se encuentra en los órganos, los vasos sanguíneos y las glándulas.

rapé Tabaco molido finamente que es inhalado o mantenido en la boca o las mejillas.

edad social Edad calculada de acuerdo con tu estilo de vida y las conexiones que tienes con los demás.

salud social Habilidad para llevarte bien con las personas que te rodean.

sodio Nutriente que ayuda a controlar la cantidad de fluidos en el cuerpo.

sistema somático Parte del sistema nervioso relacionada con las acciones que tú controlas.

especialista Profesional del cuidado de la salud que está capacitado para tratar una categoría especial de pacientes o un problema de salud específico.

espermatozoides Células reproductoras masculinas.

médula espinal Largo conjunto de neuronas que transmiten mensajes entre el cerebro y todas las otras partes del cuerpo.

accesorios deportivos Ropa para deportes y equipo de seguridad.

torcedura Daño al ligamento que conecta huesos con articulaciones.

energía La habilidad de poder realizar y mantener una actividad por largos periodos de tiempo.

estereotipo Creencia sobre las personas que pertenecen a un grupo determinado.

estimulante Droga que acelera las funciones del cuerpo.

fuerza Capacidad que tienen los músculos para ejercer una fuerza.

English

Strep throat Sore throat caused by streptococcal bacteria.

Stress The body's response to real or imagined dangers or other life events.

Stress fracture A small fracture caused by repeated strain on a bone.

Stress management Identifying sources of stress and learning how to handle them in ways that promote good mental/emotional health.

Stressor Anything that causes stress.

Stroke A serious condition that occurs when an artery of the brain breaks or becomes blocked.

Subcutaneous (suhb kyoo TAY nee uhs) layer A layer of fat under the skin.

Substance abuse Using illegal or harmful drugs, including any use of alcohol while under the legal drinking age.

Suicide The act of killing oneself on purpose.

Sunscreen A cream or lotion that filters out some UV rays.

Sympathetic (simp uh THET ik) Aware of how you may be feeling at a given moment.

Syphilis (SIH fuh luhs) A bacterial STD that can affect many parts of the body.

Systemic circulation Oxygen-rich blood travels to the cells and picks up waste products.

Español

amigdalitis estreptocócica Dolor de garganta causado por bacterias estreptocócicas.

estrés Reacción del cuerpo hacia peligros reales o imaginarios u otros eventos en la vida.

fractura por sobrecarga Fractura pequeña causada por una torcedura repetitiva del mismo hueso.

controlar el estrés Identificación de lo que causa el estrés y aprender cómo reaccionar a ello de manera que permita mantener la buena salud mental y emocional.

factor estresante Todo lo que provoca el estrés.

apoplejía Afección seria que se produce cuando una arteria en el cerebro se rompe o se obstruye.

capa subcutánea Capa de grasa debajo de la piel.

abuso de sustancias Consumo de drogas ilegales o nocivas, incluso el consumo del alcohol en cualquiera de sus formas antes de la edad legal para beber.

suicidio Matarse intencionalmente.

bloqueador solar Crema o loción que filtra algunos rayos UV.

empático Estar consciente que como te puedes estar sintiendo en un momento indicado.

sífilis Infección de transmisión sexual, causada por una bacteria, que puede afectar muchas partes del cuerpo.

circulación sistémica Circulación que lleva sangre rica en oxígeno a todos los tejidos del cuerpo, menos a los pulmones.

T

Tar A thick, dark liquid that forms when tobacco burns.

Target audience A group of people for which a product is intended.

Target heart rate The number of heartbeats per minute that you should aim for during moderate-to-vigorous aerobic activity to help your circulatory system the most.

Tartar (TAR tuhr) Hardened plaque that hurts gum health.

Tendon A type of connecting tissue that joins muscles to bones and muscles to muscles.

Tendonitis Painful swelling of a tendon caused by overuse.

alquitrán Líquido espeso y oscuro que forma el tabaco al quemarse.

publico objetivo Grupo de gente al cual es dirigido un product específico.

ritmo cardiaco objetivo Número de latidos del corazón, por minuto, que una persona debe tratar de alcanzar durante una actividad de intensidad moderada a vigorosa, para obtener lo máximo posible de beneficio para el aparato circulatorio.

sarro Placa endurecida que daña la salud de las encías.

tendón Tipo de tejido conjuntivo que une un músculo a otro o un músculo a un hueso.

tendinitis Hinchazón dolorosa de un tendon causada por el uso excesivo.

Glossary/Glosario

English

Testes The pair of glands that produce sperm.

THC The main active chemical in marijuana.

Therapy Professional counseling.

Third-degree burn A very serious burn in which all the layers of skin are damaged.

Time management Strategies for using time efficiently.

Tinnitus A constant ringing in the ears.

Tissue A group of similar cells that do a particular job.

Toddler A child between the ages of one and three.

Tolerance (TAHL er ence) 1. The ability to accept other people as they are. 2. The body's need for larger and larger amounts of a drug to produce the same effect.

Tornado A whirling, funnel-shaped windstorm that drops from storm clouds to the ground.

Trachea (TRAY kee uh) A passageway in your throat that takes air into and out of your lungs.

Trans fatty acids A kind of fat formed when hydrogen is added to vegetable oil during processing.

Traumatic brain injury A condition caused by the brain being jarred and striking the inside of the skull.

Trichomoniasis (TREE koh moh NI ah sis) An STD caused by the protozoan Trichomonas vaginalis.

Tuberculosis (TB) (too ber kyuh LOH sis) A bacterial disease that usually affects the lungs.

Tumor (TOO mer) A group of abnormal cells that forms a mass.

Type 1 diabetes A condition in which the immune system attacks insulin-producing cells in the pancreas.

Type 2 diabetes A condition in which the body cannot effectively use the insulin it produces.

Español

testículos Las dos glándulas que producen los espermatozoides.

THC Principal componente químico activo de la marihuana.

terapia Consejo profesional.

quemadura de tercer grado Quemadura muy seria en que todas las capas de la piel quedan dañadas.

organización del tiempo Estrategias para usar el tiempo eficazmente.

acúifeno Sonido constante en el oído.

tejido Grupo de células similares que tienen una función en particular.

niño que empieza a andar Niño entre uno y tres años.

tolerancia 1. Habilidad de aceptar a otras personas de la forma que son. 2. Necesidad del cuerpo de mayores cantidades de una droga para obtener el mismo efecto.

tornado Tormenta en forma de torbellino, que gira en grandes círculos y que cae del cielo a la tierra.

tráquea Pasaje en la garganta que lleva el aire de y hacia los pulmones.

ácido graso Forma de grasa formada cuando el hidrógeno es adherido al aceite vegetal durante el procesamiento.

lesión cerebral traumática Condición causada por irritación de cerebro y es notable en el interior del cráneo.

trichomoniasis Enfermedad de transmisión sexual causada por los protozoarios Trichomonas vaginalis.

tuberculosis Enfermedad causada por bacterias que generalmente afecta a los pulmones.

tumor Grupo de células anormales que forma una masa.

diabetes tipo 1 Afección por la cual el sistema inmunológico ataca las células productoras de insulina en el páncreas.

diabetes tipo 2 Afección que se caracteriza por la inhabilidad del cuerpo para usar de manera eficaz la insulina que produce.

U

U.S. Department of Agriculture (USDA) A department of the Federal government providing leadership on food, agriculture, nutrition, and other topics.

Ulcer (UHL ser) An open sore in the stomach lining.

Departamento de Agricultura de los Estados Unidos (USDA) Un departamento del Gobierno Federal de liderazgo en los alimentos, la agricultura, la nutrición y otros temas.

úlcera Llaga abierta en el forro del estómago

English

Ultraviolet (UV) rays Invisible form of radiation that can enter skin cells and change their structure.

Underweight Less than the appropriate weight for gender, height, age, body frame, and growth pattern.

Unit price The cost per unit of weight or volume.

Universal precautions Actions taken to prevent the spread of disease by treating all blood as if it were contaminated.

Unsaturated fat Fat that usually remains liquid at room temperature.

Uterus (YOO tuh ruhs) A pear-shaped organ, located within the pelvis, in which the developing baby is nourished and protected.

Vaccine (vak SEEN) A preparation of dead or weakened pathogens that is introduced into the body to cause an immune response.

Values The beliefs that guide the way a person lives.

Vector An organism, such as an insect, that transmits a pathogen.

Veins Blood vessels that carry blood from all parts of the body back to the heart.

Verbal communication Expressing feelings, thoughts, or experiences with words, either by speaking or writing.

Violence An act of physical force resulting in injury or abuse.

Viruses (VY ruh suhz) The smallest and simplest pathogens.

Vitamins (VY tuh muhns) Compounds that help to regulate body processes.

Warm-up Gentle exercises that get heart muscles ready for moderate-to-vigorous activity.

Warranty A company's or a store's written agreement to repair a product or refund your money if the product does not function properly.

Weather emergency A dangerous situation brought on by changes in the atmosphere.

Español

rayos ultravioletas Forma invisible de radiación que puede penetrar células de la piel y cambia su estructura.

bajo peso Por debajo del peso apropiado de acuerdo con el sexo, estatura, edad, estructura corporal y ritmo de crecimiento.

precio por unidad Costo por unidad de peso o volumen.

precauciones universales Medidas para prevenir la propagación de enfermedades al tratar toda la sangre como si estuviera contaminada.

grasas no saturdas Grasas que son líquidas a la temperatura del ambiente.

útero Órgano en forma de pera en el cual se nutre un bebé en desarrollo.

vacuna Fórmula compuesta por gérmenes patógenos Muertos o debilitados que es introducida en el cuerpo para causar una reacción inmune.

valores Creencias que guían la forma en la cual vive una persona.

vector Organismo, por ejemplo un insecto, que transmite un agente patógeno.

vena Tipo de vaso sanguíneo que lleva la sangre de todas partes del cuerpo de regreso al corazón.

comunicación verbal Expresar sentimientos, pensamientos, o experiencias a través de palabras, ya sea hablando o escribiendo.

violencia Acto de fuerza física que resulta en un daño o abuso.

virus Patógenos más simples y pequeños.

vitaminas Compuestos que ayudan a regular los procesos del cuerpo.

precalentamiento Ejercicio suave que prepara a los músculos cardíacos para estar listos para actividad física moderada o fuerte.

garantía comercial Promesa escrita de un fabricante o una tienda de reparar un producto o devolver el dinero al comprador, si el producto no funciona debidamente.

emergencia meteorológica Situación peligrosa debido a cambios en la atmósfera.

Glossary/Glosario

English

Wellness A state of well-being or balanced health over a long period of time.

Win-win solution An agreement that gives each party something they want.

Withdrawal A series of painful physical and mental symptoms that a person experiences when he or she stops using an addictive substance.

Y

Youth court A special school program where teens decide punishments for other teens for bullying and other problem behaviors.

Z

Zero tolerance policy A policy that makes no exceptions for anybody for any reason.

Español

bienestar Mantener una salud balanceada por un largo período de tiempo.

solución mutuamente beneficiosa Acuerdo o resultado que da algo de lo que quieren a cada lado.

síndrome de abstinencia Series de síntomas físicos y mentales dolorosos por los cuales una persona pasa cuando para el uso de una sustancia adictiva.

corte juvenil Programa escolar especial en el cual adolescents deciden el castigo para otros adolescentes por intimidar y otros problemas de comportamiento.

política de cero tolerancia Normativa en que no hay excepciones para nadie por ninguna razón.

Index

Index

Index

Index

Index